AF686257

Animal Reproduction: Semen Quality Assessment, Volume II

Guest Editor

Anna Wysokińska

MDPI

Basel • Beijing • Wuhan • Barcelona • Belgrade • Novi Sad • Cluj • Manchester

Guest Editor
Anna Wysokińska
Institute of Animal Science
and Fisheries
Faculty of Agricultural
Sciences
University of Siedlce
Siedlce
Poland

Editorial Office
MDPI AG
Grosspeteranlage 5
4052 Basel, Switzerland

This is a reprint of the Special Issue, published open access by the journal *Animals* (ISSN 2076-2615), freely accessible at: https://www.mdpi.com/journal/animals/special_issues/AYC8038B0X.

For citation purposes, cite each article independently as indicated on the article page online and as indicated below:

Lastname, A.A.; Lastname, B.B. Article Title. *Journal Name* **Year**, *Volume Number*, Page Range.

ISBN 978-3-7258-3783-0 (Hbk)
ISBN 978-3-7258-3784-7 (PDF)
https://doi.org/10.3390/books978-3-7258-3784-7

© 2025 by the authors. Articles in this book are Open Access and distributed under the Creative Commons Attribution (CC BY) license. The book as a whole is distributed by MDPI under the terms and conditions of the Creative Commons Attribution-NonCommercial-NoDerivs (CC BY-NC-ND) license (https://creativecommons.org/licenses/by-nc-nd/4.0/).

Contents

About the Editor

Anna Wysokińska

Anna Wysokińska obtained her PhD and DSc in agricultural sciences with a focus on animal science. She is currently a professor at the University of Siedlce in the Faculty of Agricultural Science, Institute of Animal Science and Fisheries. Her scientific activity mainly focuses on issues related to animal reproduction, particularly andrological diagnostics and assessing the impact of genetic and environmental factors on the effectiveness of the use of breeders of various species. She is the author and co-author of numerous scientific papers, book chapters, and implementation instructions. She has reviewed many scientific articles for renowned international journals. She has reviewed and supervised master's, engineering, and doctoral theses. She has received numerous awards and distinctions for her scientific, teaching, and organizational achievements, including the Bronze Medal for Long Service, the Medal of the Commission of National Education, and the 'Merit for Agriculture' Badge of Honour. She is also a member of the Polish Society of Animal Production.

Preface

The assessment of semen quality plays a key role in determining the reproductive performance of males. It measures the functionality of the sperm used in assisted reproduction as well as in natural mating. This reprint of 'Animal Reproduction: Semen Quality Assessment, Volume II' aims to present the latest scientific achievements in the use of advanced techniques used to assess animal semen, including the precise analysis of sperm structures and the physiological processes taking place in them at various stages of fertilization. Spermatozoa are among the most diverse types of cells and are highly sensitive to the effects of various factors. Moreover, they exhibit a high degree of morphological and functional variation. Suitable semen diagnostics is extremely important to improving semen preservation techniques. Therefore, the challenges that arise in this area have prompted researchers to undertake measures aimed at introducing new techniques and analyses that can predict sperm quality and improve semen preservation protocols, with the ultimate goal of maximizing the efficiency of assisted reproductive techniques (ARTs). This reprint was prepared with the assistance of recognized experts in the field of reproduction, whose research and observations are contributing significantly to improving the effective implementation of ARTs in various groups of animals. It provides a comprehensive review of the current advances in semen diagnostics, while simultaneously underscoring promising directions for future research. I hope that these contributions will inspire those involved in animal reproduction to continually update their knowledge and seek innovative solutions.

As Guest Editor, I would like to thank all the researchers for their contributions to this Special Issue: the reviewers who donated their time and expertise to evaluate and provide valuable feedback to the authors of each article, and the Editors for their substantive support and extensive input into editing this Special Issue.

Anna Wysokińska
Guest Editor

animals

MDPI

Editorial

Animal Reproduction: Semen Quality Assessment, Volume II

Anna Wysokińska

Institute of Animal Science and Fisheries, Faculty of Agricultural Sciences, University of Siedlce, 08-110 Siedlce, Poland; anna.wysokinska@uws.edu.pl; Tel.: +48-25-6431388

1. Introduction

Semen assessment is an important aspect of semen diagnostics used in assisted reproductive technology (ART). The most common ART technique is artificial insemination (AI), which has many advantages over natural mating. Artificial insemination technology allows the use of the best males in breeding, increases the possibility of fertilizing a larger number of females, prevents the spread of certain reproductive diseases, solves problems related to crossbreeding and mating, and enables the transport of semen over long distances [1,2]. The effectiveness of artificial insemination is closely related to the quality of the semen. Semen quality is a term that describes the probability that sperm in the ejaculate will be functional both in artificial and in natural mating [3].

Many factors affect semen quality, including semen preservation methods, type and composition of the extender, frequency of semen collection, and the individual source of the ejaculate [4,5]. As sperm quality is of key importance for the success of artificial insemination, various preservation methods are used to maintain the fertilization capacity of the sperm, such as storage of semen in liquid form and cryopreservation [6–8]. Sperm preservation is very important to prolong sperm survival time, increase effective semen volume, and utilize males with the best reproductive potential. Changes in sperm structure may take place during semen preservation. These changes result from the specific structure of the sperm cell membrane. The sperm cell membrane contains large amounts of polyunsaturated fatty acids (PUFA), which are susceptible to free radicals, such as ROS attacks, shortening the survival time of sperm [9,10]. The sperm cell membrane is a structure that plays a significant role in the connection of sperm with the oocyte. An intact sperm cell membrane is a condition for the proper functioning of cells and the passage of individual stages necessary for oocyte fertilization [11]. The main cause of sperm damage during cryopreservation is oxidative stress, caused primarily by elevated ROS levels [12].

Assessment of sperm function plays an important role in predicting optimal fertilization outcomes. Semen quality is influenced by numerous factors, so precise knowledge of the mechanisms influencing sperm formation is essential for accurate determination of fertilization capacity. Andrology is continually introducing new advances in the laboratory evaluation of semen. There is increasing evidence for the clinical importance of specialized tests for evaluating sperm quality. Andrological diagnostics is a continually developing area of research whose main role is to assess the reproductive potential of males. The aim of this special edition is to present the latest scientific achievements in the application of modern andrological diagnostic techniques for the evaluation of sperm quality in various animals, in order to improve their reproductive performance.

Received: 20 February 2025
Revised: 26 February 2025
Accepted: 27 February 2025
Published: 1 March 2025

Citation: Wysokińska, A. Animal Reproduction: Semen Quality Assessment, Volume II. *Animals* **2025**, *15*, 709. https://doi.org/10.3390/ani15050709

Copyright: © 2025 by the author. Licensee MDPI, Basel, Switzerland. This article is an open access article distributed under the terms and conditions of the Creative Commons Attribution (CC BY) license (https://creativecommons.org/licenses/by/4.0/).

2. Animal Reproduction: Semen Quality Assessment—An Overview of Published Articles

The extender used for semen preservation is of key importance in maintaining optimal quality traits of sperm during storage [Zhang et al., Contribution 1]. Various components of the extender provide nutrition for the sperm, maintain the stability of the preserved environment, and prolong sperm survival time. A study using sheep of the Chinese Hu breed assessed the effect of various extenders on semen quality. Ejaculates diluted using four extenders with different components were assessed for sperm motility parameters, acrosome integrity, cell membrane integrity, and reactive oxygen species (ROS). The authors showed that sperm quality was influenced by the type of extender used. Specifically, the extender containing Tris, fructose, citric acid, and egg yolk improved the quality of Hu sheep semen during storage at 4 °C. The components of the extender were shown to be very important for maintaining optimal fertilization capacity. The effect of the composition and concentration of extender components on the quality of preserved semen has been confirmed in a study using goat semen [Li et al., Contribution 2], which showed that the glucose concentration in the extender affected sperm motility. Following incubation of the semen with various glucose concentrations, sperm motility (progressive motility and straight-line velocity), adenosine triphosphate (ATP) levels, and mitochondrial membrane potential (MMP) were shown to be significantly higher in the environment with low glucose content. The findings indicated that sperm regulate the energy metabolism pathway on the basis of changes in metabolic substrates and that low glucose levels activate oxidative phosphorylation (OXPHOS) in the mitochondria in order to maintain normal sperm functions. Low glucose conditions promote transcription and translation in mitochondria and activate mitochondrial OXPHOS to provide energy to the sperm. In addition, the LKB1/AMPK pathway is activated in order to maintain energy homeostasis and inhibit oxidative damage induced by ferroptosis. For this reason, the straight-line velocity of sperm following the use of an extender with a low glucose level is a new and important factor improving the effectiveness of artificial insemination. Hence, the dilution of semen with a low-glucose extender may be an effective and inexpensive method for improving the straight-line velocity of sperm within the female reproductive system, thus facilitating fertilization of the ovum. Another study showed that the addition of 50 mol/L FA (ferulic acid) to goat semen extender improves the quality of goat semen preserved at 17 °C [Zhang et al., Contribution 3]. The addition of 50 mol/L FA to the extender increased sperm motility, cell membrane integrity, and acrosome integrity in comparison with the control group (without the addition of FA), while at the same time reducing the peroxidation rate. These findings indicate that ferulic acid included in an extender at an appropriate concentration can mitigate oxidative damage during storage of goat semen in liquid form and improve the quality of semen preservation.

Despite the advances of recent decades, sperm damage induced by semen storage remains a common and nearly unavoidable side effect of procedures for handling and preserving semen. Cryopreservation in particular decreases sperm quality. A key role is played by the chemical cryoprotectants in the extender, which protect sperm viability in low-temperature conditions [Taheri-Khas et al., Contribution 4]. Therefore, studies are conducted to improve the quality of the cryopreservation process by perfecting the composition of extenders. An example is research conducted by Zhu et al. [Contribution 5] aimed at assessing the effect of resveratrol and its concentration on ram sperm quality following cryopreservation. The study showed that 50 μM of resveratrol added to the extender used to freeze semen can effectively mitigate the decline in sperm motility and acrosome integrity and the damage to membrane integrity taking place during cryopreservation of ram semen, while at the same time preserving the high mitochondrial activity of the sperm.

Moreover, the addition of resveratrol was shown to activate phosphorylation of 5′ adenosine monophosphate-activated protein kinase (AMPK) and expression of sirtuin 1 (SIRT1), which reduces ROS production. It also enhanced the antioxidant defense system of sperm (e.g., glutathione (GSH) content and the activity of glutathione synthase (GPx), superoxide dismutase (SOD), and catalase (CAT)) and reduced apoptosis and DNA damage.

The type of extender used in semen cryopreservation affects the quality traits of sperm and in vitro fertilization rates [Pérez-Durand et al., Contribution 6]. A study using semen collected from the vas deferens of llamas (*Lama glama*) compared the effect of three semen extenders before and after cryopreservation on sperm quality parameters and rates of in vitro fertilization of llama oocytes [Pérez-Durand et al., Contribution 6]. Differences in semen characteristics following the use of the three different extenders were shown for acrosome integrity, sperm viability, membrane permeability, and sperm motility, determined before and after cryopreservation. This indicated that the extender used for cryopreservation is a factor determining rates of in vitro fertilization using sperm collected from the vas deferens of llamas. According to the authors, further research is needed to understand the effect of various extenders in order to improve rates of in vitro fertilization using sperm samples collected from the vas deferens of llamas.

Selection of the optimal method of sperm preservation is an important factor in animal breeding [Neuman et al., Contribution 7]. A study conducted on the semen of red deer analyzed the quality of sperm stored in liquid form and in the epididymides for six days at 5 °C. Sperm samples were evaluated for motility, viability, morphology, antioxidant enzyme activity (SOD, GPx and CAT), and lipid peroxidation (based on the content of malondialdehyde). Sperm stored in liquid form were shown to have greater motility and viability and better morphology and antioxidant status than sperm stored in the epididymis. The authors concluded that for short-term preservation, storage of red deer semen in liquid form is better than storage in the epididymis.

The correct dilution factor is of key importance for the survival of sperm [Zhang et al., Contribution 8]. A dilution factor that is too high or too low can reduce sperm quality and, thus, the effectiveness of semen preservation, as demonstrated for the semen of Hu sheep. Zhang et al. [Contribution 8] analyzed the effect of various dilution methods and dilution factors on sperm motility parameters and the functional integrity of sperm following cryopreservation. The authors showed that two-stage dilution (1:3 and 1:2) can improve the effectiveness of preservation of Hu sheep semen.

Semen quality is influenced by numerous genetic and environmental factors. In the case of fish breeding, temperature and hormonal stimulation play a key role [Taheri-Khas et al., Contribution 4]. The study showed that despite the administration of nearly identical doses of hormones in groups of females and males, temperature changes had differing effects on reproductive parameters. The best results in goldfish (*Carassius auratus*) breeding were obtained at 22 °C, as higher ovulation rates and better sperm quality were observed in comparison to high (28 °C) and low (16 °C) temperatures.

An important factor pointed out by Žura Žaja et al. [Contribution 9] is anthropogenic radiofrequency electromagnetic radiation (RF-EMR), which poses a potential risk to animal health, including the semen quality of boars. Two-hour exposure of the semen of breeding boars to RF-EMR at a frequency of 2500 MHz resulted in a reduction in progressive sperm motility and in the proportion of the subpopulation of sperm with a more elongated head and a larger midpiece outline. For efficient pig production and breeding, it is extremely important to determine the effect of this type of radiation on semen quality and fertilization of the sow. The authors suggest that further research is needed to investigate the impact of RF-EMR on the semen of other domesticated animal species, especially those undergoing artificial insemination procedures.

An important factor affecting reproduction in dogs is the prevalence of aerobic bacteria and mycoplasmas [Domrazek et al., Contribution 10]. By identifying and characterizing the bacteria present in dog ejaculates, we can determine potential sources of contamination and infection, which can affect semen quality. Domrazek et al. [Contribution 10] showed that *Mycoplasma* spp. is commonly present in dog ejaculates, but that semen quality is not correlated with the presence of *Mycoplasma* spp. in dogs. The authors suggest that there may be undescribed species of canine mycoplasmas that can only be identified using advanced diagnostic techniques. An understanding of the microbial composition of semen and its influence on fertility parameters is essential to the development of effective strategies for optimizing breeding outcomes in dogs. Another study using dog semen assessed the potential effects of bacteria on semen quality in dogs [Sorkytė et al., Contribution 11]. Bacteriospermia was shown to significantly affect the quality of dog semen, as indicated by the link between a higher bacterial count in the semen and reduced semen quality parameters. Not all bacteria present in the dog semen were shown to have a negative impact on its quality. Samples with *Corynebacterium* spp. were associated with a decreased bacterial load, which also resulted in better quality parameters. Other bacteria, however, were more often linked to poorer semen quality. Of particular interest in that study was the identification of β-haemolytic *Escherichia coli* as the most pathogenic bacteria for semen parameters in dogs.

Issues associated with identifying mechanisms ensuring reproduction in lizards in order to implement assisted reproductive technology were taken up by [Sánchez-Rivera et al., Contribution 12]. Given the lack of research on sperm physiology in lizards, the authors attempted to establish whether lizard sperm undergo capacitation, like mammalian sperm. The study showed that certain changes associated with sperm capacitation, such as changes in the type of movement or acrosome damage, take place after two hours of semen incubation with the capacitation medium. According to authors, there is a need to establish whether capacitation is necessary for lizard sperm to acquire fertilization competence and whether this process is crucial for improving the success of ART techniques in this group of animals.

3. Conclusions

This Special Issue presents several innovative studies aimed at improving semen preservation technology, various aspects of sperm analysis and preservation in different species, and also physiological processes taking place within sperm structures. The studies are very promising and provide valuable information on the biology of reproduction in mammals, reptiles, and fish. This is a continually growing area of research in which new techniques and methods are being implemented, inspiring improvements in animal reproduction. Due to the many aspects of reproduction that have yet to be explained, there is a need for further studies with the ultimate goal of maximizing the efficiency of ART.

Funding: This research received no external funding.

Conflicts of Interest: The author declares no conflict of interest.

List of Contributions

1. Zhang, L.; Wang, Y.; Sun, X.; Kang, Y.; Sohail, T.; Wang, J.; Li, Y. Effects of Different Diluents on Semen Quality of Hu Ram Stored at 4 °C. *Animals* **2023**, *13*, 2823. https://doi.org/10.3390/ani13182823.
2. Li, Y.; Zhang, G.; Wen, F.; Xian, M.; Guo, S.; Zhang, X.; Feng, X.; Hu, Z.; Hu, J. Glucose Starvation Inhibits Ferroptosis by Activating the LKB1/AMPK Signaling Pathway and Promotes the High Speed Linear Motility of Dairy Goat Sperm. *Animals* **2023**, *13*, 1442. https://doi.org/10.3390/ani13091442.

3. Zhang, F.; Han, S.; Zhang, N.; Chai, J.; Xiong, Q. Effect of Ferulic Acid on Semen Quality of Goat Bucks during Liquid Storage at 17 °C. *Animals* **2023**, *13*, 2469. https://doi.org/10.3390/ani13152469.
4. Taheri-Khas, Z.; Gharzi, A.; Vaissi, S.; Heshmatzad, P.; Kalhori, Z. Hormone-Driven Temperature Optimization for Elevated Reproduction in Goldfish (*Carassius auratus*) under Laboratory Conditions. *Animals* **2024**, *14*, 2701. https://doi.org/10.3390/ani14182701.
5. Zhu, Z.; Zhao, H.; Cui, H.; Adetunji, A.O.; Min, L. Resveratrol Improves the Frozen-Thawed Ram Sperm Quality. *Animals* **2023**, *13*, 3887. https://doi.org/10.3390/ani13243887.
6. Pérez-Durand, M.G.; Bustamante, C.W.; Machaca, P.P.; García, W.; Condori, E.A.; Macedo, R.; Fernández, E.; Manrique, Y.P.; Gutiérrez-Reinoso, M.A.; Perez-Guerra, U.H.; et al. Effect of Three Semen Extenders on Sperm Quality and In Vitro Fertilization Rates of Fresh and Cryopreserved Sperm Collected from Llama *(Lama glama)* Vas Deferens. *Animals* **2024**, *14*, 1573. https://doi.org/10.3390/ani14111573.
7. Neuman, N.M.; Orzołek, A.; Steiner-Bogdaszewska, Ż.; Dziekońska, A. Changes in the Morphology and Antioxidant Status of European Red Deer Sperm Stored in the Epididymides and in a Liquid State. *Animals* **2024**, *14*, 1653. https://doi.org/10.3390/ani14111653.
8. Zhang, L.; Wang, X.; Jiang, C.; Sohail, T.; Sun, Y.; Sun, X.; Wang, J.; Li, Y. Effect of Different Dilution Methods and Ratios of Ram Semen on Sperm Parameters after Cryopreservation. *Animals* **2024**, *14*, 907. https://doi.org/10.3390/ani14060907.
9. Žaja, I.Ž.; Vince, S.; Butković, I.; Senaši, K.; Milas, N.P.; Malarić, K.; Lojkić, M.; Folnožić, I.; Tur, S.M.; Kreszinger, M.; et al. The Distribution of Boars Spermatozoa in Morphometrically Distinct Subpopulations after In Vitro Exposure to Radiofrequency Electromagnetic Radiation at 2500 MHz and Their Motility. *Animals* **2024**, *14*, 828. https://doi.org/10.3390/ani14060828.
10. Domrazek, K.; Konieczny, P.; Majka, M.; Czopowicz, M.; Jurka, P. The Impact of Microorganisms on Canine Semen Quality. *Animals* **2024**, *14*, 1267. https://doi.org/10.3390/ani14091267.
11. Sorkytė, Š.; Šiugždinienė, R.; Virgailis, M.; Vaičiulienė, G.; Wysokińska, A.; Wójcik, E.; Matusevičius, P.; Rekešiūtė, A.; Sutkevičienė, N. The Interaction between Canine Semen Bacteria and Semen Quality Parameters. *Animals* **2024**, *14*, 2151. https://doi.org/10.3390/ani14152151.
12. Sánchez-Rivera, U.Á.; Cruz-Cano, N.B.; Medrano, A.; Álvarez-Rodríguez, C.; Martínez-Torres, M. Sperm Incubation in Biggers–Whitten–Whittingham Medium Induces Capacitation-Related Changes in the Lizard Sceloporus torquatus. *Animals* **2024**, *14*, 1388. https://doi.org/10.3390/ani14091388.

References

1. Avanzi, B.R.; Farrás, M.C.; Melo, C.M.; Alvarenga, M.A.; Dell'Aqua, J.A.; Medeiros, A.S.L.; Araújo, G.H.M.; Papa, F.O. Efficiency of different cooling and storage systems for maintaining equine semen viability in a hot environment. *Anim. Reprod. Sci.* **2006**, *94*, 152–154. [CrossRef]
2. Macan, R.C.; Camargo, C.E.; Zielinski, B.L.; Cardoso, N.; Lara, N.; Bergstein-Galan, T.G.; Weiss, R.R.; Kozicki, L.E. Timed artificial insemination in crossbred mares: Reproductive efficiency and costs. *Reprod. Domest. Anim.* **2021**, *56*, 459–466. [CrossRef] [PubMed]
3. Morrell, J.M. Effect of colloid centrifugation on boar sperm quality during storage and function in in vitro fertilization. *Theriogenology* **2019**, *137*, 122–126. [CrossRef] [PubMed]
4. Yeste, M. Recent advances in boar sperm cryopreservation: State of the art and current perspectives. *Reprod. Domest. Anim.* **2015**, *50*, 71–79. [CrossRef] [PubMed]
5. García-Vázquez, F.A.; Mellagi, A.; Ulguim, R.R.; Hernández-Caravaca, I.; Llamas-López, P.J.; Bortolozzo, F.P. Post-cervical artificial insemination in porcine: The technique that came to stay. *Theriogenology* **2019**, *129*, 37–45. [CrossRef] [PubMed]
6. Salamon, S.; Maxwell, W.M. Storage of ram semen. *Anim. Reprod. Sci.* **2000**, *62*, 77–111. [CrossRef] [PubMed]
7. Yeste, M. State-of-the-art of boar sperm preservation in liquid and frozen state. *Anim. Reprod.* **2017**, *14*, 69–81. [CrossRef]
8. Wiebke, M.; Hensel, B.; Nitsche-Melkus, E.; Jung, M.; Schulze, M. Cooled storage of semen from livestock animals (part I): Boar, bull, and stallion. *Anim. Reprod. Sci.* **2022**, *246*, 106822. [CrossRef] [PubMed]
9. Maneesh, M.; Jayalekshmi, H. Role of reactive oxygen species and antioxidants on pathophysiology of male reproduction. *Indian J. Clin. Biochem.* **2006**, *21*, 80–89. [CrossRef] [PubMed]
10. Peris, S.I.; Bilodeau, J.F.; Dufour, M.; Bailey, J.L. Impact of cryopreservation and reactive oxygen species on DNA integrity, lipid peroxidation, and functional parameters in ram sperm. *Mol. Reprod. Dev.* **2007**, *74*, 878–892. [CrossRef] [PubMed]

11. Sutkeviciene, N.; Riskeviciene, V.; Januskauskas, A.; Zilinskas, H.; Andersson, M. Assessment of sperm quality traits in relation to fertility in boar semen. *Acta Vet. Scand.* **2009**, *51*, 53. [CrossRef] [PubMed]
12. Khosravizadeh, Z.; Khodamoradi, K.; Rashidi, Z.; Jahromi, M.; Shiri, E.; Salehi, E.; Talebi, A. Sperm cryopreservation and DNA methylation: Possible implications for ART success and the health of offspring. *J. Assist. Reprod. Genet.* **2022**, *39*, 1815–1824. [CrossRef] [PubMed]

Disclaimer/Publisher's Note: The statements, opinions and data contained in all publications are solely those of the individual author(s) and contributor(s) and not of MDPI and/or the editor(s). MDPI and/or the editor(s) disclaim responsibility for any injury to people or property resulting from any ideas, methods, instructions or products referred to in the content.

animals

MDPI

Article

Hormone-Driven Temperature Optimization for Elevated Reproduction in Goldfish (*Carassius auratus*) under Laboratory Conditions

Zeynab Taheri-Khas [1], Ahmad Gharzi [1,*], Somaye Vaissi [1,*], Pouria Heshmatzad [2] and Zahra Kalhori [1]

[1] Department of Biology, Faculty of Science, Razi University, Kermanshah 6714414971, Iran; zeynab.taheri93@gmail.com (Z.T.-K.); zahrakalhori1@yahoo.com (Z.K.)

[2] Department of Fisheries, Faculty of Fisheries and Environmental Sciences, Gorgan University of Agricultural Sciences and Natural Resources, Gorgan 4918943464, Iran; pheshmatzad@gmail.com

* Correspondence: ahgharzi@yahoo.com (A.G.); s.vaissi@razi.ac.ir (S.V.)

Simple Summary: This study investigated the use of hormones and temperature control to improve breeding success. Injecting Ovaprim significantly increased egg and sperm production. However, temperature played a critical role. A medium temperature (around 22 °C) produced the best results, with more eggs, faster egg release, and healthier sperm. Both low and high temperatures negatively impacted sperm quality and larval fish survival. Using extender E4 (15% DMSO) for cryopreservation improved fertilization rates. Overall, the study highlights the importance of precise hormone control and temperature management for successful goldfish reproduction, benefiting for fish farming.

Abstract: This study investigates the efficacy of hormone-induced artificial reproduction in goldfish (*Carassius auratus*) under controlled temperatures. Ovaprim injections significantly enhanced ovulation and sperm production compared to controls. Medium temperature (22 °C) produced the highest ovulation rates, fastest ovulation timing, and optimal sperm quality (motility and morphology) compared to high (28 °C) and low (16 °C) temperature groups. The low-temperature group exhibited reduced sperm motility duration and higher rates of sperm and larvae damage. The sperm volume of the high-temperature group was higher, but their post-injection survival rates were lower. Furthermore, the lowest spawning rate and low egg quality were noted in the high temperature. Cryopreservation using extender E4 (15% DMSO) exhibited superior post-thaw sperm motility and achieved higher fertilization rates. Fertilization rates, embryo development, and larval survival were all highest at the medium temperature. Larvae hatched from fresh sperm at medium temperature exhibited faster growth and fewer deformities. These findings suggest that hormone stimulation coupled with a medium temperature regimen is critical for successful artificial reproduction in goldfish. Cryopreservation with extender E4 holds promise for sperm banking; however, further optimization is necessary to improve fertilization success with thawed sperm. Future research could explore the influence of temperature on sperm physiology and refine cryopreservation protocols to enhance fertilization rates.

Keywords: temperature; motility; in vitro fertilization; cryopreservation; morphology

Citation: Taheri-Khas, Z.; Gharzi, A.; Vaissi, S.; Heshmatzad, P.; Kalhori, Z. Hormone-Driven Temperature Optimization for Elevated Reproduction in Goldfish (*Carassius auratus*) under Laboratory Conditions. *Animals* **2024**, *14*, 2701. https://doi.org/10.3390/ani14182701

Academic Editor: Anna Wysokińska

Received: 9 April 2024
Revised: 3 May 2024
Accepted: 17 May 2024
Published: 18 September 2024

Copyright: © 2024 by the authors. Licensee MDPI, Basel, Switzerland. This article is an open access article distributed under the terms and conditions of the Creative Commons Attribution (CC BY) license (https://creativecommons.org/licenses/by/4.0/).

1. Introduction

Reproduction of fish under controlled conditions necessitates specific environmental factors, including temperature, in addition to hormonal agent stimulation [1]. The synergy between these two factors enables the attainment of gamete maturation and the production of high-quality gametes and larvae in captivity [2]. It is noteworthy that optimal temperature ranges can vary among species and may be influenced by factors such as age, size, and reproductive state [3]. In the realm of aquaculture, the meticulous control and maintenance of appropriate water temperatures are not only crucial for hormonal regulation but also have a profound impact on overall growth and health. Achieving the right temperature ensures optimal growth and minimizes the risk of stress and disease [4]. Beyond its direct effects on growth and reproduction, temperature can also influence various facets of fish physiology and behavior, including metabolism, immune function, and swimming performance [5].

The goldfish (*Carassius auratus*), classified within the family Cyprinidae, serves as a significant subject for studies in reproductive biology. This is due to their relative ease of breeding and maintenance in captivity, as well as their well-defined reproductive cycle with distinct stages of gamete maturation and spawning, which can be easily observed and manipulated [6]. Goldfish is indeed a popular ornamental fish worldwide, including in Iran. Many individuals across different regions have taken up goldfish breeding as a venture, and it has proven to be a lucrative business for employers. Successful goldfish breeding requires knowledge of proper care, water conditions, and breeding techniques [7]. Natural reproduction for this species primarily occurs in the spring; however, under carefully controlled conditions, it exhibits the potential for reproduction during other seasons as well [8]. However, broodstocks often encounter reproductive failure during the off-season due to the lack of natural stimuli necessary for fish reproduction [9]. To address this issue, hormone therapy, facilitated by changes in the endocrine system specifically, the brain-pituitary-gonadal axis emerges as an efficient method for inducing sexual maturity and obtaining a significant quantity of gametes under controlled conditions [10]. Indeed, the application of hormonal stimulation in cyprinids has been shown to elevate the percentage of ovulation in females, facilitate the synchronization of spawning, and augment the number of eggs produced by a female [11,12]. In recent years, human chorionic gonadotropin hormone (HCG) and gonadotropin-releasing hormone (GnRH) have been successfully employed in the artificial reproduction of fish [13]. Ovaprim is a liquid peptide preparation consisting of a gonadotropin-releasing hormone analog and a dopamine inhibitor [14]. This hormone stimulates the release of mature oocytes or eggs in females and mature sperm in males [15]. Ovaprim is especially beneficial for species that encounter difficulties in initiating spontaneous spawning in captivity [16].

Sperm motility and morphology play pivotal roles in influencing the fertilization rate in fish [17]. In species with external fertilization, such as many fish, spermatozoa are released into the marine or freshwater environment, which not only serves as a challenging milieu but also provides signals that govern their motility [18]. Temperature emerges as a paramount environmental factor influencing sperm motility [19]. Aquatic organisms exhibit sensitivity to rapid temperature fluctuations, and deviations from the optimal range induce stress, impacting motility [20], viability [21], morphology [22], DNA [23], and ultimately, the reproductive process [24]. Sperm motility experiences an increase, to a certain extent, with rising temperatures [25]. However, exceeding a specific temperature threshold results in thermal shock, leading to the loss of sperm motility [26]. For instance, in *Acipenser baeri*, sperm motility decreases as the temperature rises from 10 to 17.5 °C [27]. In some species, like salmonids, swimming duration has demonstrated an inverse exponential relationship with water temperature [28]. Additionally, various studies have reported that lower temperatures bolster spermatozoon motility [29]. Understanding the optimal temperature conditions for fertilization holds the potential to enhance breeding programs and augment the success rate of artificial insemination [30].

Cryopreservation is a pivotal technique in the realm of aquaculture [31], offering the capacity for the long-term storage of sperm, eggs, and embryos, thereby providing a valuable resource for breeding programs [32]. The advantages of cryopreservation are manifold, encompassing the preservation of genetic diversity [33], heightened breeding program efficiency [34], and mitigated risks associated with live animal transportation [35,36]. However, cryopreservation is not exempt from its set of challenges. The freezing and thawing processes can inflict damage upon biological material, diminishing its viability and fertility [37]. To mitigate such damage, meticulous control over freezing protocols and the utilization of appropriate cryoprotectants are imperative. By prudently managing the associated risks and challenges, researchers and breeders stand to harness the benefits of cryopreservation, thereby enhancing the overall health and productivity of the fish population [38]. The success of cryopreservation is undoubtedly influenced by factors such as extender composition, cryoprotectant concentration, and freezing procedures [39]. The selection of extenders for freezing fish sperm is species-specific. Although saline and sugar solutions are commonly employed for cryopreserving several fish species [40,41], further investigations are warranted to explore sperm viability under freezing conditions, as well as the impact of diluents and cryoprotectants on sperm longevity [31].

Given the established efficacy of various hormonal interventions in optimizing fish reproduction and breeding, even outside their natural reproductive seasons, and recognizing the paramount significance of establishing precise environmental conditions, particularly temperature, this research aims to achieve the following objectives: (1) Determining the precise temperature range for optimizing hormone-induced reproductive performance. (2) Meticulously evaluating the impact of varying temperature regimes on spawning and sperm quality. (3) Focusing on improving cryopreservation protocols-based outcomes, specifically sperm motility percentage and fertilization with thawed sperm. (4) Examining the developmental progression of larvae under distinct temperature conditions. This comprehensive research approach aims to provide profound scientific insights into the field of artificial fish reproduction within controlled environments. Ultimately, it aims to advance our understanding of the intricate interplay between hormones, temperature fluctuations, and reproductive success in aquatic organisms.

2. Materials and Methods

2.1. Sample Procurement and Breeder Preparation

A total of 198 goldfish (*Carassius auratus*), averaging 2 years old, were obtained from an ornamental fish breeding center for this study. The fish consisted of 108 females and 90 males. Females averaged 35.12 g (±1.46) in weight and 13.42 cm (±1.64) in total length. Males averaged 33.26 g (±1.72) in weight and 11.71 cm (±0.99) in total length. All fish were individually housed in separate 100 L glass tanks with gentle aeration. Both males and females were individually and randomly housed in separate 100 L glass tanks equipped with gentle aeration. Fish were held in tanks containing continuously aerated (24 h) purified water treated with an aquarium filter system. Weekly, 80% of the water was renewed [42]. For the initial 7 days, all groups experienced identical lighting and temperature conditions (21–22 °C). The total test period lasted 16 days, with a consistent photoperiod of 10 h of light followed by 14 h of darkness [2]. Fish were fed a diet of carp pellets and mealworms throughout the experiment [43]. Water quality parameters (temperature, pH, and osmolality) were monitored daily to ensure they remained within optimal levels for all treatment groups [44]. To minimize potential contamination of sperm during collection, fish were fasted for two days prior to hormone injection [45].

2.2. Temperature Regimens and Conditions

Goldfish exhibit a wide range of temperature tolerance, surviving between 5 and 35 °C [46]. This study investigated the effects of temperature on hormone-induced reproduction. Three temperature groups were established: low (LT, 16 ± 1 °C), medium (MT, 22 ± 1 °C), and high (HT, 28 ± 1 °C) using aquarium heaters (Aquaria HT 300, IRAN AQUARIA, Tehran, Iran, corporation and research facility). Each group included 15 males and 18 females, alongside a control group without hormone injection (18 females, 15 males) at each temperature. After 24 h of temperature acclimation, all fish except the controls received hormone injections [47,48].

2.3. Hormone Injection Procedure and Preparation

Fish were anesthetized using a clove oil solution, a commonly used fish anesthetic, following established protocols (0.15 mL/liter) [49] The dosage of Ovaprim was determined following the manufacturer's instructions [50]. For females, a single dose ranging from 15 to 20 μL of Ovaprim was administered, while males received a single dose of 7 to 12 μL of Ovaprim. Prior to injection, the dosage was calculated based on the weight of the fish, and normal saline was added to achieve a volume of 100 μL (approximately 0.5 mL kg^{-1} for females and 0.25 mL kg^{-1} for males) for the injection [51]. Injections were performed intramuscularly using a 1 mL syringe and a 29-gauge needle [52]. Male and female fish, which had been exposed to controlled temperature groups for 24 h, were transferred to the respective temperature groups following hormone injection.

2.4. Sperm Collection Procedure

Sperm were collected from each male using a micropipette. Before sperm collection, the genital area of each male was dried with paper towels to prevent contamination from water, urine, and feces [53]. The collected milt was placed into 2 mL macro-tubes, and its volume, measured in microliters (μL) [54]. To ensure the collection of all possible sperm, the abdomen of each fish was stripped. Following sperm collection, these macro-tubes were immediately placed on ice [55].

2.5. Sperm Motility

Fresh sperm quality was evaluated based on motility. A 10 μL aliquot of sperm was diluted 1:100 with distilled water in a micro-tube [55]. A 10 μL subsample was then placed on a slide and covered with a coverslip. Sperm motility was assessed by counting motile sperm among 100 cells under a 400× magnification Zeiss AXIO Scope A1 microscope (Carl Zeiss, Oberkochen, Germany). Five replicates were performed per sample to minimize observer bias [56,57]. An experienced observer filmed sperm motility under the microscope for later evaluation. Motility duration was measured from the initial sperm movement after adding water to the sample until at least 70% of the sperm became immotile [58]. A chronometer was used to record this duration [58,59].

2.6. Sperm Morphology

The Diff-Quik staining method was employed for sperm morphology evaluation [60]. In this process, for fresh sperm samples obtained in each temperature treatment were prepared at a 1:100 ratio, distributed on a thin slide, and stained after drying [61]. For the frozen sperm sample, the stored sperms were thawed after one week on steam at 37 degrees [62], and after preparing the smear, they were stained with the indicated color. The lengths of the head, tail, and total length were measured. A total head length of 200 cells per slide (one slide per male) for male sperm in each temperature treatment were measured on a light microscope, at 40× magnification. Characteristics of all samples were evaluated by the same technician [63].

2.7. Oocyte Collection

Fish were checked for ovulation every thirty minutes after hormone injection by gently applying abdominal pressure. When ovulation was detected, the ovulated females were dried with a towel and weighed. The sampling was performed through stripping, which involves a gentle abdominal massage in the cephalocaudal direction [64]. The oocytes were collected in a dry beaker and then weighed with an accuracy of ±0.1 g to estimate the number of oocytes produced per gram of female [65].

2.8. Fresh Sperm-Egg Fertilization

Each female's ovulated eggs in each temperature treatment were manually stripped and collected individually into a dry beaker. After the oocyte stripping, a subjective (macroscopic) quality check was conducted, focusing on size uniformity, yellow color, and the absence of blood. Subsequently, sperm from each male was carefully collected using a micropipette. To enhance the quality of fertilization, sperm collected from 2–3 randomly selected males [2], were used. The number of eggs produced by each female was estimated based on egg mass weights and the number of eggs per gram of egg weight. According to the egg samples, there were approximately 900 eggs per gram. For the fertilization trial, in the MT and LT treatments, eggs extracted from each female sample were divided into several containers (each container containing 100 eggs) and mixed with the sperm of males (50 μL) from the same temperature treatment [66]. However, in the HT treatment, while a very small number of eggs were obtained, their quality was poor and unsuitable for fertilization. Therefore, no fertilization trial was conducted for the HT group. Followed by the addition of 700 mL of water to activate the sperm in the MT and LT groups. The sperm and eggs were allowed to contact each other for two minutes. Then, they were incubated in a container at the temperatures specified for MT and LT. To prevent the growth of microorganisms, 0.1 mL of methylene blue was added to the water in both MT and LT groups, and 50% of the water was replenished daily [67].

2.9. *Fertilization Evaluation*

The percentage of ovulation was calculated by dividing the number of fish producing eggs by the total number of fish in each treatment. Individual females were considered as replicates and were assigned a value of 1 for ovulation or 0 for no ovulation, respectively [68]. Relative fecundity was calculated in terms of eggs per gram of female body weight [69]. Latency time was defined as the time elapsed between female fish injection and egg stripping [70]. Furthermore, the egg survival rate up to the eyed-egg stage was used to estimate fertilization quality [71]. The percentage of eyed eggs was calculated by dividing the total number of eyed eggs by the total number of provided eggs and then multiplying the result by 100 [72].

2.10. *Sperm Cryopreservation*

Extenders were prepared using sugars such as glucose, fructose, and sucrose, along with chemicals like NaCl, KCl, and NaOH (further details can be found in Table 1). Before use, all extenders were adjusted to a pH of 7 and stored at 4 °C [73]. For each temperature group, sperm was diluted to a final concentration of 1:10 (*v*/*v*) in extenders containing the cryoprotectant dimethyl sulfoxide (DMSO) at concentrations of 15% [74]. Immediately after mixing sperm with extenders and cryoprotectants, a micropipette aliquot of 100 μL was placed into 0.25 mL cryostraws (IMV, FR). These straws were then subjected to 4 °C for 5 min, followed by exposure to liquid nitrogen (LN2) vapor at a height of 5 cm for 10 min, and finally plunged into LN2 [75,76]. Sperm specimens were thawed seven days later by immersing them in 35 °C vapor water for 5 s [77].

Table 1. The compositions of various extenders and cryoprotectants were used in the experiment.

Extenders (E)	NaCl (g)	KCl (g)	$C_6H_5O_7Na_3$, $2H_2O$ (g)	Glucose (g)	Fructose (g)	Sucrose (g)	NaOH * (μL)	Antibiotics ** (mL)	Distilled Water (mL)	Cryoprotectant DMSO (%)
E1	0.17	0.23	-	1.06	0.9	-	12	0.5	100	10
E2	0.34	-	-	-	-	3.43	21	0.5	100	10
E3	-	0.46	-	-	1.93	-	16	0.5	100	10
E4	0.4		0.8	2.05			-	0.5	100	15

*: NaOH solution (NaOH 1 g + distilled water 100 mL). **: Antibiotic 12,000 Unit/mL penicillin.

2.11. *Sperm Motility Post Thminawing*

To evaluate the quality of frozen sperm in each temperature treatment, and choosing the best diluent for fertilization with fresh eggs, after storage in liquid nitrogen dewar for 7 days, straws were thawed by swirling them in a 35 °C water bath for 7–10 s [77]. Then, 10 μL of it was placed on a glass slide coverslip, and its motility and duration of motility were checked with a light microscope 40×. Six replicate samples were used for each treatment [41].

2.12. Egg Fertilization Using Thawed Sperm

Following selection of the optimal extender based on post-thaw sperm motility, a fertilization test was conducted using thawed sperm. Similar to the fresh sperm fertilization process, eggs were collected from each temperature treatment and fertilized with sperm that had been frozen at each corresponding temperature. For thawed sperm fertilization, one or two thawed freezing straws were rapidly thawed by swirling them in a 35 °C water bath for 7–10 s. The thawed sperm was then carefully poured over the eggs [77]. The sealed end of the straw was then quickly cut with scissors, allowing the partially thawed milt-extender mixture to flow onto a Petri dish [77]. The thawed sperm was poured onto the eggs in the Petri dish. Subsequently, the dishes were incubated at the same temperature as the LT and MT groups. The percentage of eyed eggs, the total length of larvae on the seventh day, and the percentage of deformed larvae were assessed in each temperature group.

2.13. Larvae Evaluation following Fertilization

Seven days after fertilization, larval development was assessed in both the MT and LT groups. Larvae from fertilizations using both fresh and thawed sperm were measured for total length. Digital photographs were taken using a camera mounted on a tripod at a fixed height (20 cm). For each photo, larvae were placed in a Petri dish positioned over gridded paper to facilitate size measurement. Larval images were then analyzed using Digimizer software (version 6) to determine total length. Additionally, the number of deformed larvae in each container was counted and expressed as a percentage of the total number [78].

2.14. Statistical Analysis

The data's normal distribution was confirmed using both the Kolmogorov–Smirnov and Shapiro–Wilk tests. Nonparametric data (motility rate and motility duration of sperm) were analyzed using the Kruskal–Wallis test, followed by the Mann–Whitney U comparisons test and Univariate Analysis of Variance. Parametric data (relative fecundity, percentage of eyed-eggs, sperm morphology, sperm head length, larvae total length, and abnormal larvae) were analyzed using a one-way ANOVA, followed by Tukey's post hoc test. The data were expressed as mean ± standard deviation (SD) for all comparisons, with $p \leq 0.05$ considered statistically significant [59].

3. Results

Sperm production (spermatogenesis) continued for up to 24 h after hormone injection in all temperature groups. However, the time it took to collect usable sperm (latency time) varied significantly. Males in the HT group were the fastest, producing sperm within 4–5 h. Those in the MT group required 7–8 h, and the low temperature (LT) group took the longest (10–11 h). The average volume of collected sperm was as follows: 366.66 ± 49.23 µL for HT, 354.16 ± 62.00 µL for MT, and 304.16 ± 68.94 µL for LT ($p \leq 0.03$). Interestingly, the percentage of fish producing sperm (spermiation) was highest in the LT and MT groups (83.33% each), followed by the HT group (69.23%). Notably, no sperm were collected from males in the control groups (without hormone injection) at any temperature.

3.1. Sperm Motility

3.1.1. Fresh Sperm

The MT group demonstrated the longest duration of sperm motility (197.5 ± 15.44 s), followed by the HT group (177.50 ± 27.01 s) and the LT group (135.00 ± 15.66 s). Regarding motility rate (%) within the first 10 s, the MT treatment had the highest percentage (97.08% ± 2.57%), followed by the HT treatment (95.83% ± 2.88%) and the LT treatment (94.58% ± 3.34%). At 30 s, HT with MT (p = 0.03) and HT with LT (p = 0.01) were significant. MT with LT were significant (p = 0.01) at 60 s. At 90 s, MT with LT (p = 0.006) and HT with LT (p = 0.009) were significant, and at 120 s, HT with LT (p = 0.00) and MT with LT (p = 0.00) were significant. The effects of temperature (F = 20.89, df = 2, p = 0.00) and time (F = 1210.33, df = 6, p = 0.00) were found to be significant, both separately and in combination (F = 3.80, df = 12, p = 0.00), on sperm motility (Figure 1).

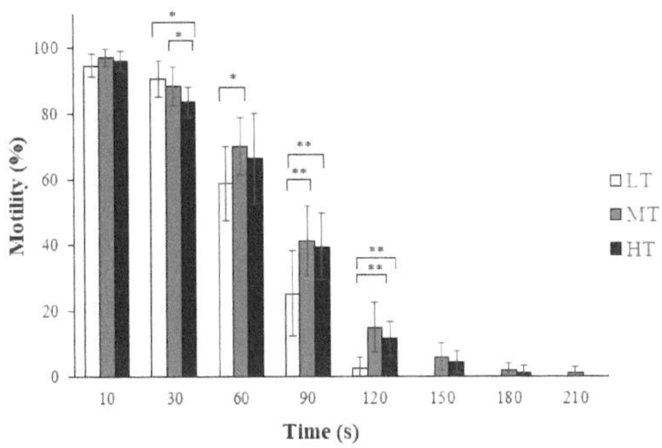

Figure 1. Percentage of goldfish (*Carassius auratus*) sperm motility at various temperatures. The temperatures are categorized as follows: 28 ± 1 °C (HT); 22 ± 1 °C (MT); and 16 ± 1 °C (LT). The values are mean ± SD. (*) p-value < 0.05; (**) p-value < 0.001.

3.1.2. Thawed Sperm

Sperm motility remained consistently high throughout one week in different temperature groups when sperm were cryopreserved in E4. However, other treatments showed average post-thaw sperm motility ranging from 0 to 50%. The MT-E4 treatment exhibited the highest motility rate (%) after thawing in E4, with a delay of 10 s (91.00% ± 4.18%), followed by HT-E4 (89.00% ± 4.18%) and LT-E4 (79.00% ± 4.18%) ($p \leq 0.00$). (Table 2). Furthermore, the MT-E4 (252.00 ± 16.43 s), LT-E4 (204.00 ± 13.41 s), and HT-E4 (180.00 ± 00 s) treatments demonstrated the longest duration of motility after thawing, respectively ($p \leq 0.00$).

Table 2. The results of artificial reproduction experiments conducted with goldfish (*Carassius auratus*) at three distinct temperature conditions. The data in the table are expressed as means with corresponding standard deviations (±SD).

Temperature	**16 ± 1 °C**	**22 ± 1 °C**	**28 ± 1 °C**	***p*-Value**
Female Weight (g)	35.15 ± 1.01	34.98 ± 2.55	35.23 ± 0.99	0.98
Ovulation (%)	66	80	13	0.001
Latency Time (h)	21–22	9–10	6–7	0.001
Relative Fecundity	46.99 ±16.84	57.84 ± 19.17	-	
Survival Rate until Eyed-Egg Stage	46.65 ± 9.98	65.80 ± 7.34	-	0.001
Percentage Deformed Larvae (%)	32.47 ± 3.09	8.10 ± 5.69	-	0.32
Total Length of Larvae (mm)	3.94 ±0.32	5.14 ± 0.29	-	0.001
Male Weight (g)	32.35 ± 1.83	34.41 ± 2.22	33.01 ± 0.56	0.37
Spermiation (%)	83.33	83.33	69.23	0.001
Sperm Motility (%)	94.58 ± 3.34	97.08 ± 2.57	95.83 ±2.88	0.001
Sperm Head Length (µm)	2.16 ± 0.25	2.19 ± 0.28	2.09 ± 0.26	0.001
Sperm Damage	33.16 ± 11.45	10.41 ± 5.61	15.66 ± 10.24	0.001
Survival Fish (%)	94.87	92.3	56.41	0.001
Post Thawing				
Sperm Motility for Extender 4 (E4) (%)	79.00 ± 4.18	91.00 ± 4.18	89.00 ± 4.18	0.001
Survival Rate until Eyed-Egg Stage	13.84 ± 1.23	40.00 ± 7.93	-	0.001
Percentage Deformed Larvae (E4) (%)	41.20 ± 9.43	17.07 ± 5.14		0.001
Total Length Larvae (mm)	3.54 ±0.11	4.78 ±0.52	-	0.001
Sperm Damage (E4)	47.91 ± 15.12	18.50 ± 6.78	27.00 ± 10.43	0.001

3.2. Sperm Morphology

3.2.1. Fresh Sperm

The analysis of sperm morphology in three different temperature groups revealed that LT group had the highest percentage of sperm damage (wrinkled and detached head, bent tail, coiled tail) (33.16 ± 11.45%). Following LT, the HT group exhibited a lower significant percentage of sperm damage (15.66 ± 10.24%). MT group had the lowest percentage of sperm damage (10.41 ± 5.61%) ($p \leq 0.00$). Additionally, measurements of sperm head length showed variations: 2.16 ± 0.25 µm in LT, 2.19 ± 0.28 µm in MT, and 2.09 ± 0.26 µm in HT ($p \geq 0.2$) (Figure 2).

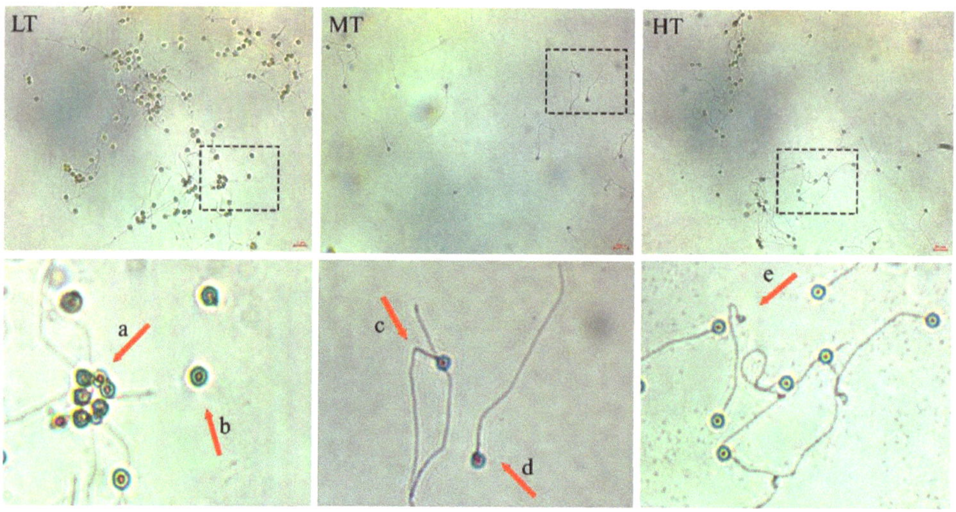

Figure 2. Sperm morphology of goldfish (*Carassius auratus*) in different temperature treatments. The temperatures are categorized as follows: 16 ± 1 °C (LT); 22 ± 1 °C (MT); and, 28 ± 1 °C (HT). (a) Head wrinkled; (b) detached head; (c) bent tail; (d) normal sperm; (e) coiled tail. (40× magnification).

3.2.2. Thawed Sperm

The analysis of sperm morphology in three different temperature groups for extender 4 revealed that LT group had the highest percentage of sperm damage (wrinkled and detached head, bent tail, coiled tail) (47.91 ± 15.12%). Following the HT group exhibited a lower significant percentage of sperm damage (27.00 ± 10.43%). MT group had the lowest percentage of sperm damage (18.50 ± 6.78%) ($p \leq 0.00$).

3.3. Fertilization Quality

3.3.1. Fresh Sperm

The highest ovulation rate was observed in the MT group (80%), followed by the LT group (66%), while the HT group exhibited the lowest rate (13%)) the eggs were released in small quantities on the aquarium floor. In the control group, where only temperature was applied without hormone injection, no spawning was observed. Among the temperature groups, the MT group had the shortest latency time for ovulation, with an average of 10 ± 1 h. In contrast, the LT group had a longer latency time, averaging 22 ± 1 h for ovulation. Relative fecundity, measured as eggs per gram of body weight (egg/g BW), was slightly higher in the MT group with an overall mean of 57.84 ± 19.17, compared to the LT group with 46.99 ± 16.84 ($p \geq 0.32$) (Table 2). The first appearance of eyed-eggs occurred earlier in the MT group, approximately 48 h after fertilization. Under the LT treatment, the first-eyed egg was observed much later, at around 120 h after fertilization. The MT group showed a significantly higher hatching rate, with 65.80 ± 7.34%, when considering hatched eyed-eggs between 48–72 h after fertilization, compared to the LT group with 46.65 ± 9.98% ($p \leq 0.004$). HT and hormone stimulation had a significant impact on fish survival rates. The HT group exhibited the lowest survival rates, with 56.41%. In contrast, the MT group showed highest survival rates, with 92.30%. The LT group had the survival rate of 94.87%.

3.3.2. Thawed Sperm

Similar to fresh sperm fertilization, the first appearance of eyed-eggs occurred 48 h after fertilization in the MT group when using E4. In contrast, in the LT group, the first-eyed egg was observed much later, at around 120 h after fertilization. When considering the percentage of eyed eggs, the MT group had the highest rate, with 40.00 ± 7.93%, followed by the LT group with 13.84 ± 1.23% ($p \leq 0.05$).

3.4. *Larvae following Fertilization*

3.4.1. Fresh Sperm

In the MT group, goldfish larvae exhibited the lowest percentage of deformed larvae after hatching (8.10 ± 5.69%), whereas the LT group had the highest percentage of deformed (32.47 ± 3.09%). On the seventh day after fertilization, the total length of the larvae was the greatest in the MT group (5.14 ± 0.29 mm), followed by the LT group (3.94 ± 0.32 mm) ($p \leq 0.00$).

3.4.2. Thawed Sperm

Concerning the MT group for Extender 4, goldfish larvae exhibited the lowest percentage of deformed larvae after hatching (17.07 ± 5.14%), whereas the LT group had Extender 4, the highest percentage of deformed larvae (41.20 ± 9.43%). The longest total length of larvae seven days after fertilization with frozen sperm was observed in the MT-E4 group (4.78 ± 0.52 mm), followed by the LT-E4 group (3.54 ± 0.11 mm). According to the ANOVA results, there were significant differences in larval length after thawing frozen sperm ($p \leq 0.00$).

4. Discussion

The results presented in this study demonstrated the intricate interplay between temperature and various aspects of fish reproduction, taking into account the pivotal influence of hormone injections. Specifically, we examined how temperature influences the fundamental processes of sperm and egg release, determining fertilization success, and subsequently affecting the development of fish larvae. Traditionally, goldfish breeding has been confined to specific seasons, with breeders typically maintaining a temperature range of 26–27 °C. However, attempts to induce breeding outside these customary seasons through temperature shocks often led to incomplete ovulation. In response to this challenge, we implemented hormone therapy, which proved to be a successful intervention.

In our study, we observed that maintaining an MT group proved to be the optimal condition for goldfish reproduction. The administration of Ovaprim hormone at 22 ± 1 °C resulted in an impressive 80% ovulation rate, signifying it as the most successful strategy for fertilization. This finding aligns with previous studies demonstrating the critical role of temperature in fish reproduction [48]. It is crucial to underscore the sensitivity of various fish species to rapid temperature changes, a factor that can significantly impact reproductive success. This underscores the importance of precise temperature management in aquaculture practices. Furthermore, the research conducted by Nowosad et al. [2], on *Barbus barbus* builds upon our understanding of optimal thermal conditions and the effectiveness of hormone interventions. Their results highlight that the use of CPH (common carp pituitary homogenate) and Ovaprim hormones yielded exceptional outcomes, achieving ovulation percentages ranging from 90% to 100%, with embryo survival rates at hatching reaching approximately 90%. This reinforces the potential benefits of hormone-based interventions in fish reproduction, albeit with species-specific considerations.

The observed outcomes from our experiment shed light on the significant influence of temperature on various aspects of goldfish reproduction. Under the MT condition, we noted a latency time of approximately 9 h and a hatching rate of 65.80%. In contrast, under the LT condition, the latency period extended to around 21 h, accompanied by a hatching rate of 46.65%. Interestingly, under the HT condition, the latency period was notably shorter, around 6 h. However, it is worth noting that our findings also reveal a significant limitation in the effectiveness of hormone injections at temperatures exceeding 27 °C, particularly concerning ovulation induction and survival rates in goldfish; as previously mentioned, a majority of the females failed to ovulate at this temperature, with only a few releasing a limited quantity of unfertilized eggs under abdominal pressure, which were of suboptimal quality in terms of color and size. Therefore, precise monitoring and strict control of both hormone levels and temperature are imperative when striving to optimize reproductive outcomes in aquaculture practices

This study showed that despite administering nearly identical hormone doses across male and female groups, the temperature variations yielded distinct effects on these reproductive parameters. This finding aligns with the research by Servili et al. [79] who emphasized the critical importance of the appropriate combination of inducing agent dosage and latency period in achieving optimal egg production in catfish *Clarias batrachus*. Similarly, a study by Dhara et al. [80] on *Clarias batrachus* highlighted that the highest rates of fertilization and hatching were achieved when fish were injected with carp pituitary gland extracts and kept at a temperature of 28 °C, with a latency period of 15 h. Our present study underscores that HT does not necessarily facilitate successful egg release post-hormone injection, and conversely, LT is unsuitable for fertilization in aquaculture due to prolonged latency times.

Moreover, a body of research has indicated that temperature fluctuations can either expedite or delay the growth and development of eggs and larvae [81]. For instance, Akatsu et al. [82] discovered enhanced growth in *Epinephelus tauvina* larvae at higher temperatures, while Kupren et al. [83] found that embryos of *Leuciscus leuciscus* and *L. idus* exhibited tolerance to temperatures up to 23 °C, albeit with reduced survival rates and an increased occurrence of deformities. In our current experiment, we observed that fertilized eggs with both fresh and frozen sperm became eyed-eggs in the MT group simultaneously, but this process was delayed in the LT temperature group. Additionally, eggs cultivated under MT treatment displayed normal growth compared to those under LT treatment, with a significantly lower percentage of abnormal larvae. These findings emphasize that the effects of temperature on fish reproduction are highly contingent on species, developmental stage, and prevailing environmental conditions.

Global warming presents a complex and multifaceted threat to fish reproduction, with cascading effects on marine ecosystems. Rising temperatures disrupt spawning cycles, particularly for species with limited geographic range [84]. Larval fish, especially vulnerable due to their sensitivity to environmental fluctuations, may face significant challenges under a changing climate. Studies suggest that temperature has a stronger influence on fish reproduction than elevated CO_2 levels [85]. The ability of fish populations to adapt to these changing conditions will be critical for their long-term survival [86,87]. However, rising temperatures act as stressors, impacting fish physiology, metabolism, and behavior, potentially compromising their capacity to cope with additional environmental challenges like ocean acidification and salinity changes [88]. Understanding the combined effects of temperature and other stressors on fish stress physiology is crucial for predicting the consequences of global warming on the persistence of fish populations.

It is well established that sperm morphology is a critical factor influencing fertilization success in fish [18]. The optimal shape and size of the sperm head are essential for successful penetration through the micropyle [89]. Although research on the effects of temperature and hormone stimulation on sperm morphology in fish is relatively limited, a study by Fenkes et al. [90] on *Salmo trutta* revealed that temperature has a significant impact on head size and tail length, which in turn affect fertilization success. Considering our findings, where we observed the least damage associated with the MT treatment, which also yielded the highest success rate compared to other treatments, it can be inferred that unfavorable temperatures not only affect sperm motility but also impact the shape and size of the sperm. This reinforces the notion that temperature plays a multifaceted role in influencing reproductive processes. The reduced percentage of sperm damage in the MT group suggests that optimal temperatures may promote better sperm quality, potentially enhancing fertilization success.

Our study highlights the pivotal role of cryoprotective chemicals and extenders in safeguarding sperm viability under low-temperature conditions. Dimethyl sulfoxide (DMSO) and carbohydrates have emerged as highly effective agents for cryopreserving sperm in freshwater species [91]. Building upon this knowledge, the authors of [41] conducted an extensive investigation in common carp, underscoring the potency of CCSE 2 and DMSO in maintaining post-thaw sperm motility at an impressive 94.53%. Traditionally, the success of fish sperm cryopreservation has been evaluated based on fertilization yield, primarily focusing on motility metrics. In our experiment, Extender 4 proved to be the optimal dilution for carp sperm, displaying the highest motility percentage, longest duration, and minimal morphological damage. This selection has paved the way for successful post-thaw fertilization. Studies by [92,93] provided valuable insights into the impact of cryopreservation on sperm morphology in zebrafish and catfish. Our findings further emphasize the importance of tailoring cryopreservation solutions to the specific requirements of each species. Extender 4 exhibited fewer morphological alterations, solidifying its suitability for carp sperm preservation. Moreover, Extender 4 maintained a remarkable motility rate of over 90% even after a week, showcasing its potential for long-term preservation of highly motility and minimally damaged sperm.

5. Conclusions

This study explored the interplay between temperature, hormones, and goldfish reproduction. Precise temperature control was crucial, with Ovaprim at 22 °C achieving an 80% ovulation rate. Synchronization of temperature and hormones is essential to avoid breeding failure. The research also identified temperature's influence on various reproductive aspects and the importance of sperm morphology in fertilization success. An optimal extender solution for carp sperm preservation was discovered, showing promise for long-term storage and breeding strategies. These findings offer valuable insights for optimizing aquaculture practices and potentially contribute to fish conservation efforts.

Author Contributions: Conceptualization, Z.T.-K., S.V., A.G., P.H. and Z.K.; data curation, Z.T.-K. and S.V.; formal analysis, Z.T.-K., A.G. and S.V.; investigation, Z.T.-K., S.V., P.H. and Z.K.; methodology, Z.T.-K., A.G., S.V., P.H. and Z.K.; project administration, S.V.; supervision, A.G. and S.V.; writing—original draft, Z.T.-K. and S.V.; writing—review and editing, S.V., Z.T.-K., A.G., P.H. and Z.K. All authors have read and agreed to the published version of the manuscript.

Funding: This research received no external funding.

Institutional Review Board Statement: All procedures in this study were conducted in accordance with the Animal Care and Use Committees of Razi University [IR.RAZI.REC.1402.064] approved protocols.

Informed Consent Statement: Informed consent was obtained from all subjects involved in the study.

Data Availability Statement: The data presented in this study are contained within the article.

Conflicts of Interest: The authors declare no conflicts of interest.

References

1. Mylonas, C.C.; Fostier, A.; Zanuy, S. Broodstock Management and Hormonal Manipulations of Fish Reproduction. *Gen. Comp Endocrinol.* **2010**, *165*, 516–534. [CrossRef]
2. Nowosad, J.; Kucharczyk, D.; Targońska, K.; Wyszomirska, E.; Chwaluczy, R.; Kupren, K. The Synergistic Effect of Temperature and Hormonal Stimulation on Spawning Efficiency of Common Barbel, *Barbus barbus* L. *Turkish J. Fish. Aquat. Sci.* **2016**, *16* 517–524.
3. Jonsson, B.; Jonsson, N. A Review of the Likely Effects of Climate Change on Anadromous Atlantic Salmon *Salmo salar* and Brown Trout *Salmo trutta*, with Particular Reference to Water Temperature and Flow. *J. Fish Biol.* **2009**, *75*, 2381–2447. [CrossRef] [PubMed]
4. Sallenave, R. *Important Water Quality Parameters in Aquaponics Systems*; College of Agricultural, Consumer and Environmental Sciences: Las Cruces, NM, USA, 2016.
5. Kim, J.-H.; Park, H.-J.; Kim, K.-W.; Hwang, I.-K.; Kim, D.-H.; Oh, C.W.; Lee, J.S.; Kang, J.-C. Growth Performance, Oxidative Stress, and Non-Specific Immune Responses in Juvenile Sablefish, Anoplopoma Fimbria, by Changes of Water Temperature and Salinity. *Fish Physiol. Biochem.* **2017**, *43*, 1421–1431. [CrossRef] [PubMed]
6. Stacey, N.; Fraser, E.J.; Sorensen, P.; Van Der Kraak, G. Milt Production in Goldfish: Regulation by Multiple Social Stimuli. *Comp. Biochem. Physiol. Part C Toxicol. Pharmacol.* **2001**, *130*, 467–476. [CrossRef]
7. Tarkhani, R.; Imanpoor, M.R.; Taghizadeh, V. Effect of Dietary 17 β-Estradiol on Serum Sex Hormones' Levels and Gamete Quality in Goldfish (*Carassius auratus*). *J. Comp. Clin. Pathol. Res.* **2012**, *2252*, 422.
8. PASMANIK, M.; CALLARD, G. V Changes in Brain Aromatase and 5α-Reductase Activities Correlate Significantly with Seasonal Reproductive Cycles in Goldfish (*Carassius auratus*). *Endocrinology* **1988**, *122*, 1349–1356. [CrossRef] [PubMed]
9. Zohar, Y.; Mylonas, C.C. Endocrine Manipulations of Spawning in Cultured Fish: From Hormones to Genes. In *Reproductive Biotechnology in Finfish Aquaculture*; Elsevier: Amsterdam, The Netherlands, 2001; pp. 99–136.
10. Ghanem, S.F. Effect of Endocrine Disrupting Chemicals Exposure on Reproduction and Endocrine Functions Using the Zebrafish Model. *Egypt. J. Aquat. Biol. Fish.* **2021**, *25*, 951–981. [CrossRef]
11. Zarski, D.; Kucharczyk, D.; Targonska, K.; Jamróz, M.; Krejszeff, S.; Mamcarz, A. Application of Ovopel and Ovaprim and Their Combinations in Controlled Reproduction of Two Reophilic Cyprinid Fish Species. *Polish J. Nat. Sci.* **2009**, *4*, 235–244. [CrossRef]
12. Mehdi, Y.; Mousavi, S.E.; Mehdi, Y.; Mousavi, S.E. A Review of the Control of Reproduction and Hormonal Manipulations in Finfish Species. *African J. Agric. Res.* **2011**, *6*, 1643–1650.
13. Abdel-Latif, H.M.R.; Shukry, M.; Saad, M.F.; Mohamed, N.A.; Nowosad, J.; Kucharczyk, D. Effects of GnRHa and HCG with or without Dopamine Receptor Antagonists on the Spawning Efficiency of African Catfish (*Clarias gariepinus*) Reared in Hatchery Conditions. *Anim. Reprod. Sci.* **2021**, *231*, 106798. [CrossRef] [PubMed]
14. Afzal, M.; Rab, A.; Akhtar, N.; Khan, M.F.; Khan, S.U.; Qayyum, M. Induced Spawning of Bighead Carp, *Aristichthys nobilis* (Richardson), by Using Different Hormones/Hormonal Analogues. *Pak. J. Zool.* **2008**, *40*, 283–287.
15. Nargesi, E.A.; Falahatkar, B.; Żarski, D.; Gorouhi, D. The Effectiveness of Ovaprim, Ovopel, and Their Combinations in Artificial Reproduction of Common Rudd *Scardinius erythrophthalmus* under Controlled Conditions. *Theriogenology.* **2023**, *199*, 114–120. [CrossRef] [PubMed]
16. Hill, J.E.; Kilgore, K.H.; Pouder, D.B.; Powell, J.F.F.; Watson, C.A.; Yanong, R.P.E. Survey of Ovaprim Use as a Spawning Aid in Ornamental Fishes in the United States as Administered through the University of Florida Tropical Aquaculture Laboratory. *N. Am. J. Aquac.* **2009**, *71*, 206–209. [CrossRef]
17. Van Look, K.J.W.; Kime, D.E. Automated Sperm Morphology Analysis in Fishes: The Effect of Mercury on Goldfish Sperm. *J. Fish Biol.* **2003**, *63*, 1020–1033. [CrossRef]
18. Cosson, J. *Fish Sperm Physiology: Structure, Factors Regulating Motility, and Motility Evaluation*; IntechOpen: London, UK, 2019; Volume 1, pp. 1–26.
19. Alavi, S.M.H.; Cosson, J. Sperm Motility in Fishes. I. Effects of Temperature and PH: A Review. *Cell Biol. Int.* **2005**, *29*, 101–110. [CrossRef] [PubMed]
20. Merino, O.; Risopatrón, J.; Valdebenito, I.; Figueroa, E.; Farías, J.G. Effect of the Temperature of Activation Medium on Fish Sperm Quality: Impact on Fertilization in Vitro in Aquaculture Practice. *Rev. Aquac.* **2023**, *15*, 434–451. [CrossRef]
21. Geffroy, B.; Sandoval-Vargas, L.; Boyer-Clavel, M.; Pérez-Atehortúa, M.; Lallement, S.; Isler, I.V. A Simulated Marine Heatwave Impacts European Sea Bass Sperm Quantity, but Not Quality. *J. Fish Biol.* **2023**, *103*, 784–789. [CrossRef]
22. Dadras, H.; Dzyuba, B.; Cosson, J.; Golpour, A.; Siddique, M.A.M.; Linhart, O. Effect of Water Temperature on the Physiology of Fish Spermatozoon Function: A Brief Review. *Aquac. Res.* **2017**, *48*, 729–740. [CrossRef]
23. Brionne, A.; Goupil, A.-S.; Kica, S.; Lareyre, J.-J.; Labbé, C.; Laurent, A. Spermatozoa Methylome and Its Sensitivity to Water Temperature in a Teleost Fish. *Sci. Total Environ.* **2023**, *892*, 164077. [CrossRef]
24. Donelson, J.M.; Munday, P.L.; McCormick, M.I.; Pankhurst, N.W.; Pankhurst, P.M. Effects of Elevated Water Temperature and Food Availability on the Reproductive Performance of a Coral Reef Fish. *Mar. Ecol. Prog. Ser.* **2010**, *401*, 233–243. [CrossRef]
25. Rurangwa, E.; Kime, D.E.; Ollevier, F.; Nash, J.P. The Measurement of Sperm Motility and Factors Affecting Sperm Quality in Cultured Fish. *Aquaculture* **2004**, *234*, 1–28. [CrossRef]
26. Geffen, A.J.; Frayer, O. Retention of Sperm Motility in Turbot, Scophthalmus Maximus L.: The Effects of Time from Activation, Thermal Shock and Adenosine Triphosphate Levels. *Aquac. Res.* **1993**, *24*, 203–209. [CrossRef]

27. Williot, P.; Kopeika, E.F.; Goncharov, B.F. Influence of Testis State, Temperature and Delay in Semen Collection on Spermatozoa Motility in the Cultured Siberian Sturgeon (*Acipenser baeri* Brandt). *Aquaculture* **2000**, *189*, 53–61. [CrossRef]
28. Lahnsteiner, F.; Mansour, N. The Effect of Temperature on Sperm Motility and Enzymatic Activity in Brown Trout Salmo Trutta, Burbot Lota Lota and Grayling Thymallus Thymallus. *J. Fish Biol.* **2012**, *81*, 197–209. [CrossRef]
29. Bombardelli, R.A.; Sanches, E.A.; Baggio, D.M.; Sykora, R.M.; de Souza, B.E.; Tessaro, L.; Piana, P.A. Effects of the Spermatozoa: Oocyte Ratio, Water Volume and Water Temperature on Artificial Fertilization and Sperm Activation of Cascudo-Preto. *Rev. Bras. Zootec.* **2013**, *42*, 1–6. [CrossRef]
30. Mehlis, M.; Bakker, T. The Influence of Ambient Water Temperature on Sperm Performance and Fertilization Success in Three-Spined Sticklebacks (*Gasterosteus aculeatus*). *Evol. Ecol.* **2014**, *28*, 655–667. [CrossRef]
31. Ekici, A.; Yamaner, G.; Demircan, M.D. Cryopreservation Studies in Aquaculture from Past to Present: Scientific Techniques and Quality Controls for Commercial Applications. In *Cryopreservation-Applications and Challenges*; IntechOpen: London, UK, 2022.
32. Hassan, M.M.; Qin, J.G.; Li, X. Sperm Cryopreservation in Oysters: A Review of Its Current Status and Potentials for Future Application in Aquaculture. *Aquaculture* **2015**, *438*, 24–32. [CrossRef]
33. Howard, J.G.; Lynch, C.; Santymire, R.M.; Marinari, P.E.; Wildt, D.E. Recovery of Gene Diversity Using Long-term Cryopreserved Spermatozoa and Artificial Insemination in the Endangered Black-footed Ferret. *Anim. Conserv.* **2016**, *19*, 102–111. [CrossRef]
34. Judycka, S.; Nynca, J.; Ciereszko, A. Opportunities and Challenges Related to the Implementation of Sperm Cryopreservation into Breeding of Salmonid Fishes. *Theriogenology* **2019**, *132*, 12–21. [CrossRef]
35. Fraser, D.J. How Well Can Captive Breeding Programs Conserve Biodiversity? A Review of Salmonids. *Evol. Appl.* **2008**, *1*, 535–586. [CrossRef]
36. Cabrita, E.; Sarasquete, C.; Martínez-Páramo, S.; Robles, V.; Beirão, J.; Pérez-Cerezales, S.; Herráez, M.P. Cryopreservation of Fish Sperm: Applications and Perspectives. *J. Appl. Ichthyol.* **2010**, *26*, 623–635. [CrossRef]
37. Torres, L.; Hu, E.; Tiersch, T.R. Cryopreservation in Fish: Current Status and Pathways to Quality Assurance and Quality Control in Repository Development. *Reprod. Fertil. Dev.* **2016**, *28*, 1105–1115. [CrossRef]
38. Diwan, A.D.; Harke, S.N.; Gopalkrishna; Panche, A.N. Cryobanking of Fish and Shellfish Egg, Embryos and Larvae: An Overview. *Front. Mar. Sci.* **2020**, *7*, 251. [CrossRef]
39. DeGraaf, J.D.; Berlinsky, D.L. Cryogenic and Refrigerated Storage of Atlantic Cod (*Gadus morhua*) and Haddock (*Melanogrammus aeglefinus*) Spermatozoa. *Aquaculture* **2004**, *234*, 527–540. [CrossRef]
40. Suquet, M.; Dreanno, C.; Fauvel, C.; Cosson, J.; Billard, R. Cryopreservation of Sperm in Marine Fish. *Aquac. Res. Orig. Artic.* **2000**, *31*, 231–243. [CrossRef]
41. Irawan, H.; Vuthiphandchai, V.; Nimrat, S. The Effect of Extenders, Cryoprotectants and Cryopreservation Methods on Common Carp (*Cyprinus carpio*) Sperm. *Anim. Reprod. Sci.* **2010**, *122*, 236–243. [CrossRef] [PubMed]
42. Minaz, M. *The Potential of Ornamental Fish Culture in Biofloc Technology with Different C/N Ratio and Multi-Criteria Decision Making Model: An Example of Goldfish (Carassius auratus)*; Research Square: Durham, NC, USA, 2024.
43. Targońska, K.; Kucharczyk, D. The Application of HCG, CPH and Ovopel in Successful Artificial Reproduction of Goldfish (*Carassius auratus auratus*) under Controlled Conditions. *Reprod. Domest. Anim.* **2011**, *46*, 651–655. [CrossRef] [PubMed]
44. Dumorne, K.; Valdebenito, I.; Contreras, P.; Rodríguez, P.U.; Risopatrón, J.; Figueroa, E.; Estévez, M.L.; Díaz, R.; Farías, J. Effect of PH, Osmolality and Temperature on Sperm Motility of Pink Cusk-Eel (*Genypterus blacodes*, (Forster, 1801)). *Aquac. Rep.* **2018**, *11*, 42–46. [CrossRef]
45. Ulikowski, D. European Catfish (*Silurus glanis* L.) Reproduction Outside of the Spawning Season. *Fish. Aquat. Life* **2004**, *12*, 121–131.
46. Ford, T.; Beitinger, T.L. Temperature Tolerance in the Goldfish, Carassius Auratus. *J. Therm. Biol.* **2005**, *30*, 147–152. [CrossRef]
47. Targońska, K.; Kucharczyk, D.; Kujawa, R.; Mamcarz, A.; Żarski, D. Controlled Reproduction of Asp, *Aspius aspius* (L.) Using Luteinizing Hormone Releasing Hormone (LHRH) Analogues with Dopamine Inhibitors. *Aquaculture* **2010**, *306*, 407–410. [CrossRef]
48. Nowosad, J.; Targońska, K.; Chwaluczyk, R.; Kaszubowski, R.; Kucharczyk, D. Effect of Temperature on the Effectiveness of Artificial Reproduction of Dace [Cyprinidae (*Leuciscus leuciscus* (L.))] under Laboratory and Field Conditions. *J. Therm. Biol.* **2014**, *45*, 62–68. [CrossRef]
49. Abdolazizi, S.; Ghaderi, E.; Naghdi, N.; Kamangar, B.B. Effects of Clove Oil as an Anesthetic on Some Hematological Parameters of *Carassius auratus*. *J. Aquac. Res. Dev.* **2011**, *2*, 108. [CrossRef]
50. Acharjee, A.; Chaube, R.; Joy, K.P. Ovaprim, a Commercial Spawning Inducer, Stimulates Gonadotropin Subunit Gene Transcriptional Activity: A Study Correlated with Plasma Steroid Profile, Ovulation and Fertilization in the Catfish Heteropneustes Fossilis. *Gen. Comp. Endocrinol.* **2017**, *251*, 66–73. [CrossRef]
51. Olumuji, O.K.; Mustapha, M.K. Induced Breeding of African Mud Catfish, Clarias Gariepinus (Burchell 1822), Using Different Doses of Normal Saline Diluted Ovaprim. *J. Aquac. Res. Dev.* **2012**, *3*, 1–4.
52. Black, B.J.; Black, M. Efficacy of Two Exogenous Hormones (GnRHa and HCG) for Induction of Spontaneous Spawning in Captive Yellowfin Bream, *Acanthopagrus australis* (Sparidae) and Influence of Sex Ratio on Spawning Success. *Aquaculture* **2013**, *416*, 105–110. [CrossRef]
53. Carolsfeld, J.; Godinho, H.P.; Zaniboni Filho, E.; Harvey, B.J. Cryopreservation of Sperm in Brazilian Migratory Fish Conservation. *J. Fish Biol.* **2003**, *63*, 472–489. [CrossRef]

54. Kása, E.; Bernáth, G.; Kollár, T.; Żarski, D.; Lujić, J.; Marinović, Z.; Bokor, Z.; Hegyi, Á.; Urbányi, B.; Vílchez, M.C. Development of Sperm Vitrification Protocols for Freshwater Fish (Eurasian perch, *Perca fluviatilis*) and Marine Fish (European Eel, *Anguilla anguilla*). *Gen. Comp. Endocrinol.* **2017**, *245*, 102–107. [CrossRef]
55. Hossen, S.; Kim, S.C.; Cho, Y.; Kho, K.H. Vital Analysis of Cryopreserved Sperm of Marbled Flounder, Pseudopleuronectes Yokohamae. *Front. Physiol.* **2021**, *12*, 696737. [CrossRef]
56. Abinawanto, A.; Yimastria, S.; Pertiwi, P. Sperm Analysis of Lukas Fish (*Puntius bramoides*): Motility, Viability and Abnormalities. In *AIP Conference Proceedings*; AIP Publishing: Depok, Indonesia, 2018.
57. Bisai, K.; Kumar, V.; Roy, A.; Parida, S.N.; Dhar, S.; Das, B.K.; Behera, B.K.; Pati, M.K. Effects of Di-(2-Ethylhexyl) Phthalate (DEHP) on Gamete Quality Parameters of Male Koi Carp (*Cyprinus carpio*). *Curr. Issues Mol. Biol.* **2023**, *45*, 7388–7403. [CrossRef]
58. Tunçelli, G.; Memiş, D. Effects of Constant and Changing Water Temperature on Sperm Quality of the Endangered Çoruh Trout, Salmo Coruhensis. *J. Appl. Aquac.* **2021**, *33*, 209–220. [CrossRef]
59. Caldas, J.S.; Godoy, L. Sperm Characterization of the Endangered Amazonian Fish Hypancistrus Zebra: Basic Knowledge for Reproduction and Conservation Strategies. *Anim. Reprod. Sci.* **2019**, *204*, 117–124. [CrossRef]
60. Tuset, V.M.; Dietrich, G.J.; Wojtczak, M.; Słowińska, M.; De Monserrat, J.; Ciereszko, A. Comparison of Three Staining Techniques for the Morphometric Study of Rainbow Trout (*Oncorhynchus mykiss*) Spermatozoa. *Theriogenology* **2008**, *69*, 1033–1038. [CrossRef]
61. Rodrigues, J.; Dos Santos, R.S.; Cordeiro, J.G.; Leite, M.; e Oliveira, H.S.T.O.; Tercya, H.; Costa, B.P.D.; do Nascimento, N.F.; Maximino, C.; de Siqueira-Silva, D.H. Seminal Characterization of the Amazonian Fire-Eye Tetra Moenkhausia Oligolepis (Günther, 1864). *Zygote* **2020**, *28*, 453–458. [CrossRef]
62. Diwan, A.D.; Ayyappan, S.; Lal, K.K.; Lakra, W.S. Cryopreservation of Fish Gametes and Embryos. *Indian J. Anim. Sci.* **2010**, *80*, 109–124.
63. Viveiros, A.T.M.; Orfão, L.H.; Nascimento, A.F.; Corrêa, F.M.; Caneppele, D. Effects of Extenders, Cryoprotectants and Freezing Methods on Sperm Quality of the Threatened Brazilian Freshwater Fish Pirapitinga-Do-Sul Brycon Opalinus (Characiformes). *Theriogenology* **2012**, *78*, 361–368. [CrossRef]
64. Chen, Y.F. Induced Ovulation and Embryonic Development of Ocellated Puffer, Takifugu Ocellatus. *J. Appl. Ichthyol.* **2005**, *21*, 136–140. [CrossRef]
65. Sanches, E.A.; Caneppele, D.; Okawara, R.Y.; Damasceno, D.Z.; Bombardelli, R.A.; Romagosa, E. Inseminating Dose and Water Volume Applied to the Artificial Fertilization of *Steindachneridion parahybae* (Steindachner, 1877)(Siluriformes: Pimelodidae): Brazilian Endangered Fish. *Neotrop. Ichthyol.* **2016**, *14*, e140158. [CrossRef]
66. Żarski, D.; Horváth, A.; Bernáth, G.; Palińska-Żarska, K.; Krejszeff, S.; Müller, T.; Kucharczyk, D. Application of Different Activating Solutions to in Vitro Fertilization of Crucian Carp, *Carassius carassius* (L.), Eggs. *Aquac. Int.* **2014**, *22*, 173–184. [CrossRef]
67. Varga, Z.M. Aquaculture and Husbandry at the Zebrafish International Resource Center. *Methods Cell Biol.* **2011**, *104*, 453–478.
68. Kristanto, A.H.; Umali, G.; Beam, R.; Dunham, R.A. Effect of Postmanufacturing Processing and Shipping of Luteinizing Hormone Releasing Hormone Analog on Induced Ovulation for Production of Channel Catfish Female × Blue Catfish Male Hybrid Fry. *N. Am. J. Aquac.* **2009**, *71*, 307–311. [CrossRef]
69. Kjesbu, O.S.; Witthames, P.R.; Solemdal, P.; Walker, M.G. Temporal Variations in the Fecundity of Arcto-Norwegian Cod (*Gadus morhua*) in Response to Natural Changes in Food and Temperature. *J. Sea Res.* **1998**, *40*, 303–321. [CrossRef]
70. Legendre, M. Seasonal Changes in Sexual Maturity and Fecundity, and HCG-Induced Breeding of the Catfish, *Heterobranchus longifilis* Val.(Clariidae), Reared in Ebrie Lagoon (Ivory Coast). *Aquaculture* **1986**, *55*, 201–213. [CrossRef]
71. Targońska, K.; Żarski, D.; Müller, T.; Krejszeff, S.; Kozłowski, K.; Demeny, F.; Urbanyi, B.; Kucharczyk, D. Controlled Reproduction of the Crucian Carp *Carassius carassius* (L.) Combining Temperature and Hormonal Treatment in Spawners. *J. Appl. Ichthyol.* **2012**, *28*, 894–899. [CrossRef]
72. Arndt, R.E.; Wagner, E.J.; Routledge, M.D. Reducing or Withholding Hydrogen Peroxide Treatment during a Critical Stage of Rainbow Trout Development: Effects on Eyed Eggs, Hatch, Deformities, and Fungal Control. *N. Am. J. Aquac.* **2001**, *63*, 161–166. [CrossRef]
73. Lahnsteiner, F.; Berger, B.; Horvath, A.; Urbányi, B.; Weismann, T. Cryopreservation of Spermatozoa in Cyprinid Fishes. *Theriogenology* **2000**, *54*, 1477–1498. [CrossRef] [PubMed]
74. Zidni, I.; Lee, Y.H.; Park, J.Y.; Lee, H.B.; Hur, J.W.; Lim, H.K. Effects of Cryoprotective Medium Composition, Dilution Ratio, and Freezing Rates on Spotted Halibut (*Verasper variegatus*) Sperm Cryopreservation. *Animals* **2020**, *10*, 2153. [CrossRef] [PubMed]
75. Ding, S.; Ge, J.; Hao, C.; Zhang, M.; Yan, W.; Xu, Z.; Pan, J.; Chen, S.; Tian, Y.; Huang, Y. Long-Term Cryopreservation of Sperm from Mandarin Fish Siniperca Chuatsi. *Anim. Reprod. Sci.* **2009**, *113*, 229–235. [CrossRef] [PubMed]
76. Tsai, S.; Chong, G.; Meng, P.; Lin, C. Sugars as Supplemental Cryoprotectants for Marine Organisms. *Rev. Aquac.* **2018**, *10*, 703–715. [CrossRef]
77. Melo, F.; Godinho, H.P. A Protocol for Cryopreservation of Spermatozoa of the Fish Brycon Orthotaenia. *Anim. Reprod.* **2018**, *3*, 380–385.
78. Brzuska, E. Artificial Spawning of Herbivorous Fish: Use of an LHRH—A to Induce Ovulation in Grass Carp *Ctenopharyngodon idella* (Valenciennes) and Silver Carp *Hypophthalmichthys molitrix* (Valenciennes). *Aquac. Res.* **1999**, *30*, 849–856. [CrossRef]
79. Servili, A.; Canario, A.V.M.; Mouchel, O.; Muñoz-Cueto, J.A. Climate Change Impacts on Fish Reproduction Are Mediated at Multiple Levels of the Brain-Pituitary-Gonad Axis. *Gen. Comp. Endocrinol.* **2020**, *291*, 113439. [CrossRef] [PubMed]

80. Dhara, K.; Saha, N.C. Controlled Breeding of Asian Catfish Clarias Batrachus Using Pituitary Gland Extracts and Ovaprim at Different Temperatures, Latency Periods and Their Early Development. *J. Aquac. Res. Dev.* **2013**, *4*, 186.
81. Ashaf-Ud-Doulah, M.; Islam, S.M.M.; Zahangir, M.M.; Islam, M.S.; Brown, C.; Shahjahan, M. Increased Water Temperature Interrupts Embryonic and Larval Development of Indian Major Carp Rohu Labeo Rohita. *Aquac. Int.* **2021**, *29*, 711–722. [CrossRef]
82. Akatsu, S.; Al-Abdul-Elah, K.M.; Teng, S.K. Effects of Salinity and Water Temperature on the Survival and Growth of Brown-spotted Grouper Larvae (*Epinephelus tauvina*, Serranidae). *J. World Maric. Soc.* **1983**, *14*, 624–635. [CrossRef]
83. Kupren, K.; Mamcarz, A.; Kucharczyk, D. Effects of Temperature on Survival, Deformations Rate and Selected Parameters of Newly Hatched Larvae of Three Rheophilic Cyprinids (*Genus leuciscus*). *Polish J. Nat. Sci.* **2010**, *25*, 299–312. [CrossRef]
84. Pankhurst, N.W.; Munday, P.L. Effects of Climate Change on Fish Reproduction and Early Life History Stages. *Mar. Freshw. Res.* **2011**, *62*, 1015–1026. [CrossRef]
85. Muhling, B.; Lindegren, M.; Clausen, L.W.; Hobday, A.; Lehodey, P. Impacts of Climate Change on Pelagic Fish and Fisheries. *Clim. Chang. Impacts Fish. Aquac. A Glob. Anal.* **2017**, *2*, 771–814.
86. Quigley, J.T.; Hinch, S.G. Effects of Rapid Experimental Temperature Increases on Acute Physiological Stress and Behaviour of Stream Dwelling Juvenile Chinook Salmon. *J. Therm. Biol.* **2006**, *31*, 429–441. [CrossRef]
87. Alfonso, S.; Gesto, M.; Sadoul, B. Temperature Increase and Its Effects on Fish Stress Physiology in the Context of Global Warming. *J. Fish Biol.* **2021**, *98*, 1496–1508. [CrossRef] [PubMed]
88. Salisbury, J.E.; Jönsson, B.F. Rapid Warming and Salinity Changes in the Gulf of Maine Alter Surface Ocean Carbonate Parameters and Hide Ocean Acidification. *Biogeochemistry* **2018**, *141*, 401–418. [CrossRef] [PubMed]
89. Ginsburg, A.S. *Fertilization of Fishes and the Problem of Polyspermy*; Translation: NOOAA and National Science Foundation; Academy of Science USSR: New York, NY, USA, 1968; p. 354.
90. Fenkes, M.; Fitzpatrick, J.L.; Shiels, H.A.; Nudds, R.L. Acclimation Temperature Changes Spermatozoa Flagella Length Relative to Head Size in Brown Trout. *Biol. Open.* **2019**, *8*, bio039461. [CrossRef] [PubMed]
91. Biswas, J.K.; Maurye, P. Aquaculture Biotechnology: Prospects and Challenges. *Prog. Biotechnol. India.* **2017**, *1*, 33–47.
92. Da Costa, B.B.; de Oliveira, D.L.; Rodrigues, R.B.; Gomes, I.C.; Streit, D.P., Jr. Morphological Abnormalities in Zebrafish Cryopreserved Sperm. *Cryobiology* **2020**, *97*, 235–237. [CrossRef]
93. Da Costa, B.B.; Marques, L.S.; Lassen, P.G.; Rodrigues, R.B.; Tais Da Rosa Silva, H.; Moreira, J.C.F.; Streit, D.P. Effects of Cysteine Supplementation on the Quality of Cryopreserved Sperm of South American Silver Catfish. *Aquac. Res.* **2020**, *51*, 455–464. [CrossRef]

Disclaimer/Publisher's Note: The statements, opinions and data contained in all publications are solely those of the individual author(s) and contributor(s) and not of MDPI and/or the editor(s). MDPI and/or the editor(s) disclaim responsibility for any injury to people or property resulting from any ideas, methods, instructions or products referred to in the content.

animals

MDPI

Article

The Interaction between Canine Semen Bacteria and Semen Quality Parameters

Šarūnė Sorkytė [1,*], Rita Šiugždinienė [2], Marius Virgailis [2], Gintarė Vaičiulienė [1], Anna Wysokińska [3], Ewa Wójcik [3], Paulius Matusevičius [4], Audronė Rekešiūtė [1] and Neringa Sutkevičienė [1]

1 Animal Reproduction Laboratory, Faculty of Veterinary Medicine, Veterinary Academy, Lithuanian University of Health Sciences, Tilžės Str. 18, LT-47181 Kaunas, Lithuania; gintare.vaiciuliene@lsmu.lt (G.V.); audrone.rekesiute@lsmu.lt (A.R.); neringa.sutkeviciene@lsmu.lt (N.S.)

2 Institute of Microbiology and Virology, Faculty of Veterinary Medicine, Veterinary Academy, Lithuanian University of Health Sciences, Tilžės Str. 18, LT-47181 Kaunas, Lithuania; rita.siugzdiniene@lsmu.lt (R.Š.); marius.virgailis@lsmu.lt (M.V.)

3 Institute of Animal Science and Fisheries, Faculty of Agricultural Sciences, University of Siedlce, Konarskiego 2, 08110 Siedlce, Poland; anna.wysokinska@uws.edu.pl (A.W.); ewa.wojcik@uph.edu.pl (E.W.)

4 Department of Animal Nutrition, Faculty of Animal Sciences, Veterinary Academy, Lithuanian University of Health Sciences, Tilžės Str. 18, LT-47181 Kaunas, Lithuania; paulius.matusevicius@lsmu.lt

* Correspondence: sarune.sorkyte@lsmu.lt

Simple Summary: In recent decades, the practice of artificial insemination has become increasingly prevalent in canine reproduction, thereby underscoring the necessity for improved semen quality assessment routines. While the bacterial microflora of the reproductive system of female dogs has been extensively studied, there is comparatively limited analysis of the male dog's reproductive tract's microbiota. In the field of andrology, a few connections have already been discovered between seminal bacteria and semen's qualitative parameters. The objective of this study was to determine whether the bacteria present in semen samples from breeding dogs can influence the semen's characteristics. The results indicated that multiple bacterial species present in semen samples from breeding dogs are associated with inferior sperm parameters. Moreover, a greater bacterial load was observed to be related with poorer semen quality in breeding dogs.

Abstract: Assessing canine semen quality helps to detect infertility in males, but identifying factors that influence canine semen quality is a complicated task. The objective of this study was the assessment of the potential influence of bacteria found in canine semen samples on the characteristics of dogs' semen. In this study, semen samples were collected manually from 30 dogs and subjected to a comprehensive examination. The results of sperm motility, concentration, viability, and morphology were statistically analysed in relation to the number of bacteria in the semen (CFUs/mL) and the seminal microbiota. Samples with an increased bacterial count per millilitre were associated with lower-quality sperm motility ($p < 0.05$). The most frequently isolated bacterial genera from the analysed semen samples were *Staphylococcus* spp. (26.0%), *Corynebacterium* spp. (17.8%), and *Streptococcus* spp. (16.4%). The presence of β-haemolytic *Escherichia coli* bacteria was linked to suboptimal semen samples, characterised by significantly reduced semen viability and a lower proportion of morphologically normal spermatozoa ($p < 0.05$). *Corynebacterium* spp. was associated with reduced bacterial load and superior semen quality ($p < 0.01$). These findings highlight the importance of bacterial cell counts and microbiota diversity in relation to various factors influencing canine semen quality, providing a more comprehensive understanding of canine reproductive well-being.

Keywords: canine; seminal bacteria; semen quality

Citation: Sorkytė, Š.; Šiugždinienė, R.; Virgailis, M.; Vaičiulienė, G.; Wysokińska, A.; Wójcik, E.; Matusevičius, P.; Rekešiūtė, A.; Sutkevičienė, N. The Interaction between Canine Semen Bacteria and Semen Quality Parameters. *Animals* **2024**, *14*, 2151. https://doi.org/10.3390/ani14152151

Academic Editor: Jesús Dorado

Received: 20 June 2024
Revised: 16 July 2024
Accepted: 23 July 2024
Published: 24 July 2024

Copyright: © 2024 by the authors. Licensee MDPI, Basel, Switzerland. This article is an open access article distributed under the terms and conditions of the Creative Commons Attribution (CC BY) license (https://creativecommons.org/licenses/by/4.0/).

1. Introduction

The assessment of canine semen's quality has become a standard procedure for the identification of infertility in dogs [1]. The evaluation of the semen's quality determines the male's fertility status and helps to identify semen that is suitable for artificial insemination (AI) and cryopreservation [2]. The assessment of semen's quality parameters has become straightforward with the advent of modern technologies and advanced research methodologies [3]. Nevertheless, identifying the specific factors that influence variations in semen's quality characteristics remains a challenging task.

The microbiota of animals can influence several physiological factors in organisms and affect their health [4]. It is noteworthy that studies investigating the microbiota in the male reproductive system of dogs and its influence on animal fertility are limited in number [4]. In contrast, research has predominantly focused on the microbiota of the bitch reproductive tract and its association with reproductive success [5]. However, seminal bacteria, which can be transferred to female dogs' reproductive tracts, have the potential to trigger infections, spontaneous abortion, or even infertility [6,7].

It is well established that canine semen is not sterile, and various opportunistic bacteria have been detected within the male dog reproductive system [8,9]. The data presented in other articles indicate a correlation between the identified bacteria and semen quality across a range of animal species, including bulls, boars, poultry, etc. [4,10–13]. In the literature, some male genital tract bacteria are described as potentially more pathogenic. However, there are a lack of published data on the seminal microbiota, including which bacteria are commensal and which are pathogenic [14]. *Escherichia coli* is more frequently associated with less desirable semen quality [13]. A high bacterial load in semen (bacteriospermia) is related to lower sperm motility, membrane and acrosome integrity alterations, and higher oxidative stress. In canines, alterations in semen quality, including teratospermia, have been linked to specific pathogenic bacterial strains, such as β-haemolytic *Streptococcus* spp. [9]. Additionally, it has been observed that canine semen samples with bacteriospermia demonstrate lower sperm viability and a higher prevalence of morphological abnormalities in their spermatozoa [8].

As recent findings have highlighted the significant role of semen microbiota in male infertility, it is becoming increasingly important to extend bacteriological studies to include a wider variety of animal species. Acknowledging that canine semen contains bacterial populations indicates the possibility that specific bacteria may impact semen quality characteristics, particularly its motility, which is closely associated with reproductive success in canines [15]. Therefore, the aim of this study was to evaluate the impact of bacteria in the semen of breeding dogs on the semen's quality.

2. Materials and Methods

2.1. Animals

Semen samples were obtained from 30 clinically healthy canines (*Canis familiaris*). The dogs exhibited a range of sizes, with the following breeds represented: German Shepherds (n = 5), Labrador Retrievers (n = 3), Salukis (n = 2), Bull Terriers (n = 2), Shih Tzus (n = 3), and several other breeds. It was confirmed that the animals in question had not received any antibiotics within the preceding month. The collection of canine semen and the external examination of the canines were conducted in accordance with EU Directive 2010/63/EU for animal experimentation.

2.2. Semen Collection and Assessment

The samples were collected at a veterinary clinic. Prior to semen collection, the clinical condition of each dog was assessed. An external examination of the dog's reproductive system was conducted to identify any signs of pain or discomfort in the reproductive organs, including signs of inflammation [1]. The detection of inflammation in dogs was done through palpation, which involved the detection of symptoms such as swelling, redness, heat, and pain. The sample collection procedure was conducted in the presence of a female in oestrus or with the use of swabs containing vaginal secretions from a female dog in oestrus. Sperm-rich fractions were collected into sterile plastic bags through digital manipulation, as previously described [16]. To prevent bacterial contamination from the penile prepuce, the tip of the glans penis was protruded from the prepuce and cleaned with sterile gauze before collection. Samples of semen for bacteriological investigation were obtained immediately after semen collection with a pipette using sterile tips. Volumes of 0.2 mL of semen were transferred into sterile 1.5 mL Snap Cap Microcentrifuge Tubes (Thermo Fisher Scientific, Hennigsdorf, Germany) and transported to the microbiology laboratory within a 15–30 min period in an insulated box with a cold pack. Semen samples, for the assessment of their quality, were delivered to the laboratory in a thermos with a temperature of 35 ± 2 °C within 30 min.

2.3. Semen Quality Assessment

The evaluation of semen quality was performed at the Laboratory of Animal Reproduction of the Lithuanian University of Health Sciences. The following parameters were analysed: semen volume, pH, sperm motility, concentration, viability, and morphological changes of spermatozoa.

The most important semen quality parameter, motility, was assessed subjectively immediately after sample collection: 10 µL of the semen sample were placed on a microscope slide (Avantor VWR®, Radnor, PA, USA), covered with a coverslip, and placed on a heated plate at 37 °C. The slide was then observed under an Eclipse 50i microscope (Nikon®, Tokyo, Japan).

The pH of the samples was measured using an electronic pH meter, the Fisherbrand accumet AB150 (Fisher scientific, Loughborough, UK). The concentration of semen was determined using an improved neubauer haemocytometer chamber (BLAUBRAND®, Wertheim, Germany) and the results were expressed as the number of spermatozoa per millilitre. The total number of sperm present in the ejaculate was calculated by multiplying the concentration per millilitre by the volume.

Semen viability was assessed through smears of the samples, which were stained with the following eosin–nigrosin dyes: Eosin G stain, 2% solution, and Nigrosin stain, 4% solution for live/dead cells (Minitube, Tiefenbach, Germany). A drop of the semen sample was placed on a slide, followed by a drop of eosin–nigrosin dye, which was then mixed. After half a minute, a smear was made, the slide was air-dried, and a microscopic evaluation was performed, assessing the staining of 100 spermatozoa. In accordance with the test principles employed, the unstained sperm were considered to be viable.

For the assessment of morphological abnormalities in the spermatozoa, a semen sample smear was stained with the SpermBlue staining kit (Microptic SL, Barcelona, Spain) in accordance with the procedure of van der Horst and Maree [17]. A total of 500 sperm were evaluated under a light microscope, the Eclipse 50i, with an oil immersion objective at 100× magnification. The following abnormalities were included in the sperm head morphological defects: macrocephaly, microcephaly, tapered head, thin, narrow head, pinhead, undeveloped sperm, and asymmetrical midpiece insertion into the head. Abnormalities of the tail included irregularities of the midpiece (e.g., a bent neck, a neck that is too thick or too thin, or an irregularly shaped neck), a coiled tail, a stump tail, and a duplicate tail. Other morphological defects encompassed droplets, acrosome defects, vacuoles, or absent tails (tailless). The information on the spermatozoa's morphology was systematically recorded, including head, tail, and other abnormalities, in the prepared data tables. Additionally, the number of morphologically normal spermatozoa was recorded.

2.4. Bacteriological Analysis of Semen

The bacteriological analysis of the semen samples was conducted at the Lithuanian University of Health Sciences, Institute of Microbiology and Virology. In brief, 10 µL of each sample was inoculated on Columbia Agar plates with 5% Sheep Blood (E&O Laboratories Ltd., Bonnybridge, Scotland, UK) using serial dilutions. The inoculated plates were incubated for 24–48 h at 37 °C under aerobic conditions. After incubation, the number of colony-forming units (CFUs) per millilitre were recorded.

The bacteria colonies were collected for further investigation. For further identification, pure bacterial cultures were inoculated onto selective agars. The morphology of the bacteria was examined by Gram staining, and the genera and families were identified by biochemical properties, including the catalase test [18,19]. The bacteria belonging to the *Staphylococcus* genus were identified to the species level using the "RapID™ STAPH PLUS System" identification systems (Thermo Fisher Scientific, Hennigsdorf, Germany) and associated computer program according to the instructions. Gram-negative bacteria were identified using the "Novacyt–Microgen Bioproducts GN-ID A+B-Combined 30 Test System" identification systems (Thermo Fisher Scientific, Hennigsdorf, Germany), along with the corresponding computer program and according to the manufacturer's instructions.

Bacteria that could not be identified under laboratory conditions were sent to the Lithuanian National Food and Veterinary Risk Assessment Institute for MALDI-TOF MS (matrix-assisted laser desorption ionization–time of flight Mass Spectrometry) (Bruker Daltonics, Bremen, Germany) biotyping.

2.5. Sample Grouping

The research subjects were grouped into three size groups: small dogs (<10 kg), n = 10; medium dogs (10–30 kg), n = 11; and large dogs (>30 kg), n = 9.

Semen samples were divided into 3 high- and 3 low-quality groups (H and L) by qualitative criteria, according to motility, viability, and morphology (Table 1). The morphology was based on the percentage of normal morphology form spermatozoa (normal spermatozoa). Samples that met all of the pre-established criteria were included in the high-quality (H QUALITY) group, while those samples that failed to meet at least one of the criteria were included in the low-quality (L QUALITY) group.

Table 1. Canine semen sample grouping criteria.

Parameter	Superior Quality Semen (H); n = 10	Inferior Quality Semen (L); n = 20
Sperm motility (%)	>75%	≤75%
Live spermatozoa (%)	≥50%	<50%
Normal spermatozoa (%)	≥70%	<70%

2.6. Data Analysis

The statistical data analysis was performed using "IBM SPSS Statistics 29.0.0.0". Descriptive statistics were calculated, and statistical tests, including the Student's *t*-test, Mann–Whitney U test, and Pearson's correlation, were used to assess differences and correlations among the samples' bacterial numbers, variable aerobic bacteria isolations, and different dog semen quality parameters. Because the normal distribution was not always fulfilled, non-parametric statistical analysis was performed. The data are presented as means (standard deviation (SD)). Results were considered statistically significant when the p-value < 0.05. Exact p-values are provided for all comparisons, except when p is below 0.001 ($p < 0.001$).

3. Results

Bacterial populations were detected in all 30 canine semen samples. In these samples, a total of 11 different bacterial families were identified. The bacterial genera associated with each family are presented in Table 2. The most frequently isolated bacterial genera from the analysed semen samples were *Staphylococcus* spp. (26.0%), *Corynebacterium* spp. (17.8%), and *Streptococcus* spp. (16.4%). The most predominantly isolated bacterial species from the *Staphylococcus* spp. genus were *S. pseudintermedius* (31.6%), *S. epidermidis* (15.8%), and *S. aureus* (15.8%). Additionally, *S. schleiferi* and *S. intermedius* were isolated, while other staphylococci species were identified but not at the species level. *Corynebacterium imitans* was the only species that could be identified from the *Corynebacterium* spp. genera with the system used. The bacterium *Streptococcus canis* constituted 50 percent of all identified streptococci, with the remaining half consisting of *S. sanguinis* and other streptococci. From *Neisseria* spp., the only species present was *Neisseria weaveri*. *Acinetobacter* spp. consisted entirely of the species *Acinetobacter haemolyticus*. Of the *Micrococcus* spp. isolates, only one was identified to the species level, as *Micrococcus luteus*, while the remainder remained unidentified at the species level. Both bacteria from the *Escherichia* spp. genus were identified as *Escherichia coli* (β-haemolytic). All bacteria from the families *Pasteurellaceae*, *Microbacteriaceae*, and *Lactobacillaceae* were identified to the species level. These included *Haemophilus* spp.—*Haemophilus haemoglobinophilus*, *Frederiksenia* spp.—*Frederiksenia canicolla*, *Pasteurella* spp.—*Pasteurella multocida*, *Canibacter* spp.—*Canibacter oris*, and *Limosilactobacillus* spp.—*Limosilactobacillus reuteri*.

Table 2. Bacterial families and genera isolated and identified from canine semen samples.

Identified Bacteria Families	Identified Bacteria Genus	Positive Semen Samples	
		N	Percentage [1], %
Staphylococcaceae	*Staphylococcus* spp.	19	63.3
Corynebacteriaceae	*Corynebacterium* spp.	13	43.3
Streptococcaceae	*Streptococcus* spp.	12	40.0
Moraxellaceae	*Moraxella* spp.	5	16.7
	Acinetobacter spp.	3	10.0
Enterococcaceae	*Enterococcus* spp.	5	16.7
Neisseriaceae	*Neisseria* spp.	4	13.3
Micrococcaceae	*Micrococcus* spp.	3	10.0
Enterobacteriaceae	*Escherichia* spp.	2	6.7
Pasteurellaceae	*Haemophilus* spp.	2	6.7
	Frederiksenia spp.	2	6.7
	Pasteurella spp.	1	3.3
Microbacteriaceae	*Canibacter* spp.	1	3.3
Lactobacillaceae	*Limosilactobacillus* spp.	1	3.3

[1] The proportion of semen samples in which bacteria were identified.

The bacterial count in the canine semen samples was not found to be significantly affected by the size (weight) of the dog ($p = 0.347$).

The bacterial contamination results for the canine semen samples were determined according to the categorisations of motility, viability, and morphology, as presented in Table 1. These results are displayed in Figure 1. In samples exhibiting optimal viability and morphological quality (blue colour), a lower bacterial number was observed, although this was not statistically significant ($p > 0.05$). Samples with lower motility showed a statistically significantly higher number of bacteria in their semen ($p = 0.035$) (Figure 1).

The association between different bacteria and the quality of the examined semen samples was evaluated by monitoring the frequencies of detection of different bacteria in samples of high and low semen quality (Figure 2). *Limosilactobacillus reuteri* was identified only in one optimal-quality semen sample. The isolation of *Corynebacterium* spp., *Neisseria* spp., *Frederiksenia canicolla*, and *Haemophilus haemoglobinophilus* was observed to occur with greater frequency in superior quality semen samples. *Canibacter oris*, *Pasteurella multocida*, *E. coli*, *A. haemolyticus*, and *Enterococcus* spp. were only detected in samples of inferior semen quality.

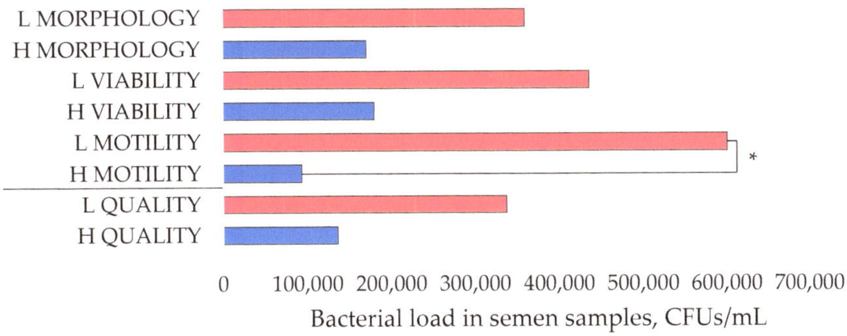

Figure 1. A diagram illustrating the bacterial contamination in semen samples, expressed as colony-forming units (CFUs) per millilitre of sample. Semen samples are classified into higher (H, blue colour) and lower (L, red colour) quality groups according to their quality parameters: viability, morphology, and motility. H VIABILITY represents samples with 50% viability and higher, while L VIABILITY represents samples with less than 50% viability. H MORPHOLOGY indicates samples with 70% morphologically normal spermatozoa and higher, and L MORPHOLOGY represents samples with less than 70% morphologically normal spermatozoa. H MOTILITY designates samples with motility higher than 75%, and L MOTILITY represents samples with 75% motility and lower. H QUALITY and L QUALITY represent general quality groups. * Significant differences between groups when p-value < 0.05.

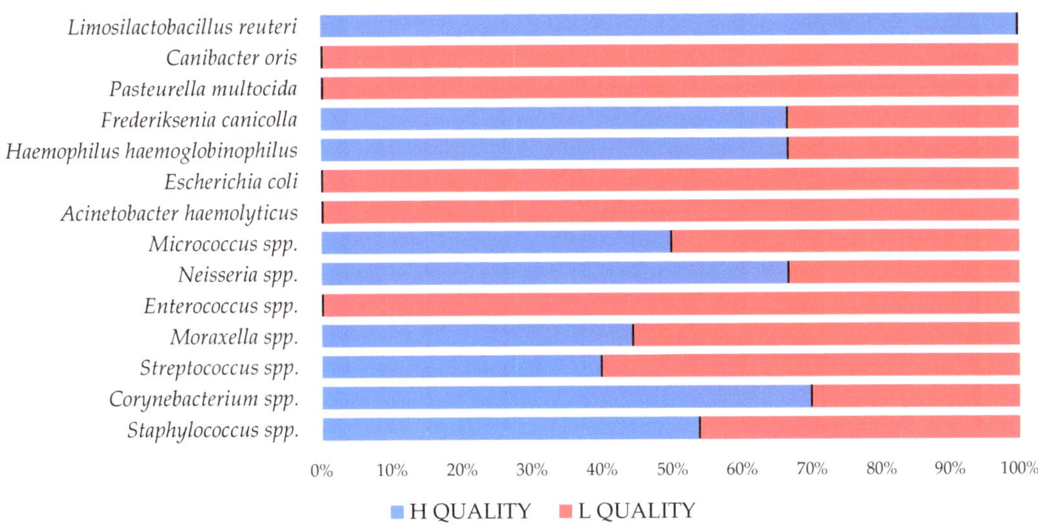

Figure 2. A diagram illustrating the percentage distribution of identified bacteria in semen samples of two quality groups, according to their viability, motility, and morphology.

The means of the parameters, including motility, viability, and detailed semen morphology, are presented in Table 3 by dividing the samples according to the presence or absence of bacteria. Samples with each detected bacterium were compared with samples where that bacterium was not detected. Several notable differences were observed. Samples containing β-haemolytic *E. coli* were associated with inferior sperm viability (p = 0.03) and a lower number of morphologically normal spermatozoa (p = 0.02). Conversely, semen samples in which *Corynebacterium* spp. were isolated had a significantly reduced number of spermatozoa other abnormalities (p = 0.045), while sperm motility was significantly higher than in other samples without this bacterium ($p = 0.006$). Moreover, the bacterial count was found to be significantly lower in these samples ($p = 0.028$).

Table 3. Comparison of semen quality parameters between samples where specific bacteria were identified and samples where those bacteria were not identified.

Identified Bacteria (Genus/Species)	In Samples	Motility, %	Viability, %	Head Pathologies, %	Tail Pathologies, %	Other Pathologies, %	CFUs per mL, 10^5
Staphylococcus spp.	Identified	66.39 (31.98)	55.35 (22.28)	7.77 (6.79)	14.53 (13.48)	19.97 (22.98)	2.52 (4.06)
	Not identified	79.55 (11.50)	52.36 (22.61)	8.64 (3.56)	8.09 (4.04)	24.59 (23.47)	2.76 (4.65)
Streptococcus spp.	Identified	76.67 (19.92)	52.91 (18.03)	7.35 (4.84)	13.38 (13.02)	20.21 (22.86)	1.85 (3.84)
	Not identified	67.65 (30.57)	55.00 (24.79)	8.62 (6.36)	11.18 (10.12)	22.79 (23.50)	3.11 (4.47)
Corynebacterium spp.	Identified	** 83.46 (4.74)	51.85 (25.52)	6.08 (2.90)	8.04 (4.86)	* 12.39 (13.76)	* 0.44 (0.96)
	Not identified	** 60.94 (32.52)	50.38 (22.79)	9.74 (6.92)	15.38 (13.83)	* 29.31 (26.21)	* 4.27 (4.96)
Escherichia coli (β-haemolytic)	Identified	75.00 (7.07)	* 21.50 (2.12)	6.10 (4.10)	6.00 (7.78)	* 70.75 (0.35)	5.05 (6.99)
	Not identified	70.74 (27.52)	* 55.33 (21.58)	8.24 (5.85)	12.54 (11.43)	* 18.09 (19.08)	2.43 (4.09)
Enterococcus spp.	Identified	51.25 (42.11)	49.00 (12.65)	10.45 (5.66)	17.25 (15.84)	26.00 (15.53)	2.44 (4.29)
	Not identified	74.60 (23.00)	53.64 (23.88)	7.72 (5.75)	11.26 (10.53)	21.04 (24.00)	2.64 (4.28)
Moraxella spp.	Identified	57.00 (43.53)	40.20 (19.61)	12.60 (11.35)	14.70 (14.83)	31.90 (29.08)	* 6.35 (5.03)
	Not identified	74.38 (21.93)	55.67 (22.57)	7.16 (3.48)	11.54 (10.66)	19.60 (21.49)	* 1.86 (3.67)
Acinetobacter haemolyticus	Identified	38.33 (40.72)	30.67 (7.77)	12.13 (5.33)	31.17 (22.43)	21.50 (16.89)	3.43 (5.69)
	Not identified	75.19 (22.74)	55.58 (22.31)	7.63 (5.67)	9.89 (7.25)	21.75 (23.72)	2.52 (4.14)
Limosilactobacillus reuteri	Identified	90.00 (0.00)	74.00 (0.00)	3.40 (0.00)	5.50 (0.00)	2.50 (0.00)	0.01 (0.00)
	Not identified	70.71 (26.90)	52.25 (22.59)	8.26 (5.75)	12.32 (11.37)	22.41 (22.97)	2.70 (4.25)
Neisseria spp.	Identified	85.00 (4.08)	58.50 (29.24)	4.80 (3.07)	7.88 (6.55)	27.00 (32.17)	0.03 (0.045)
	Not identified	69.20 (28.13)	52.12 (21.92)	8.62 (5.90)	12.76 (11.77)	20.88 (21.79)	3.01 (4.39)
Micrococcus spp.	Identified	61.67 (44.81)	37.33 (13.05)	9.40 (0.87)	23.33 (20.50)	16.50 (12.99)	4.98 (0.82)
	Not identified	72.50 (24.95)	54.81 (22.87)	7.95 (6.03)	10.79 (9.54)	22.33 (23.84)	2.84 (4.37)

The data are represented as means (SD). * Significant differences between groups when p-value < 0.05; ** p-value < 0.01.

4. Discussion

The results of our study demonstrated that none of the canine semen samples were sterile, and 11 distinct bacterial families were identified. In the analysis of 30 healthy canine semen samples, the most frequently detected bacteria were identified as *Staphylococcus* spp. (26.0%), *Corynebacterium* spp. (17.8%), and *Streptococcus* spp. (16.4%). In smaller quantities, the *Moraxella* spp., *Enterococcus* spp., *Pastuarellaceae*, *Neisseria* spp., *Microbacteriaceae*, *Acinetobacter haemolyticus*, *E. coli*, *Canibacter oris*, and *Lactobacillus reuteri* families, genera, and species were detected. In 2022, a related study conducted by S. Agudelo-Yepes et al. revealed relevant results about nonsterile dog semen samples. The study demonstrated that bacterial populations were present in 100% of the examined dog semen samples [8].

In the previously mentioned study on the microbiota of canine semen, the presence of the following bacteria was detected in healthy canine semen samples: *Staphylococcus* spp., *Streptococcus* spp., *E. coli*, *Klebsiella* spp., *Neisseria gonorrhoeae*, *Pseudomonas* spp., and *Chlamydia trachomatis* [8]. Other research has revealed a broader variety of bacteria in canine semen [9]. However, none of these studies identified the presence of *Canibacter oris*. *C. oris* bacteria are strongly associated with the canine oral bacterial microflora and are more frequently described as pathogens in dog bite wounds [20]. Consequently, the detection of this bacterium in semen indicates that the oral microbiota of dogs can enter the canine reproductive tract, and bacteria from the mouth can be identified in the seminal microbiota of dogs. Knowing that bacteria can enter a dog's reproductive tract from its mouth, it is possible that bacteria can also travel in the opposite direction, from the male reproductive tract to the oral cavity. Scientists have observed that several bacterial species may be transmitted from the oral flora of dogs to their owners through close physical contact, as described in this article [21]. Overall, there is a possibility that pathogenic bacteria discovered in the male dog reproductive system could, over time, be transported to the dog's oral flora and subsequently transmitted as zoonotic bacteria to the dog owner.

In our research, samples with a higher number of bacteria were associated with lower motility in semen samples. A previous study conducted at the Columbian university (Universidad de Antioquia) has discovered related results. The study found that bacteriospermia in dog semen negatively affect sperm motility and spermatozoa morphology [8]. The negative influence of seminal bacteria can be explained as follows: several bacteria can directly attach to spermatozoa, immobilise them, and induce agglutination, autoimmune reactions, or oxidative stress, and some bacteria may even excrete toxins [4,22]. The present study aimed to analyse the effects of different bacteria on the parameters of semen quality. The results revealed significant differences. In the present study, potentially pathogenic seminal bacteria previously associated with poor semen quality in other species were detected. Samples containing β-haemolytic *Escherichia coli* exhibited reduced viability and a greater prevalence of other sperm morphological pathologies. To the best of our knowledge, this bacterium has not previously been associated with changes in the quality parameters of canine semen. However, related results have been observed in other animal species research, where *E. coli* was associated with inferior semen samples of boars [23,24]. In our study, for dog semen samples with β-haemolytic *E. coli*, the average viability was two times lower than in samples without this bacterium. This could be attributed to the α-haemolysin produced by *E. coli*, which is a potent cytotoxin capable of damaging the integrity of the sperm membrane and forming pores in the cell membrane [22]. In an earlier study, a statistically significant negative correlation was observed between the presence of *E. coli* in semen used for gilt insemination and litter size [24]. Nevertheless, studies comparing the microbiota of dog semen with the number of offspring produced by females, according to our knowledge, have not yet been conducted.

In microbiology field, some authors claim that a few strains of *Staphylococcus* spp. can produce substances believed to be like the staphylococcal enterotoxins SAF (sperm agglutination factor) and SIF (sperm immobilization factor), which can immobilize sperm and promote their agglutination [22]. This could be a contributing factor to the decrease in sample motility observed with *Staphylococcus* spp. in our research. *Enterococcus* spp. is described as secreting the toxin β-haemolysin, which also has the ability to immobilize sperm [25]. Additionally, some bacteria can form biofilms, which induce sperm agglutination and decreases motility [22].

In our study, samples in which bacteria from the *Corynebacterium* spp., *Neisseria* spp., and *Streptococcus* spp. genera or the bacterium *L. reuteri* were identified had a higher motility than samples in which these bacteria were not detected. A statistically significant difference was observed for *Corynebacterium* spp. Additionally, in samples where *Corynebacterium* spp. was isolated, statistically significantly fewer other abnormalities were found in comparison to samples without this bacterium. Moreover, these samples were associated with lower bacterial counts in semen. Thereby, the assumption can be made that possibly the *Corynebacterium* spp. found in our study samples can inhibit the proliferation of other microbes. The results indicate that bacteria belonging to the *Corynebacterium* spp., *Neisseria* spp., *Streptococcus* spp., and *Lactobacillus* spp. genera, found in this research, do not negatively impact sperm motility. It can be assumed that these bacteria represent a normal component of the seminal microflora. The presence of such bacteria may be indicative of the absence of competing species that have a more negative effect on semen. It can be hypothesized that these bacteria may act as commensals, protecting the dog's semen from invasion by more pathogenic bacteria.

In our research, *Limosilactobacillus reuteri* was found to be associated with an excellent-quality semen sample. However, due to the fact that this bacterium was identified in only one semen sample, the result was not statistically significant. In other articles, the bacterium *L. reuteri* is described as a probiotic strain capable of excreting antimicrobial molecules, inhibiting the growth of pathogenic microbes, and having a strong anti-inflammatory effect. As previously demonstrated, the administration of *L. reuteri* as an oral supplement to canines has the potential to enhance the qualitative parameters of semen [26]. It can be considered that a similar mechanism of action is present in canine semen, whereby the bacterium directly reduces inflammation or the growth of pathogenic microbes. Nevertheless further in vitro studies are required to confirm this assumption.

5. Conclusions

It has been demonstrated that bacteriospermia has a significant impact on the quality of canine semen. This is evidenced by an association between a higher number of bacteria (CFUs) in semen and decreased seminal qualitative parameters. It is important to note that not all bacteria present in canine semen have a detrimental impact on its quality. Rather different bacteria may be associated with alterations in distinct qualitative parameters. A number of bacterial species, including *Corynebacterium* spp., *Neisseria* spp., *Streptococcus* spp., and *Limosilactobacillus reuteri*, were not found to have a negative impact on semen quality. In contrast, the samples with *Corynebacterium* spp. were linked to a reduced bacterial load, which also resulted in superior qualitative parameters. However, other bacteria were more frequently associated with inferior semen quality. Of particular interest in this study was the identification of β-haemolytic *Escherichia coli* as the most pathogenic bacteria for canine semen parameters.

Author Contributions: Conceptualization, N.S., Š.S., M.V. and R.Š.; methodology N.S. and R.Š.; investigation, N.S., A.R., E.W., A.W. and Š.S.; resources, N.S., R.Š. and M.V.; writing—original draft preparation, Š.S. and N.S.; writing—review and editing, M.V., G.V., P.M., E.W. and A.W. All authors have read and agreed to the published version of the manuscript.

Funding: This research received no external funding.

Institutional Review Board Statement: For our study, we received the semen samples as biological material for further examination at the laboratories of the Lithuanian University of Health Sciences. At the time study was conducted, our institution did not require bioethics approval for research involving only animal biological materials. The collection of semen and assessment of the physiological status of the dogs were performed by a licensed veterinarian as part of routine procedures for breeding animals in a veterinary clinic. This was conducted in accordance with the provisions of the Law on Animal Welfare and Protection of the Republic of Lithuania, No XI-2271.

Informed Consent Statement: Not applicable.

Data Availability Statement: The data presented in this study are available within the article.

Conflicts of Interest: The authors declare no conflicts of interest.

References

1. Kolster, K.A. Evaluation of Canine Sperm and Management of Semen Disorders. *N. Am. Small Anim. Pract.* **2018**, *48*, 533–545. [CrossRef] [PubMed]
2. Schäfer-Somi, S.; Colombo, M.; Luvoni, G.C. Canine Spermatozoa—Predictability of Cryotolerance. *Animals* **2022**, *12*, 733. [CrossRef] [PubMed]
3. Robert, M.A.; Jayaprakash, G.; Pawshe, M.; Tamilmani, T.; Sathiyabarathi, M. Collection and evaluation of canine semen—A review. *Int. J. Sci. Environ. Technol.* **2016**, *5*, 1586–1595.
4. Rowe, M.; Veerus, L.; Trosvik, P.; Buckling, A.; Pizzari, T. The Reproductive Microbiome: An Emerging Driver of Sexual Selection, Sexual Conflict, Mating Systems, and Reproductive Isolation. *Trends Ecol. Evol.* **2020**, *35*, 220–234. [CrossRef] [PubMed]
5. Pereira, A.M.; Clemente, A. Dogs' Microbiome from Tip to Toe. *Top. Companion Anim. Med.* **2021**, *45*, 100584. [CrossRef]
6. England, G.C.W.; Rijsselaere, T.; Campbell, A.; Moxon, R.; Freeman, S.L. Normal and abnormal response to sperm deposition in female dogs: A review and new hypotheses for endometritis. *Theriogenology* **2021**, *159*, 176–183. [CrossRef]
7. Antunes, J.M.A.P.; Freire, D.A.C.; Oliveira, I.V.P.M.; Moura, G.H.F.; Demoner, L.C.; Ferreira, H.I.P. Infectious Causes of Abortion, Stillbirth and Neonatal Death in Bitches. *Vet. Clin. Small Anim.* **2012**, *42*, 501–513. [CrossRef]
8. Agudelo-Yepes, S.; Puerta-Suárez, J.; Carrillo-Gonzalez, D.F.; Cardona-Maya, W.D. Bacteriospermia assessment and its relationship with conventional seminal parameters in stud dogs ejaculates (*Canis familiaris*). *J. Hell. Vet. Med. Soc.* **2023**, *73*, 4785–4794. [CrossRef]
9. Goericke-Pesch, S.; Weiss, R.; Wehrend, A. Bacteriological findings in different fractions of canine ejaculates showing normospermia, teratozoospermia or azoospermia. *Aust. Vet. J.* **2011**, *89*, 318–322. [CrossRef]

10. Koziol, J.H.; Sheets, T.; Wickware, C.L.; Johnson, T.A. Composition and diversity of the seminal microbiota in bulls and its association with semen parameters. *Theriogenology* **2022**, *182*, 17–25. [CrossRef]
11. Ďuračka, M.; Belić, L.; Tokárová, K.; Žiarovská, J.; Kačániová, M.; Lukáč, N.; Tvrdá, E. Bacterial communities in bovine ejaculates and their impact on the semen quality. *Syst. Biol. Reprod. Med.* **2021**, *67*, 438–449. [CrossRef] [PubMed]
12. Lenický, M.; Slanina, T.; Kačániová, M.; Galovičová, L.; Petrovičová, M.; Ďuračka, M.; Benko, F.; Kováč, J.; Tvrdá, E. Identification of Bacterial Profiles and Their Interactions with Selected Quality, Oxidative, and Immunological Parameters of Turkey Semen. *Animals* **2021**, *11*, 1771. [CrossRef] [PubMed]
13. Tvrdá, E.; Petrovičová, M.; Benko, F.; Ďuračka, M.; Kováč, J.; Slanina, T.; Galovičová, L.; Žiarovská, J.; Kačániová, M. Seminal Bacterioflora of Two Rooster Lines: Characterization, Antibiotic Resistance Patterns and Possible Impact on Semen Quality. *Antibiotics* **2023**, *12*, 336. [CrossRef] [PubMed]
14. Domrazek, K.; Kaszak, I.; Kanafa, S.; Sacharczuk, M.; Jurka, P. The influence of Mycoplasma species on human and canine semen quality: A review. *Asian J. Androl.* **2023**, *25*, 29–37. [CrossRef] [PubMed]
15. Tesi, M.; Sabatini, C.; Vannozzi, I.; Di Petta, G.; Panzani, D.; Camillo, F.; Rota, A. Variables affecting semen quality and its relation to fertility in the dog: A retrospective study. *Theriogenology* **2018**, *118*, 34–39. [CrossRef] [PubMed]
16. Kutzler, M.A. Semen collection in the dog. *Theriogenology* **2005**, *64*, 747. [CrossRef]
17. van der Horst, G.; Maree, L. SpermBlue®: A new universal stain for human and animal sperm which is also amenable to automated sperm morphology analysis. *Biotech. Histochem.* **2010**, *84*, 299–308. [CrossRef]
18. Paray, A.A.; Singh, M.; Amin Mir, M. Gram Staining: A Brief Review. *Int. J. Res. Rev.* **2023**, *10*, 336–341. [CrossRef]
19. Talaiekhozani, A.; Sanaz, A.; Mohanadoss, P. Guidelines for Quick Application of Biochemical Tests to Identify Unknown Bacteria. *Acc. Biotechnol. Res.* **2015**, *2*, 65–82. [CrossRef]
20. Hof, H.; Bode, K.; von Stillfried, F. Canibacter oris—A fairly unknown pathogenic agent of bite wound infections. *GMS Infect. Dis.* **2021**, *9*, Doc01. [CrossRef]
21. Oh, C.; Lee, K.; Cheong, Y.; Lee, S.; Park, S.; Song, C.; Choi, I.; Lee, J. Comparison of the Oral Microbiomes of Canines and Their Owners Using Next-Generation Sequencing. *PLoS ONE* **2015**, *10*, 0131468. [CrossRef]
22. Tvrdá, E.; Ďuračka, M.; Benko, F.; Lukáč, N. Bacteriospermia—A formidable player in male subfertility. *Open Life Sci.* **2022**, *17*, 1001–1029. [CrossRef] [PubMed]
23. Dalmutt, A.C.; Moreno, L.Z.; Gomes, V.T.M.; Cunha, M.P.V.; Barbosa, M.R.F.; Sato, M.I.Z.; Knöbl, T.; Pedroso, A.C.; Moreno, A.M. Characterization of bacterial contaminants of boar semen: Identification by MALDI-TOF mass spectrometry and antimicrobial susceptibility profiling. *J. Appl. Anim. Res.* **2020**, *48*, 559–565. [CrossRef]
24. Maroto Martín, L.O.; Muñoz, E.C.; De Cupere, F.; Van Driessche, E.; Echemendia-Blanco, D.; Rodríguez, J.M.M.; Beeckmans, S. Bacterial contamination of boar semen affects the litter size. *Anim. Reprod. Sci.* **2010**, *120*, 95–104. [CrossRef] [PubMed]
25. Zhang, F.; Dai, J.; Chen, T. Role of Lactobacillus in Female Infertility Via Modulating Sperm Agglutination and Immobilization. *Front. Cell Infect. Microbiol.* **2021**, *10*, 620529. [CrossRef]
26. Mahiddine, F.Y.; You, I.; Park, H.; Kim, M.J. Commensal Lactobacilli Enhance Sperm Qualitative Parameters in Dogs. *Front. Vet. Sci.* **2022**, *9*, 238–246.

Disclaimer/Publisher's Note: The statements, opinions and data contained in all publications are solely those of the individual author(s) and contributor(s) and not of MDPI and/or the editor(s). MDPI and/or the editor(s) disclaim responsibility for any injury to people or property resulting from any ideas, methods, instructions or products referred to in the content.

animals

MDPI

Article

Changes in the Morphology and Antioxidant Status of European Red Deer Sperm Stored in the Epididymides and in a Liquid State

Nicoletta M. Neuman [1], Aleksandra Orzołek [1], Żaneta Steiner-Bogdaszewska [2] and Anna Dziekońska [1,*]

[1] Department of Animal Biochemistry and Biotechnology, University of Warmia and Mazury in Olsztyn, Oczapowskiego 5, 10-719 Olsztyn, Poland; nicoletta.neuman@uwm.edu.pl (N.M.N.); aleksandra.deszczka@uwm.edu.pl (A.O.)

[2] Witold Stefański Institute of Parasitology of the Polish Academy of Sciences, Research Station in Kosewo Górne, 11-700 Mrągowo, Poland; kosewopan@kosewopan.pl

* Correspondence: a.dziekonska@uwm.edu.pl

Simple Summary: The choice of the optimal sperm preservation method is an important consideration in animal breeding. Stored semen can be used for reproductive purposes to introduce new genotypes and prevent inbreeding, which poses a considerable problem in cervid farms. The aim of this study was to evaluate and compare the effect of storage time and storage method (liquid state/epididymides) on the motility, morphology, and antioxidant status of European red deer sperm stored at 5 °C for up to six days (D0-D6). Sperm samples were assessed for motility, viability, morphology, activity of antioxidant enzymes (superoxide dismutase, SOD; glutathione peroxidase, GPx; catalase, CAT), and lipid peroxidation (based on malondialdehyde, MDA, content). Significant differences between storage variants were noted on D2 in sperm morphology; on D4 in the percentage of progressively motile sperm, MDA content, and SOD and GPx activity; and on D6 in the percentage of motile and viable spermatozoa. Sperm motility, viability, and antioxidant status are more effectively preserved during liquid storage than epididymal storage. Morphological and functional abnormalities of sperm were observed earlier during epididymal storage, which suggests that spermatozoa can be stored for shorter periods of time in the epididymides than in a liquid state.

Abstract: The aim of this study was to evaluate the motility, morphology, and antioxidant status of European red deer sperm stored in a liquid state (variant I) and in the epididymides (variant II). Spermatozoa were harvested post-mortem from the cauda epididymis. Sperm samples in both variants were stored for up to six days (D6) at 5 °C. Spermatozoa were assessed for motility, viability, morphology, activity of antioxidant enzymes (superoxide dismutase, SOD; glutathione peroxidase, GPx; catalase, CAT), and lipid peroxidation (malondialdehyde, MDA, content). Sperm samples were analyzed on storage days 0, 2, 4, and 6 (D0-D6). Storage time and storage method significantly ($p \leq 0.05$) influenced the examined variables. On D2, a decrease in motility and acrosomal integrity was observed in both storage variants, whereas a decrease in viability and an increase in MDA content were noted in spermatozoa stored in the epididymides. On D4, higher values of SOD and GPx activity and MDA content were noted in variant I than in variant II. Catalase activity was very low. GPx is the key enzyme that participates in the reduction of hydrogen peroxide in sperm cells. Spermatozoa stored in a liquid state were characterized by higher motility and viability, improved morphology and antioxidant status than those stored in the epididymides; therefore, liquid storage is more recommended for short-term preservation of epididymal spermatozoa.

Keywords: epididymal sperm; morphology; antioxidant status; storage; European red deer

Citation: Neuman, N.M.; Orzołek, A.; Steiner-Bogdaszewska, Ż.; Dziekońska, A. Changes in the Morphology and Antioxidant Status of European Red Deer Sperm Stored in the Epididymides and in a Liquid State. *Animals* **2024**, *14*, 1653. https://doi.org/10.3390/ani14111653

Academic Editor: Eva Bussalleu

Received: 29 April 2024
Revised: 25 May 2024
Accepted: 29 May 2024
Published: 31 May 2024

Copyright: © 2024 by the authors. Licensee MDPI, Basel, Switzerland. This article is an open access article distributed under the terms and conditions of the Creative Commons Attribution (CC BY) license (https://creativecommons.org/licenses/by/4.0/).

1. Introduction

The knowledge about the semen of free-living animals and effective semen preservation methods is limited and should be expanded. The existing sperm storage methods have been developed based on our knowledge of livestock species. Deer sperm is preserved mostly with the use of methods designed for domesticated ruminants [1], and ejaculated semen is recommended for preservation. However, semen samples are difficult to obtain from cervids due to their unique behavior, and sperm is often collected post-mortem from the epididymides [2,3]. The epididymal sperm of cervids, similarly to ejaculated sperm, can be stored in a liquid state [4–6], in a frozen state [7,8], and in the epididymides [9,10]. Of these methods, cryopreservation is the most commonly used. However, it is not always recommended for economic and practical reasons. Moreover, the fertilizing capacity of cryopreserved sperm is reduced, compared with fresh sperm and sperm stored for a short period of time in a liquid state [1]. For this reason, short-term sperm storage may be a better method to protect sperm quality than cryopreservation. Epididymal sperm harvested from deer in their natural environment is a source of valuable genetic material for reproductive purposes [11,12]. Their use in assisted reproduction techniques can reduce the phenomenon of inbreeding in limited-population deer farms and contribute to significant breeding progress [2]. However, spermatozoa cannot always be collected from hunter-harvested animals directly after culling, and in some cases, sperm is stored in the epididymides for many hours. Previous studies have shown that prolonged storage in the epididymides can adversely affect sperm functions and their suitability for reproduction [9,13].

The quality of stored spermatozoa can be undermined by various factors, including oxidative stress. Oxidative stress increases lipid peroxidation and disrupts the activity of antioxidant enzymes [14,15]. Oxidative stress also causes oxidative damage to cell organelles and increases the production of reactive oxygen species (ROS) [14,16,17]. Spermatozoa stored at low temperatures are exposed to cold shock. Cold shock enhances ROS production in sperm cells and increases their susceptibility to lipid peroxidation [18].

In healthy organisms, antioxidant defense systems offer protection against the harmful effects of ROS [19,20]. The key antioxidant enzymes include superoxide dismutase (SOD), glutathione peroxidase (GPx), and catalase (CAT) [16,17,19,21]. Sperm cells rely on the antioxidant enzymes in the epididymides and their own antioxidant capacity to mitigate the harmful consequences of ROS during transport through the epididymides, storage, and fertilization [22].

Increased lipid peroxidation and impaired antioxidant status during storage can affect sperm viability, morphology, and motility [23,24]. Sperm morphology is an important consideration during sperm assessment [25]. Spermatozoa with a normal morphology have a higher fertilization potential [26]. Morphological defects, in particular, head defects, significantly undermine sperm's ability to fuse with an egg cell [25]. Changes in the morphology and antioxidant status of stored cervid epididymal spermatozoa have not been analyzed to date.

The aim of this study was to evaluate the influence effect of storage time on the motility, morphology, and antioxidant status of spermatozoa stored in a liquid state and in the epididymides at a temperature of 5 °C. Furthermore, the effect of the storage variant (liquid storage vs. storage in the epididymides) on sperm parameters was determined on subsequent days of storage (D2, D4, D6). Sperm motility, viability, sperm morphology (including morphological defects), the activity of antioxidant enzymes (SOD, GPx, CAT), and lipid peroxidation (based on malondialdehyde, MDA, content) were assessed in spermatozoa stored for up to six days.

2. Materials and Methods

2.1. Collection of Epididymal Sperm

Testicles and epididymides stored in scrotal sacs were collected from 32 European red deer (*Cervus elaphus elaphus*) stags that were culled during legal hunts in the breeding season

(September–October). European red deer were hunter-harvested in the Nowe Ramuki Forest District (Warmian-Masurian Voivodeship, Poland) according to hunting and wildlife management rules. Testicles and epididymides were collected within 5 h after the hunt and were transported to the laboratory of the Department of Animal Biochemistry and Biotechnology of the University of Warmia and Mazury in Olsztyn.

Sperm samples for analyses were obtained from the cauda epididymis by multiple incisions made with a scalpel and were collected in Eppendorf tubes [5]. During a preliminary assessment of sperm samples, sperm concentrations were determined using a Bürker chamber (Equimed-Medical Instruments, Cracow, Poland). In each sample, sperm motility was assessed subjectively under a microscope [5].

2.2. Sperm Storage in a Liquid State (Variant I)

Sperm collected from the cauda epididymis of each animal were stored in a liquid state. After standard evaluation, each sperm sample was diluted in Salomon's extender [4] (to a concentration of 100 × 10^6 spermatozoa/mL). The samples were then incubated at room temperature for two hours and stored for up to six days in a refrigerator at 5 °C.

2.3. Storage of Spermatozoa in the Epididymis (Variant II)

The second testicle with the epididymis was stored in scrotal sacs in the refrigerator for up to six days. In this variant, testicles with epididymides were randomly divided into three groups that were stored in the refrigerator for two days (D2, n = 11), four days (D4, n = 11), and six days (D6, n = 10). After refrigeration, sperm samples were collected, subjected to a standard evaluation, and diluted in Salomon's extender (100 × 10^6 spermatozoa/mL).

2.4. Sperm Analysis

Sperm motility, viability, morphology, antioxidant status, and lipid peroxidation were determined after dilution of sperm samples (D0) and on subsequent days of storage (D2, D4, and D6).

2.4.1. Assessment of Sperm Motility

Before the analysis, sperm samples were diluted with Dulbecco's phosphate-buffered saline (DPBS; Gibco, Grand Island, NY, USA) in a 1:4 ratio (to obtain a concentration of 20–30 × 10^6 spermatozoa/mL). The prepared samples were heated for approximately 5 min at 37 °C in a thermoblock (Thermo Block TDR-120, Göttingen, Germany). An aliquot (5 µL) of each sample was placed in a Makler counting chamber (Sefi-Medical Instruments Ltd., Haifa, Israel) preheated to 37 °C. Every sample was assessed in the CASA system and the Hamilton Thorne IVOS v. 12.3 sperm analyzer (Hamilton Thorne Bioscience, Beverley, MA, USA).

The evaluation involved software settings that were recommended by the manufacturer for analyses of gazelle/deer sperm: frame acquired—60, frame rate—60 Hz, minimum cell contrast—60, minimum cell size—5 pixels, straightness threshold—80%, low VAP cut-off—21.9 µm/s, low VSL cut-off—6 µm/s. Total motility (TMOT, %) and progressive motility (PMOT, %) were defined according to Hamilton Thorne requirements (VAP > 75 µm/s, STR > 80%).

2.4.2. Sperm Viability Analysis

Sperm viability was assessed using the Muse® Cell Analyzer (Luminex Corporation, Austin, TX, USA). Sperm samples were prepared for the analysis with the use of the Muse Count & Viability Kit (Luminex Corporation, Austin, TX, USA), according to the manufacturer's recommendations. The reagent contains two dyes. The first dye penetrates the cell membrane and stains DNA, and nucleated cells are separated from non-nucleated cells and impurities. The second dye contains 7-aminoactinomycin D (7-AAD), which stains only cells with damaged cell membranes. This combination of dyes is used to differentiate between populations of live and dead cells. Before analysis, each sperm sample was diluted with

DPBS to a concentration of 1×10^6–1×10^7 cells/mL. An aliquot of 20 μL was combined with 380 μL of the Muse Count & Viability Kit reagent and incubated for 5 min at room temperature. Sperm viability was determined with the use of the Muse® Cell Analyzer.

2.4.3. Assessment of Sperm Morphology

The morphology of epididymal sperm was assessed by staining smears with the Giemsa method [27]. To prepare smear slides, 10 μL of the sperm sample (previously diluted with DPBS solution in a 1:4 ratio) was applied to a glass slide, spread over the entire slide, and left to dry at room temperature. Dried smears were fixed in a formalin buffer for 15 min, rinsed with water, and allowed to dry. The smears were then stained with Giemsa's solution for 180 min. Smear slides were rinsed with distilled water and dried. In each sample, 200 sperm cells were counted under a phase contrast microscope at ×1000 magnification with oil immersion. During the analysis, normal spermatozoa; spermatozoa with intact acrosomes; spermatozoa with head, tail, and midpiece defects; and spermatozoa with proximal and distal cytoplasmic droplets were counted (Figure 1).

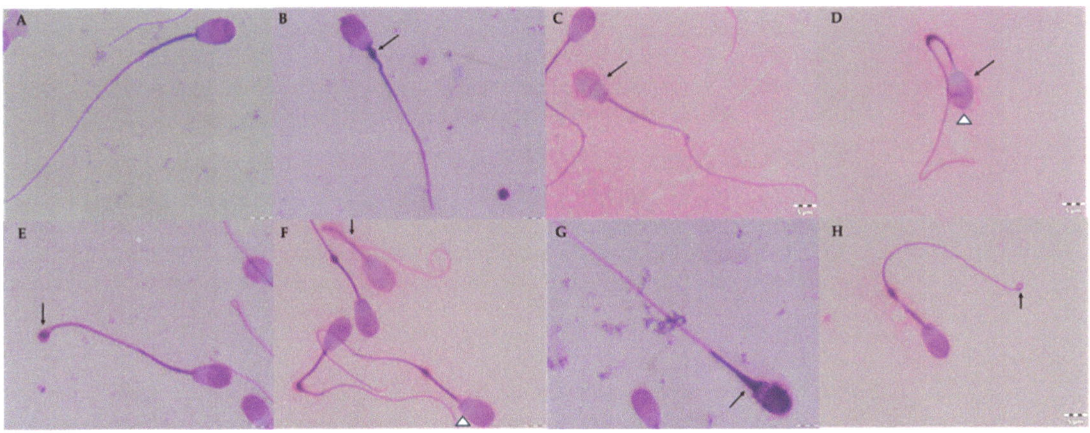

Figure 1. Morphological defects in European red deer epididymal spermatozoa stored in a liquid state and in the epididymides at 5 °C. (**A**)—normal spermatozoa; (**B**)—spermatozoa with proximal cytoplasmic droplets (↑); (**C**)—spermatozoa with head defects (↑); (**D**)—spermatozoa with head defects (↑) and damaged acrosomes (Δ); (**E**)—spermatozoa with distal cytoplasmic droplets (↑); (**F**)—spermatozoa with curled tails (↑), spermatozoa with head defects (Δ); (**G**)—spermatozoa with midpiece defects (↑); (**H**)—spermatozoa with a single bent tail (↑). Scale bars = 5 μm.

2.5. *Activity of Antioxidant Enzymes and Lipid Peroxidation*

On designated days, 0.5 mL aliquots of diluted sperm samples were collected in separate Eppendorf tubes. Each sample was diluted with 0.5 mL of 0.85% NaCl to a concentration of 50×10^6 spermatozoa/mL. The samples were rinsed twice with 0.85% NaCl (3000× *g*, 5 min, 10 °C). The supernatant was discarded, and the precipitate was suspended in 1 mL of 0.85% NaCl. The prepared samples were frozen at −80 °C and stored for further analyses.

Sperm sediments were homogenized in a FastPrep®-24 apparatus (MP Biomedicals, Santa Ana, CA, USA) before the determination of antioxidant enzyme activity and MDA content. To obtain a clear supernatant, the samples were centrifuged at 12,000× *g* for 15 min at 10 °C, and the supernatant was transferred to clean Eppendorf tubes and stored at −80 °C for analysis.

2.5.1. Superoxide Dismutase Activity

Superoxide dismutase activity was assessed with a Beckman Coulter DU 800 spectrophotometer (Beckman Coulter INC., Fullerton, CA, USA). The samples were prepared for analysis using the Ransod kit (Randox Laboratories, Crumlin, UK), according to the instructions provided by the manufacturer. SOD activity was measured at a wavelength of 505 nm. One unit (U) of SOD was defined as the amount of the enzyme that caused 50% inhibition of 2-(4-iodophenyl)-3-(4-nitrophenol)-5 phenyltetrazolium chloride (INT) reduction at 37 °C and pH 7.0. The results were expressed as U/10^6 spermatozoa (spz).

2.5.2. Glutathione Peroxidase Activity

Glutathione peroxidase activity was measured spectrophotometrically using a commercial Ransel assay (Randox Laboratories, Crumlin, UK) according to the manufacturer's instructions. GPx activity was measured at a wavelength of 340 nm. One GPx unit was defined as the amount of the enzyme that catalyzes the oxidation of 1 µM of NADPH per minute at 37 °C and pH 7.2. The results were expressed as U/10^6 spz.

2.5.3. Catalase Activity

Catalase activity was measured using a commercial Catalase Assay Kit (Sigma-Aldrich Co., Saint Louis, MI, USA) according to the manufacturer's instructions. Catalase activity was measured by determining the amount of H_2O_2 remaining after the reaction catalyzed by CAT. CAT activity was measured spectrophotometrically at 520 nm. One CAT unit was defined as the amount of the enzyme that decomposed 1 µM of H_2O_2 per minute at 25 °C and pH 7.0. The results were expressed as U/10^6 spz.

2.5.4. Malondialdehyde (MDA) Content

Lipid peroxidation in sperm cell membranes was evaluated by spectrophotometric determination of MDA content with a commercial BIOXYTECH® MDA-586TM Assay Kit (OxisResearch, Burlingame, CA, USA) according to the manufacturer's instructions. The measurement was performed at a wavelength of 586 nm. The results were expressed as µM MDA/10^6 spz.

2.6. Statistical Analysis

Data were processed statistically in Statistica v. 12.5 (StatSoft Incorporation, Tulsa OK, USA). The normality of data distribution was checked using the Shapiro–Wilk test. The results were analyzed with the use of non-parametric tests because the number of samples (n) differed across groups, and some variables did not have a normal distribution. Two independent samples were compared using the Mann–Whitney U test to determine the presence of significant differences between storage times in each variant and between storage variants on D2, D4, and D6. The results were presented as means ± SEM.

In addition, the presence of significant correlations ($p \leq 0.05$, $p \leq 0.01$, $p \leq 0.001$) between variables in each storage variant and storage time was determined by calculating Spearman's rank correlation coefficient. The results of the correlation analysis were presented for the sixth day of storage (Supplementary Materials).

3. Results

3.1. The Effect of Storage Time on the Motility, Viability, Morphology, and Antioxidant Status of Spermatozoa Stored in a Liquid State and in the Epididymides

Storage time significantly influenced the motility and viability of spermatozoa stored in a liquid state (variant I) and in the epididymides (variant II) (Figure 2). The percentage of motile sperm (TMOT) decreased significantly in both variants already on the second day of storage. The percentage of progressively motile spermatozoa (PMOT) and the percentage of spermatozoa with integral plasma membranes (Viability) also decreased significantly on D2 in sperm stored in the epididymides. In sperm stored in a liquid state, a significant decrease in PMOT and Viability was observed only on D4. However, on D6, TMOT and Viability exceeded 70%, and PMOT approximated 24% in variant I, whereas in variant II, TMOT and Viability decreased to approximately 30–40%, and PMOT decreased below 10%.

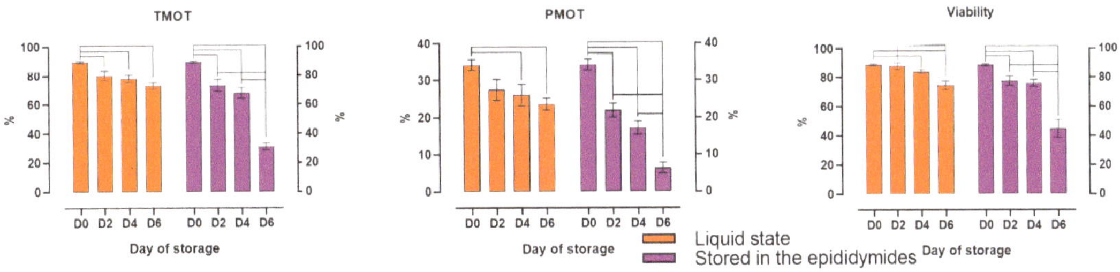

Figure 2. The effect of storage time on the motility and viability of spermatozoa stored in a liquid state and in the epididymides at 5 °C. TMOT, total motility; PMOT, progressive motility; Viability, live spermatozoa. The mean (±SEM) values of stored epididymal spermatozoa are presented. The results are significant at $p \leq 0.05$.

The changes in the morphology of spermatozoa stored in a liquid state and in the epididymides for six days are presented in Figure 3. The percentage of spermatozoa with normal morphology (MOR) was highest on D2 in samples stored in a liquid state. The values of MOR decreased gradually over time in both variants. In variant I, significant differences in MOR values were noted between D0 and D2, and between D0 and D2 vs. D4 and D6. In variant II, MOR decreased significantly to around 45% only on D6, compared with around 55% in variant I. Significant changes in acrosome integrity were also observed during storage. In both storage variants, the percentage of spermatozoa with normal apical ridge acrosomes (NAR) was lowest on D2 and D6 relative to D0. The percentage of spermatozoa with head (HD), midpiece (MD), and tail defects (TD) increased over time. In both variants, significant differences ($p \leq 0.05$) in HD and MD were noted between D0 and D6. A significant increase ($p \leq 0.05$) in HD and MD was observed on D6 in spermatozoa stored in the epididymides relative to those stored in a liquid state. The percentage of spermatozoa with tail defects (TD) also increased over time in both variants. However, significant differences ($p \leq 0.05$) in TD were noted only in variant I between D0 and D2, and between D2 vs. D4 and D6. The percentage of spermatozoa with proximal cytoplasmic droplets was highest on D2 in both variants, but significant differences ($p \leq 0.05$) in this parameter were observed only in variant II between D0 and D2. The percentage of spermatozoa with distal cytoplasmic droplets also decreased over time, in particular in variant II, where this parameter decreased from around 48% on D0 to around 20% on D6. In both variants, significant differences ($p \leq 0.05$) in this parameter were observed between D0 and D2.

The changes in the activity of antioxidant enzymes and lipid peroxidation in spermatozoa stored in a liquid state and in the epididymides for six days are presented in Figure 4. Significant changes in SOD activity were observed over time, subject to storage variant. In samples stored in a liquid state, SOD activity was highest on D2 and D4. In spermatozoa stored in the epididymides, SOD activity was highest on D0 and decreased on successive days of storage. Significant differences ($p \leq 0.05$) in SOD activity were noted between D0 and D6 in both variants. In variant I, GPx activity and MDA content were highest on D4 and lowest on D6. Significant differences ($p \leq 0.05$) in the values of both parameters were observed between D0 and D4 in variant I. In variant II, GPx activity and MDA content were highest on D2 and lowest on D6. Significant differences ($p \leq 0.05$) in MDA content were noted between D0 and D6 in variant II. CAT activity was lowest on D4 in both variants, and significant differences ($p \leq 0.05$) in this parameter were observed only between D0 and D4.

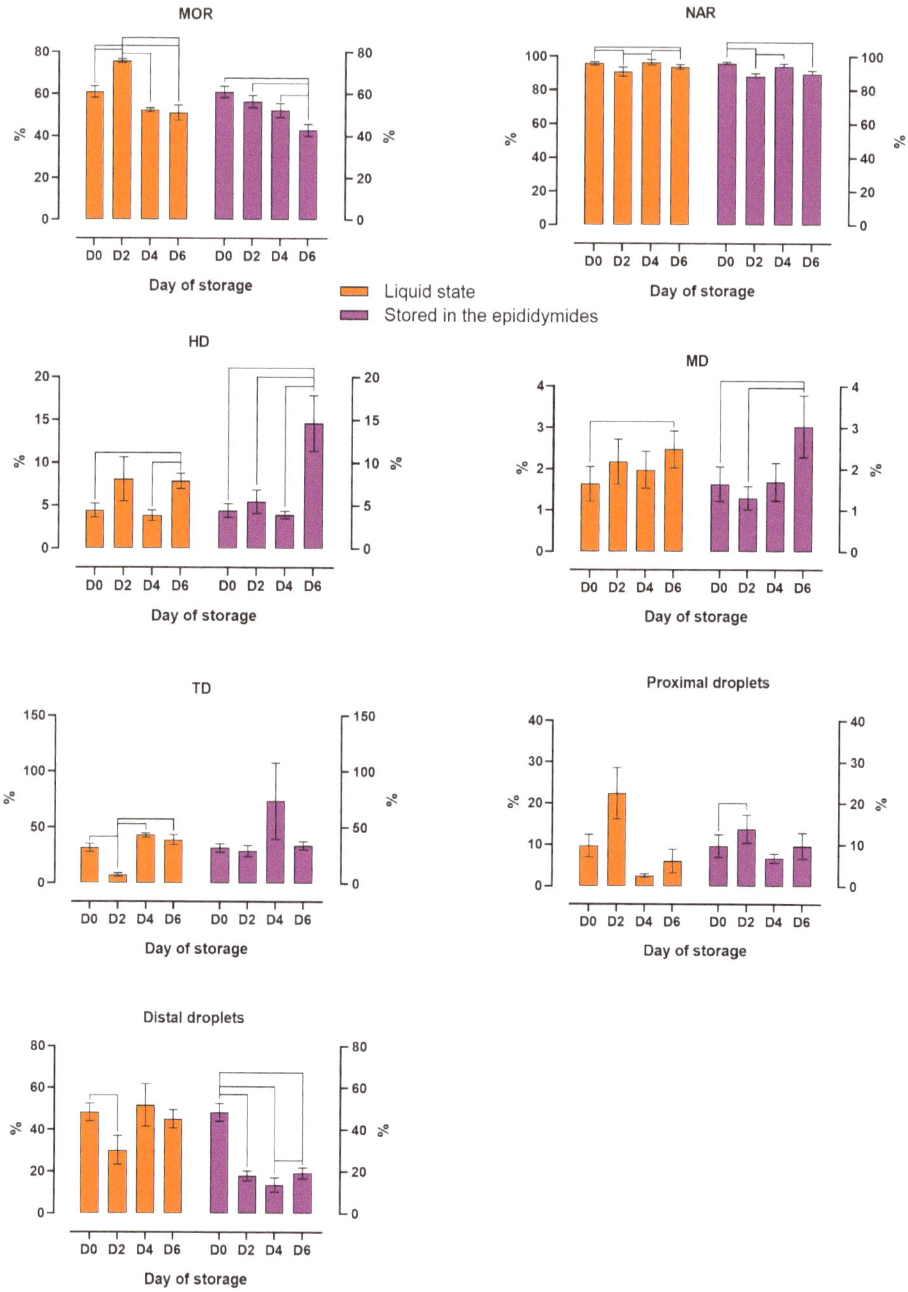

Figure 3. The effect of storage time on the morphology of spermatozoa stored in a liquid state and in the epididymides at 5 °C. MOR, normal sperm; NAR, normal apical ridge acrosomes; HD, head defects; MD, midpiece defects; TD, tail defects; proximal droplets; distal droplets. The mean (±SEM) values of stored epididymal sperm are presented. The results are significant at $p \leq 0.05$.

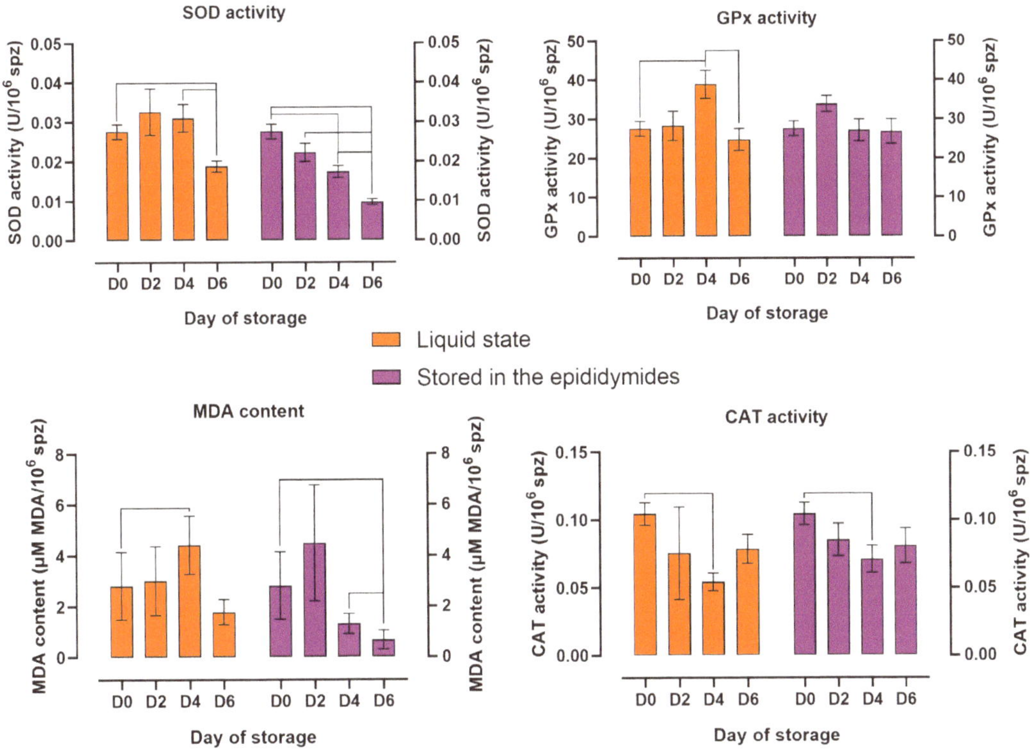

Figure 4. The effect of storage time on the activity of antioxidant enzymes and lipid peroxidation in spermatozoa stored in a liquid state and in the epididymides at 5 °C. SOD activity, superoxide dismutase activity; GPx activity, glutathione peroxidase activity; MDA content, malondialdehyde content; CAT activity, catalase activity. The mean (±SEM) values of stored epididymal sperm are presented. The results are significant at $p \leq 0.05$.

3.2. The Effect of Storage Variant on Sperm Motility, Viability, Morphology, and Antioxidant Status

The storage variant had a significant influence on the analyzed variables, depending on storage time (Table 1). The earliest significant differences between the compared variants were noted on D2 in MOR values, which can be attributed to higher TD values in variant II than in variant I. On D4, a significant decrease ($p \leq 0.05$) in PMOT, Viability, Distal Droplets, SOD and GPx activity, and MDA content, and a significant increase ($p \leq 0.05$) in Proximal Droplets were noted in epididymal sperm relative to sperm stored in a liquid state. On the last day of storage (D6), a further significant decrease in PMOT, Viability, Distal Droplets, and SOD activity, a decrease in TMOT values, and an increase in HD values were noted in variant II relative to variant I.

Table 1. The effect of storage variant (liquid state vs. storage in the epididymides) on the motility, viability, morphology, and antioxidant status of European red deer spermatozoa stored at 5 °C.

Sperm Quality Parameters	Day of Storage					
	D2 n = 11		D4 n = 11		D6 n = 10	
	Variant I	Variant II	Variant I	Variant II	Variant I	Variant II
TMOT	80.00 ± 3.16	73.17 ± 4.11	78.14 ± 2.66	68.00 ± 3.66	73.00 ± 2.26 [a]	31.00 ± 5.37 [b]
PMOT	27.40 ± 2.79	21.92 ± 1.86	26.00 ± 2.90 [a]	17.17 ± 1.77 [b]	23.58 ± 1.59 [a]	6.30 ± 1.48 [b]
Viability	87.80 ± 22.23	77.56 ± 3.26	84.02 ± 1.27 [a]	76.05 ± 2.37 [b]	74.61 ± 2.84 [a]	44.69 ± 6.22 [b]
MOR	75.74 ± 0.88 [a]	56.47 ± 2.86 [b]	52.35 ± 0.89	52.27 ± 3.34	51.08 ± 3.58	42.85 ± 2.84
NAR	90.94 ± 2.83	88.21 ± 1.86	96.73 ± 1.56	94.54 ± 1.48	93.93 ± 1.51	89.75 ± 2.09
HD	8.08 ± 2.58	5.47 ± 1.35	42.97 ± 1.98	3.92 ± 0.42	7.85 ± 0.88 [a]	14.63 ± 3.23 [b]
MD	2.18 ± 0.55	1.30 ± 0.29	1.98 ± 0.45	1.69 ± 0.46	2.48 ± 0.45	3.04 ± 0.74
TD	7.48 ± 1.65 [a]	28.99 ± 5.03 [b]	42.97 ± 1.98	73.60 ± 33.94	38.89 ± 4.62	33.80 ± 3.63
Prox Drop	22.41 ± 6.21	13.87 ± 3.40	2.55 ± 0.45 [a]	6.78 ± 1.04 [b]	6.11 ± 2.85	9.73 ± 3.09
Dist Drop	30.22 ± 6.91	18.05 ± 2.28	51.75 ± 10.03 [a]	13.65 ± 3.39 [b]	45.22 ± 4.32 [a]	19.28 ± 2.55 [b]
SOD activity	0.03 ± 0.01	0.02 ± 0.01	0.03 ± 0.01 [a]	0.02 ± 0.01 [b]	0.02 ± 0.01 [a]	0.01 ± 0.01 [b]
GPx activity	28.32 ± 3.75	33.88 ± 2.07	38.90 ± 3.54 [a]	27.17 ± 2.78 [b]	24.75 ± 2.80	26.85 ± 3.12
MDA content	3.02 ± 1.35	4.50 ± 2.28	4.43 ± 1.14 [a]	1.32 ± 0.40 [b]	1.79 ± 0.50	0.67 ± 0.37
CAT activity	0.08 ± 0.03	0.09 ± 0.01	0.05 ± 0.01	0.07 ± 0.01	0.08 ± 0.01	0.08 ± 0.01

TMOT, total motility (%); PMOT, progressive motility (%); Viability, live spermatozoa (%); MOR, normal sperm (%); NAR, normal apical ridge acrosomes (%); HD, head defects (%); MD, midpiece defects (%); TD, tail defects (%); Prox Drop, Proximal Droplets (%); Dist Drop, Distal Droplets (%); SOD activity, superoxide dismutase activity (U/10^6 spz); GPx activity, glutathione peroxidase activity (U/10^6 spz); MDA content, malondialdehyde content (µm MDA/10^6 spz); CAT activity, catalase activity (U/10^6 spz). The mean (±SEM) values of stored epididymal sperm are presented. Values marked with different letters ([a,b]) denote significant differences between storage variants on the same day of storage at $p \leq 0.05$.

4. Discussion

This is the first study to analyze the influence of storage time and short-term storage method on changes in the morphology and antioxidant status of European red deer epididymal sperm. Previous research was conducted on other animal species [18,28–30], whereas very few studies examined cervids and reported only on morphological changes [31,32].

Sperm motility and viability are crucial for successful fertilization, and these parameters were also analyzed in the present study. Motility is a key determinant of sperm quality [33] and suitability for assisted reproductive technologies (ART). The integrity of the acrosomal membrane is essential for sperm functions, including motility [34]. In the current study, the motility of spermatozoa stored in a liquid state and in the epididymides was similar to that noted by other researchers [3,6,10,13,35]. The motility and viability of epididymal spermatozoa stored in a liquid state were maintained at a satisfactory level (above 70%) until the sixth day of storage, which could indicate that their functional capacity was retained. This finding corroborates the results of our previous study, which demonstrated that epididymal sperm stored in a liquid state for up to 10 days remained useful for reproductive purposes [6]. In spermatozoa stored in the epididymides, comparable levels of motility and viability were observed only until the second day of storage, and a significant decrease in these parameters (to approx. 31% and 44%, respectively) was noted on the sixth day of storage. Motility and viability are closely correlated parameters [5], and this observation was confirmed in this study in both storage variants (Tables S1 and S2, Figure S1).

Storage induces many age-related changes that disrupt metabolic processes, increase lipid peroxidation, and cause damage to sperm structures [14,28,36,37]. In the present study, the percentage of spermatozoa with normal morphology decreased over time, but the prevalence of morphological defects differed across storage variants. The percentage of sperm with tail defects and, on the last day of storage (D6), with midpiece and head defects was significantly higher in spermatozoa stored in the epididymides than in a liquid state. These changes can undermine sperm motility and viability [34], which was observed on

D2 in spermatozoa stored in the epididymides, but only on D4 in sperm stored in a liquid state. Interestingly, a greater decrease in the percentage of spermatozoa with cytoplasmic droplets (in particular distal droplets) was noted in sperm stored in the epididymides than in a liquid state. Cytoplasmic droplets are characteristic of immature sperm cells, and they are commonly found in epididymal sperm [38–40]. Cytoplasmic droplets are shed during maturation, and the results of the present study indicate that epididymal storage supported sperm maturation. During liquid storage, the decrease in the percentage of spermatozoa with cytoplasmic droplets was correlated with changes in motility, and similar observations had been previously made in the epididymal sperm of mice [41], which suggests that cytoplasmic droplets play an important role in motility regulation [38]. Other researchers found that cytoplasmic droplets are transient organelles that provide the necessary energy for the maturation of epididymal spermatozoa [25,41].

Disruptions in metabolic processes, such as lipid peroxidation, induce irreversible damage to sperm structures [14]. Stored spermatozoa become more susceptible to oxidative damage caused by ROS. Reactive oxygen species are produced in the mitochondria and plasma membranes of spermatozoa, and their generation is controlled by the antioxidant system composed of enzymatic and non-enzymatic antioxidants [42,43]. The enzymatic system of many animal species is composed mainly of GPx, CAT, and SOD [14,22,42]. SOD is considered the key enzyme in the antioxidant defense system of spermatozoa [21,44].

In the current study, SOD activity in spermatozoa stored in a liquid state decreased over time, and similar results were reported in the spermatozoa of rams [45], cattle, and buffalo bulls during liquid storage [18]. In sperm stored in the epididymides, a significant decrease in SOD activity was reported much earlier (on D2). In addition, on D4, SOD activity was higher in spermatozoa stored in a liquid state than in the epididymides, which could suggest that the antioxidant capacity of sperm cells is more effectively preserved during liquid storage. Interestingly, GPx activity in the epididymal storage variant remained fairly constant during the entire storage period, whereas in sperm cells stored in a liquid state, the GPx activity was highest on D4. In contrast, GPx activity in cattle and buffalo bull spermatozoa continued to decrease on successive days of storage [18].

The content of MDA, a common biomarker of lipid peroxidation, increased during storage. This observation suggests that peroxidation occurred much earlier during epididymal storage (MDA content peaked on D2) than liquid storage (D4). The increase in MDA content was accompanied by an increase in GPx activity. The activity of this antioxidant enzyme probably increased to counteract excessive ROS production [14,46,47].

Lipid peroxidation is triggered by ROS, and ROS generation is controlled by SOD, GPx, and CAT. SOD catalyzes the dismutation of superoxide radicals (O^{2-}) to O_2 and H_2O_2 [22,48,49]. The generated H_2O_2 is decomposed by CAT to oxygen and water. In addition, H_2O_2 is neutralized by GPx and glutathione reductase [22,49,50].

In the present study, CAT activity was low in both storage variants, which could suggest that H_2O_2 in deer spermatozoa is reduced mainly by GPx. This observation corroborates the previous findings of Drevet et. al. [20]. Similarly to other antioxidant enzymes, CAT activity decreased over time. Low CAT activity could also suggest that deer sperm contain trace amounts of this enzyme. However, further research involving more sensitive tests and molecular techniques is needed to validate this hypothesis since CAT was not detected in the spermatozoa of many animal species, including cattle, dogs, and boars [22,51,52]. Several authors reported extremely low levels of CAT activity in human and rat spermatozoa [53,54]. In turn, CAT was detected in stallion sperm [29].

The antioxidant status of deer spermatozoa has not been investigated to date, and the present study provides novel information that could be used to develop new methods for storing epididymal spermatozoa.

5. Conclusions

The study demonstrated that the storage time of up to 6 days significantly affects the quality of the epididymal spermatozoa of European red deer stored in a liquid state and in the epididymides. However, the motility, viability, morphology, and antioxidant status of deer spermatozoa are more effectively preserved during liquid storage than epididymal storage. Lipid peroxidation, morphological defects, and a decrease in SOD activity occurred earlier in spermatozoa stored in the epididymides than in a liquid state, which exerted a negative effect on sperm viability and motility. Low CAT activity in epididymal spermatozoa suggests that hydrogen peroxide in sperm cells is reduced mainly by GPx. Based on the present findings, liquid storage is more recommended for the short-term preservation of deer spermatozoa than epididymal storage. The results of this study also indicate that sperm stored in the epididymides at 5 °C for up to 4 days could be suitable for reproduction. Further comparative studies involving in vitro fertilization or artificial insemination are recommended to confirm these findings.

Supplementary Materials: The following supporting information can be downloaded at: https://www.mdpi.com/article/10.3390/ani14111653/s1, Figure S1: Correlation coefficients describing the relationships between the quality parameters of European red deer epididymal spermatozoa stored in a liquid state (A) and in the epididymides (B) at 5 °C for 6 days. Table S1: Correlation coefficients describing the relationships between the quality parameters of European red deer epididymal spermatozoa stored in a liquid state at 5 °C for 6 days. Table S2: Correlation coefficients describing the relationships between the quality parameters of European red deer spermatozoa stored in the epididymides at 5 °C for 6 days.

Author Contributions: Conceptualization, N.M.N. and A.D.; methodology, N.M.N. and A.D.; analysis, N.M.N. and A.O.; resources, Ż.S.-B.; writing—original draft preparation, N.M.N.; writing—review and editing, A.D. All authors have read and agreed to the published version of the manuscript.

Funding: Funded by the Minister of Science under "the Regional Initiative of Excellence Program" (RID/SP/0025/2024/01) and the University of Warmia and Mazury in Olsztyn (project No. 11.610.003-110).

Institutional Review Board Statement: A hunting license was obtained for the needs of this study. The animals were hunter-harvested in accordance with hunting and game management regulations stipulated in Section 10 of the Polish Hunting Law of 13 October 1995 and European Union regulations.

Informed Consent Statement: Not applicable.

Data Availability Statement: The data presented in this study are available on request from the corresponding author.

Acknowledgments: The samples were obtained under a scientific cooperation agreement concluded between the University of Warmia and Mazury in Olsztyn and Nowe Ramuki Forest District. The authors would like to thank the employees of Nowe Ramuki Forest Inspectorate, especially K. Zdun and T. Dzietowiecki, as well as the hunters for their help in collecting samples.

Conflicts of Interest: The authors declare no conflicts of interest.

References

1. Asher, G.W.; Berg, D.K.; Evans, G. Storage of semen and artificial insemination in deer. *Anim. Reprod. Sci.* **2000**, *62*, 195–211. [CrossRef] [PubMed]
2. Garde, J.J.; Martínez-Pastor, F.; Gomendio, M.; Malo, A.F.; Soler, A.J.; Fernández-Santos, M.R.; Esteso, M.C.; García, A.J.; Anel, L.; Roldán, E.R.S. The application of reproductive technologies to natural populations of red deer. *Reprod. Domest. Anim.* **2006**, *41*, 93–102. [CrossRef] [PubMed]
3. Dziekońska, A.; Neuman, N.M.; Burdal, K.K.; Wiszniewska-Łaszczych, A.; Bogdaszewski, M. The Effect of Different Extenders on the Quality Characteristics of European Red Deer Epididymal Sperm Stored at 5 °C. *Animals* **2022**, *12*, 2669. [CrossRef] [PubMed]
4. Fernández-Santos, M.R.; Esteso, M.C.; Soler, A.J.; Montoro, V.; Garde, J.J. Effects of egg yolk and cooling rate on the survival of refrigerated red deer (*Cervus elaphus hispanicus*) epididymal spermatozoa. *Reprod. Domest. Anim.* **2006**, *41*, 114–118. [CrossRef] [PubMed]
5. Dziekońska, A.; Niedźwiecka, E.; Niklewska, M.E.; Koziorowska-Gilun, M.; Kordan, W. Viability longevity and quality of epididymal sperm stored in the liquid state of European red deer (*Cervus elaphus elaphus*). *Anim. Reprod. Sci.* **2020**, *213*, 106269. [CrossRef]

6. Dziekońska, A.; Koziorowska-Gilun, M.; Kordan, W.; Neuman, N.M.; Kotlarczyk, A.M.; Korzekwa, A.J. The Quality and Fertilizing Potential of Red Deer (*Cervus elaphus* L.) Epididymal Spermatozoa Stored in a Liquid State. *Int. J. Mol. Sci.* **2022**, *23*, 14591. [CrossRef] [PubMed]
7. Martinez-Pastor, F.; Anel, L.; Guerra, C.; Alvarez, M.; Soler, A.J.; Garde, J.J.; Chamorro, C.; da Paz, P. Seminal plasma improves cryopreservation of Iberian red deer epididymal sperm. *Theriogenology* **2006**, *66*, 1847–1856. [CrossRef] [PubMed]
8. Fernández-Santos, M.R.; Martinez-Pastor, F.; Garcia-Macias, V.; Esteso, M.C.; Soler, A.J.; da Paz, P.; Anel, L.; Garde, J.J. Extender osmolality and sugar supplementation exert a complex effect on the cryopreservation of Iberian red deer (*Cervus elaphus hispanicus*) epididymal spermatozoa. *Theriogenology* **2007**, *67*, 738–753. [CrossRef] [PubMed]
9. Fernández-Santos, M.R.; Dominguez-Rebolledo, A.E.; Esteso, M.C.; Garde, J.J.; Martínez-Pastor, F. Refrigerated Storage of Red Deer Epididymal Spermatozoa in the Epididymis, Diluted and with Vitamin C Supplementation. *Reprod. Domest. Anim.* **2009**, *44*, 212–220. [CrossRef]
10. Neuman, N.M.; Dziekońska, A.; Orzołek, A.; Gilun, P. A comparison of the biological properties of European red deer (*Cervus elaphus elaphus*) spermatozoa stored in the epididymides and in a liquid state at 5 °C. *Anim. Reprod. Sci.* **2024**, *260*, 107387. [CrossRef]
11. Morrow, C.J.; Penfold, L.M.; Wolfe, B.A. Artificial insemination in deer and non-domestic bovids. *Theriogenology* **2009**, *71*, 149–165. [CrossRef] [PubMed]
12. Comizzoli, P.; Mauget, R.; Mermillod, P. Assessment of in vitro fertility of deer spermatozoa by heterologous IVF with zona-free bovine oocytes. *Theriogenology* **2001**, *56*, 261–274. [CrossRef] [PubMed]
13. Soler, A.J.; Esteso, M.C.; Fernández-Santos, M.R.; Garde, J.J. Characteristics of Iberian red deer (*Cervus elaphus hispanicus*) spermatozoa cryopreserved after storage at 5 °C in the epididymis for several days. *Theriogenology* **2005**, *64*, 1503–1517. [CrossRef] [PubMed]
14. Partyka, A.; Łukaszewicz, E.; Niżański, W. Effect of cryopreservation on sperm parameters, lipid peroxidation and antioxidant enzymes activity in fowl semen. *Theriogenology* **2012**, *77*, 1497–1504. [CrossRef] [PubMed]
15. Cassani, P.; Beconi, M.T.; O'Flaherty, C. Relationship between total superoxide dismutase activity with lipid peroxidation, dynamics and morphological parameters in canine semen. *Anim. Reprod. Sci.* **2005**, *86*, 163–173. [CrossRef]
16. Aitken, R.J.; Roman, S.D. Antioxidants systems and oxidative stress in the testes. In *Molecular Mechanisms in Spermatogenesis, Advances in Experimental Medicine and Biology*; Cheng, Y.C., Ed.; Landes Bioscience: Austin, TX, USA; Springer Science Business Media: Boston, MA, USA, 2008; Volume 636, pp. 154–171.
17. Chabory, E.; Damon, C.; Lenoir, A.; Henry-Berger, J.; Vernet, P.; Cadet, R.; Saez, F.; Drevet, J.R. Mammalian glutathione peroxidases control acquisition and maintenance of spermatozoa integrity. *J. Anim. Sci.* **2010**, *88*, 1321–1331. [CrossRef] [PubMed]
18. Nair, S.J.; Brar, A.S.; Ahuja, C.S.; Sangha, S.P.S.; Chaudhary, K.C. A comparative study on lipid peroxidation, activities of antioxidant enzymes and viability of cattle and buffalo bull spermatozoa during storage at refrigeration temperature. *Anim. Reprod. Sci.* **2006**, *96*, 21–29. [CrossRef] [PubMed]
19. Koziorowska-Gilun, M.; Fraser, L.; Gilun, P.; Koziorowski, M.; Kordan, W. Activity of antioxidant enzymes and their mRNA expression in different reproductive tract tissues of the male roe deer (*Capreolus capreolus*) during the pre-rut and rut seasons. *Small Rumin. Res.* **2015**, *129*, 97–103. [CrossRef]
20. Drevet, J.R. The antioxidant glutathione peroxidase family and spermatozoa: A complex story. *Mol. Cell. Endocrinol.* **2006**, *250*, 70–79. [CrossRef]
21. Bartosz, G. *Druga Twarz Tlenu*; Wydawnictwo Naukowe PWN: Warszawa, Polska, 2013; pp. 87–91.
22. Koziorowska-Gilun, M.; Koziorowski, M.; Fraser, L.; Strzeżek, J. Antioxidant Defence System of Boar Cauda Epididymidal Spermatozoa and Reproductive Tract Fluids. *Reprod. Domest. Anim.* **2011**, *46*, 527–533. [CrossRef]
23. Agarwal, A.; Virk, G.; Ong, C.; du Plessis, S.S. Effect of oxidative stress on male reproduction. *World J. Men's Health* **2014**, *32*, 1–17. [CrossRef] [PubMed]
24. Domínguez-Rebolledo, A.E.; Fernández-Santos, M.R.; Bisbal, A.; Ros-Santaella, J.L.; Ramón, M.; Carmona, M.; Martínez-Pastor, F.; Garde, J.J. Improving the effect of incubation and oxidative stress on thawed spermatozoa from red deer by using different antioxidant treatments. *Reprod. Fertil. Dev.* **2010**, *22*, 856–870. [CrossRef] [PubMed]
25. Kozioł, K.; Koziorowski, M. Morphological defects of epididymal spermatozoa in male roe deer (*Capreolus capreolus*) during the reproductive season. *Pol. J. Vet. Sci.* **2015**, *18*, 565–572. [CrossRef] [PubMed]
26. Malo, A.F.; Garde, J.J.; Soler, A.J.; García, A.J.; Gomendio, M.; Roldan, E. Male fertility in natural populations of red deer is determined by sperm velocity and the proportion of normal spermatozoa. *Biol. Reprod.* **2005**, *72*, 822–829. [CrossRef] [PubMed]
27. Watson, P.F. Use of a Giemsa stain to detect changes in acrosomes of frozen ram spermatozoa. *Vet. Rec.* **1975**, *97*, 12–15. [CrossRef] [PubMed]
28. Wysokińska, A.; Szablicka, D.; Dziekońska, A.; Wójcik, E. Analysis of changes in the morphological structures of sperm during preservation of liquid boar semen in two different seasons of the year. *Anim. Reprod. Sci.* **2023**, *256*, 107297. [CrossRef] [PubMed]
29. Kankofer, M.; Kolm, G.; Aurich, J.; Aurich, C. Activity of glutathione peroxidase, superoxide dismutase and catalase and lipid peroxidation intensity in stallion semen during storage at 5 degrees C. *Theriogenology* **2005**, *63*, 1354–1365. [CrossRef] [PubMed]
30. Perumal, P. Effect of Superoxide Dismutase on Semen Parameters and Antioxidant Enzyme Activities of Liquid Stored (5 °C) Mithun (*Bos frontalis*) Semen. *J. Anim.* **2014**, *2014*, 821954. [CrossRef]

31. Soler, A.J.; Pérez-Guzmán, M.D.; Garde, J.J. Storage of red deer epididymides for four days at 5 degrees C: Effects on sperm motility, viability, and morphological integrity. *J. Exp. Zool. Part A Comp. Exp. Biol.* **2003**, *295*, 188–199. [CrossRef]
32. Hishinuma, M.; Suzuki, K.; Sekine, J. Recovery and cryopreservation of sika deer (*Cervus nippon*) spermatozoa from epididymides stored at 4 degrees C. *Theriogenology* **2003**, *59*, 813–820. [CrossRef]
33. Dott, H.M.; Foster, G.C. The estimation of sperm motility in semen, on membrane slide, by measuring the area change frequency with an image analysing computer. *J. Reprod. Fertil.* **1979**, *55*, 161–166. [CrossRef] [PubMed]
34. Love, C.C. Relationship between sperm motility, morphology and the fertility of stallions. *Theriogenology* **2011**, *76*, 547–557. [CrossRef] [PubMed]
35. Martínez-Pastor, F.; Diaz-Corujo, A.R.; Anel, E.; Herraez, P.; Anel, L.; de Paz, P. Post mortem time and season alter subpopulation characteristics of Iberian red deer epididymal sperm. *Theriogenology* **2005**, *64*, 958–974. [CrossRef]
36. Conejo-Nava, J.; Fierro, R.; Gutierrez, C.G.; Betancourt, M. Membrane status and in vitro capacitation of porcine sperm preserved in long-term extender at 16 °C. *Arch. Androl.* **2003**, *49*, 287–295. [CrossRef]
37. Neild, D.M.; Brouwers, J.F.H.M.; Colenbrander, B.; Agüero, A.; Gadella, B.M. Lipid peroxide formation in relation to membrane stability of fresh and frozen thawed stallion spermatozoa. *Mol. Reprod. Dev.* **2005**, *72*, 230–238. [CrossRef] [PubMed]
38. Ros-Santaella, J.L.; Domínguez-Rebolledo, A.E.; Garde, J.J. Sperm flagellum volume determines freezability in red deer spermatozoa. *PLoS ONE* **2014**, *9*, e112382. [CrossRef]
39. Xu, H.; Yuan, S.-Q.; Zheng, Z.-H.; Yan, W. The cytoplasmic droplet may be indicative of sperm motility and normal spermiogenesis. *Asian J. Androl.* **2013**, *15*, 799–805. [CrossRef]
40. Cooper, T.G.; Yeung, C.-H.; Fetic, S.; Sobhani, A.; Nieschlag, E. Cytoplasmic droplets are normal structures of human sperm but are not well preserved by routine procedures for assessing sperm morphology. *Hum. Reprod.* **2004**, *19*, 2283–2288. [CrossRef]
41. Yuan, S.; Zheng, H.; Zheng, Z.; Yan, W. Proteomic Analyses Reveal a Role of Cytoplasmic Droplets as an Energy Source during Epididymal Sperm Maturation. *PLoS ONE* **2013**, *8*, e77466. [CrossRef]
42. Marchlewicz, M.; Szypulska-Koziarska, D.; Grzegrzółka, A.; Kruk, J.; Duchnik, E.; Wiszniewska, B. Ochrona przed stresem oksydacyjnym w męskim układzie płciowym. *Pomeranian J. Life Sci.* **2016**, *62*, 44–52. [CrossRef]
43. Agarwal, A.; Saleh, R.A.; Bedaiwy, M.A. Role of reactive oxygen species in the pathophysiology of human reproduction. *Fertil. Steril.* **2003**, *79*, 829–843. [CrossRef] [PubMed]
44. Alvarez, J.G.; Touchstone, J.C.; Blasco, L.; Storey, B.T. Spontaneous Lipid Peroxidation and Production of Hydrogen Peroxide and Superoxide in Human Spermatozoa Superoxide Dismutase as Major Enzyme Protectant against Oxygen Toxicity. *J. Androl.* **1987**, *8*, 338–348. [CrossRef]
45. Hashem, E.Z.; Haddad, R.; Eslami, M. Evaluation of ram semen enrichment with oleic acid on different spermatozoa parameters during low temperature liquid storage. *Small Rumin. Res.* **2017**, *150*, 30–39. [CrossRef]
46. Coyan, K.; Başpınar, N.; Bucak, M.N.; Akalin, P.P.; Ataman, M.B.; Omür, A.D.; Güngör, S.; Küçükgünay, S.; Ozkalp, B.; Sarıözkan, S. Influence of methionine and dithioerythritol on sperm motility, lipid peroxidation and antioxidant capacities during liquid storage of ram semen. *Res. Vet. Sci.* **2010**, *89*, 426–431. [CrossRef] [PubMed]
47. Bayomy, M.F.F.; El-Nabi, S.E.H.; El Kassas, T.A.; Attia, Z.I.; Saeed, A.M.; Taha, H.S.A.; Alagawany, M.; Galosi, L.; Biagini, L.; El-Kassas, S. Extender Supplementation with Glutathione (GSH) and Taurine Improves In Vitro Sperm Quality and Antioxidant Status of New Zealand Rabbits during Chilled Storage for up to 72 hours. *Vet. Med. Int.* **2023**, *2023*, 8339591. [CrossRef] [PubMed]
48. Koziorowska-Gilun, M.; Szurnicka, M.; Dziekonska, A.; Kordan, W.; Giżejewski, Z.; Filipowicz, K. Characteristics of Antioxidant Systems of Yellow Fraction of Red Deer's (*Cervus elaphus* L.) Semen During the Rutting Period. *Reprod. Domest. Anim.* **2016**, *51*, 248–254. [CrossRef]
49. Gotowiecka, M.; Niżański, W.; Partyka, A.; Strzeżek, R.; Koziorowska-Gilun, M. Mechanizm obrony antyoksydacyjnej i oddziaływanie reaktywnych form tlenu na plemniki samców zwierząt domowych. *Med. Wet.* **2015**, *71*, 418–422.
50. Aitken, R.J.; Buckingham, D.W.; Carreras, A.; Irvine, D.S. Superoxide dismutase in human sperm suspensions: Relationship with cellular composition, oxidative stress, and sperm function. *Free Radic. Biol. Med.* **1996**, *21*, 495–504. [CrossRef] [PubMed]
51. Bilodeau, J.F.; Chatterjee, S.; Sirard, M.A.; Gagnon, C. Levels of antioxidant defenses are decreased in bovine spermatozoa after a cycle of freezing and thawing. *Mol. Reprod. Dev.* **2000**, *55*, 282–288. [CrossRef]
52. Strzezek, R.; Koziorowska-Gilun, M.; Kowalówka, M.; Strzezek, J. Characteristics of antioxidant system in dog semen. *Pol. J. Vet. Sci.* **2009**, *12*, 55–60.
53. Jeulin, C.; Soufir, J.C.; Weber, P.; Laval-Martin, D.; Calvayrac, R. Catalase activity in human spermatozoa and seminal plasma. *Gamete Res.* **1989**, *24*, 185–196. [CrossRef] [PubMed]
54. Tramer, F.; Rocco, F.; Micali, F.; Sandri, G.; Panfili, E. Antioxidant systems in rat epididymal spermatozoa. *Biol. Reprod.* **1998**, *59*, 753–758. [CrossRef] [PubMed]

Disclaimer/Publisher's Note: The statements, opinions and data contained in all publications are solely those of the individual author(s) and contributor(s) and not of MDPI and/or the editor(s). MDPI and/or the editor(s) disclaim responsibility for any injury to people or property resulting from any ideas, methods, instructions or products referred to in the content.

animals

Article

Effect of Three Semen Extenders on Sperm Quality and In Vitro Fertilization Rates of Fresh and Cryopreserved Sperm Collected from Llama (*Lama glama*) Vas Deferens

Manuel G. Pérez-Durand [1], Carlos W. Bustamante [2], Pedro P. Machaca [3], Wilber García [4], Eloy A. Condori [2], Rassiel Macedo [2], Eliseo Fernández [1], Yan P. Manrique [1], Miguel A. Gutiérrez-Reinoso [5], Uri H. Perez-Guerra [1,*] and Manuel García-Herreros [6,*]

[1] Facultad de Medicina Veterinária y Zootecnia, Universidad Nacional del Altiplano, Puno 21001, Peru; mgperez@unap.edu.pe (M.G.P.-D.); epfernandez@unap.edu.pe (E.F.); yan.manrique@unap.edu.pe (Y.P.M.)
[2] Facultad de Ciencias Agrárias, Universidad Nacional San Antonio Abad del Cusco, Cusco 08000, Peru; carlos.bustamante@unsaac.edu.pe (C.W.B.); eloy.condorich@unsaac.edu.pe (E.A.C.); rassiel.macedo@unsaac.edu.pe (R.M.)
[3] Facultad de Ciencias Agropecuárias, Escuela Profesional de Medicina Veterinária y Zootécnia, Universidad Jorge Basadre Grohmann, Tacna 23000, Peru; pedro.mac20@gmail.com
[4] Facultad de Medicina Veterinária, Instituto Veterinário de Investigaciones Tropicales y de Altura, Universidad Nacional Mayor de San Marcos, Maranganí 08258, Peru; wgarciav20@unmsm.edu.pe
[5] Facultad de Ciencias Agropecuarias y Recursos Naturales, Carrera de Medicina Veterinaria, Universidad Técnica de Cotopaxi (UTC), Latacunga 050150, Ecuador; miguel.gutierrez@utc.edu.ec
[6] National Institute for Agricultural and Veterinary Research (INIAV), 2005-424 Santarém, Portugal
* Correspondence: uperez@unap.edu.pe (U.H.P.-G.); herrerosgm@gmail.com (M.G.-H.); Fax: +351-24-3767 (M.G.-H.)

Citation: Pérez-Durand, M.G.; Bustamante, C.W.; Machaca, P.P.; García, W.; Condori, E.A.; Macedo, R.; Fernández, E.; Manrique, Y.P.; Gutiérrez-Reinoso, M.A.; Perez-Guerra, U.H.; et al. Effect of Three Semen Extenders on Sperm Quality and In Vitro Fertilization Rates of Fresh and Cryopreserved Sperm Collected from Llama (*Lama glama*) Vas Deferens. *Animals* **2024**, *14*, 1573. https://doi.org/10.3390/ani14111573

Academic Editor: Anna Wysokińska

Received: 14 April 2024
Revised: 19 May 2024
Accepted: 20 May 2024
Published: 25 May 2024

Copyright: © 2024 by the authors. Licensee MDPI, Basel, Switzerland. This article is an open access article distributed under the terms and conditions of the Creative Commons Attribution (CC BY) license (https://creativecommons.org/licenses/by/4.0/).

Simple Summary: Simple Summary: There are very few reports regarding Assisted Reproductive Technologies in South American camelids, including llamas (*Lama glama*). The sperm quality and in vitro fertilization rates from fresh and cryopreserved sperm collected from vas deferens using different preservation extenders were investigated in llama males as a model for its application in other South American camelids. The obtained results from our research showed that the preservation extender was a determining factor in significantly improving the in vitro fertilization rates when using fresh and cryopreserved sperm samples obtained from the vas deferens in llamas (*Lama glama*).

Abstract: The advances in Assisted Reproductive Technologies (ARTs) applied in South American camelid species are still scarce. The aim of this study was to compare the effects of three semen extenders, before and after the cryopreservation of spermatozoa obtained from the vas deferens, on sperm quality parameters and in vitro fertilization rates of llama (*Lama glama*) oocytes. Mature fertile llama males (*Lama glama*; $n = 6$; age: 48–60 mo.; BCS: ~2.7) were included in the study. Sperm samples were collected from each male using the surgical technique of the vas deferens deviation. Then, the sperm samples were pooled and diluted with the Tris-EY, Andromed®, or BioxCell® extender in order to subsequently carry out the sperm cryopreservation process. The sperm quality assessment related to each extender was performed before and after cryopreservation with regard to sperm morphological abnormalities, acrosome integrity, sperm viability, membrane permeability, and sperm motility traits. Moreover, in vitro fertilization (IVF) procedures were carried out to evaluate the in vitro fertility of the cryopreserved sperm samples using each extender. Overall, significant differences were observed before and after cryopreservation regarding acrosome integrity, sperm viability, membrane permeability, and sperm motility traits among the extenders used, where Tris-EY and Andromed® were better than BioxCell® ($p < 0.05$); however, no differences were observed regarding the sperm morphological abnormalities among extenders ($p > 0.05$). Moreover, multiple differences were observed with regard to the velocity and linearity kinematic parameters obtained by computerized analysis before and after the cryopreservation process, irrespective of the extender used ($p < 0.05$). Finally, differences were observed regarding the in vitro fertilization rates among the different extender-derived samples ($p < 0.05$). In conclusion, the sperm quality using Tris-EY and Andromed® was better before and after cryopreservation compared to that using BioxCell® Although the number of fertilized oocytes obtained after the IVF process between Tris-EY and Andromed® was similar, Andromed®-derived samples showed the best sperm quality results before and after cryopreservation. This indicates that the cryopreservation extender is a determining factor in significantly improving in vitro fertilization rates when using sperm samples obtained from vas deferens in llama (*Lama glama*) males.

Keywords: llamas; sperm quality; cryopreservation; vas deferens; in vitro fertilization

1. Introduction

Semen collection and sperm processing in South American camelids have important limitations due to the particular species-specific semen characteristics, which are frequently viscous, stringy, and sometimes foamy, making it difficult for in vitro processing and its application in different sperm quality determination procedures [1–5]. For these reasons, the application of artificial insemination (AI) in South American camelids is not as widely used as in other domestic species [6,7]. Sperm preservation in South American camelids has several advantages related to the transference of genetic material at low costs, and this fact could improve the development of artificial insemination in different camelid species [8]. However, the development of this biotechnology in camelids includes the study of various procedures, such as semen collection, handling, and cryopreservation [9–12].

Sperm extenders are usually based on biological compounds whose function has been, among others, to avoid sudden pH changes, to improve the energy source, to increase the tolerance to chilling and cryopreservation, and to avoid damage or oxidative stress during the cryopreservation process [13]. The use of commercial extenders has been characterized by ease of preparation (avoiding the addition of egg yolk) and avoiding additional specific equipment for sperm sample processing [14]. Another advantage of commercial extenders is that they allow repeatable results [15]. On the other hand, the most commonly used non-commercial extender in South American camelids has been the Tris (hydroxymethyl aminomethane) based extender supplemented with egg yolk as a sperm membrane protector and glycerol as a cryoprotectant during the cryopreservation process [16,17]. The routine assessment of sperm viability has been characterized by determining motility, membrane integrity, and acrosome integrity, among others, including several DNA evaluation techniques [18–20]. In camelids, several extenders containing animal- or plant-origin proteins have been investigated previously in fresh, refrigerated, or cryopreserved sperm [1,3,6,11,14,15,18–20]. Tris-EY is an egg yolk-based extender that has been commonly used in camelids [1,14,15,20], whereas BioxCell® is a soybean lecithin-based extender that has only been used in alpacas [1]. On the other hand, Andromed® is another soy lecithin-based extender that has only been used in fresh and refrigerated sperm samples from alpacas and llamas [3,6,11,15]. However, no studies have been carried out on vas deferens-derived sperm samples and the effects of these extenders on in vitro fertility based on IVF assays. Therefore, it would be necessary to carry out more in-depth studies on fresh and cryopreserved sperm samples recovered from the vas deferens in llamas and the impact of the Tris-EY, Andromed®, and BioxCell® extender on sperm quality and IVF results.

Recently, the evaluation of sperm motility using computerized-aided sperm analysis systems has been applied in South American camelid species, increasing the objectivity of sperm kinematic parameters [21]. The use of new methodologies for the analysis of sperm quality of little-known species has become necessary, as is the case in South American camelids, always with the aim of using the new information obtained for application in different reproductive biotechnologies, such as in vitro fertilization (IVF) [22]. The use of new Assisted Reproductive Technologies (ARTs) in South American camelids, such as IVF, allows new information to be obtained regarding the gamete and embryo physiology for the subsequent application of developing ARTs in different South American camelid species [22–24]. However, these types of technologies require a series of processes, such as oocyte retrieval, oocyte maturation, sperm capacitation, in vitro fertilization, and in vitro embryo culture, among others [25]. Currently, these types of technologies are being used by different researchers in order to develop more successful protocols, although most of the trials carried out have been considered experimental in order to generate a suitable protocol for application in different South American camelid species [26]. Recently, our research team has obtained morula and blastocyst percentages that vary between 17.4% and 16.5%, respectively, at high altitude environmental conditions, using oocytes obtained from ovaries collected in slaughterhouses, which have been recovered using follicular aspiration techniques similar to those used in cattle [27,28].

Thus, the main objective of the present study is to compare the effect of different extenders on sperm quality traits during the cryopreservation process in order to determine the potential use of cryopreserved sperm samples recovered from vas deferens in in vitro fertilization procedures in llamas (*Lama glama*) as a model to study other South American camelid species.

2. Materials and Methods

2.1. Ethical Statement

The present research was conducted according to the guidelines of the Declaration of Helsinki and following the Code of Ethics for animal experiments, as reflected in the ARRIVE guidelines available at http://www.nc3rs.org.uk/ARRIVEchecklist (accessed on 1 December 2023). This study was approved by the Bioethics Committee for the use of experimental animals at the Universidad Nacional del Altiplano (Puno, Perú; Approval Date: 17 January 2020, Code Number: DE-003866-2019).

2.2. Experiment Location, Reagents, and Media

This study was conducted in the Laboratory of Animal Reproduction of the Faculty of Veterinary Medicine and Animal Husbandry at the Universidad Nacional del Altiplano Puno (Perú) (latitude: $-15.82435°$ W 15°49′28″ S; longitude: $-70.01573°$/70°0′57″ at 3827 m.a.s.l.). All reagents and media used in the experiments, unless otherwise stated, were of analytical grade and purchased from Sigma-Aldrich Chemical Company (St. Louis, MO, USA).

2.3. Animals and Procedure of Sperm Sample Collection from Vas Deferens

Six healthy reproductively mature fertile and genetically heterogeneous llama (*Lama glama*) males ($n = 6$; Age: 48–60 mo.; B.W.: 120–140 kg.; B.C.S.: ~2.7) were used [29]. All animals were bred in the Faculty of Veterinary Medicine and Animal Husbandry and kept in isolated enclosures to avoid herd conditioning effects on semen quality. The males were maintained under the same nutritional (standard diet, natural pastures, and water ad libitum) and environmental conditions (outdoor access) during all the experiments. Two sperm samples per male (a total of twelve samples) were collected and pooled per day to obtain homogeneous samples. These sperm samples were collected from each male on consecutive weeks (twice a week) over three months (January, February, and March) using the surgical technique of the deviation of the vas deferens described by Perez et al. [30]. The first collection of sperm samples was performed two months after the surgical procedure. The surgical procedure of the vas deferens consisted of several steps. First of all, the males were fasted for 24 h, and then they were administered 0.1 mg/kg B.W. of acepromazine maleate. All males were placed in a dorsal recumbency position, and the surgical field (inguinal region) was prepared using local anesthesia. Then, a small cut (4 cm) was made above the penis, and the vas deferens (right and left side) were located and dissected (length: 7 cm). The dissected vas deferens were redirected underneath the subcutaneous tissue and fixed to the skin of the femoral region, protected with a temporary patch. The llama males were previously immobilized, knocked down, and placed in lateral recumbency. Then, prior to each sperm sample collection, the fistulas (externalized vas deferens) were cleaned using distilled water. In order to obtain the sperm samples, it was necessary to rub (linear friction) the vas deferens with the fingertips towards the exit of the fistula (located on the inner thighs). The spermatozoa (spz) were obtained as the sample drops appeared at the edge of the fistula (~10 µL per vas deferens) using a micropipette (Micropipette, Boeco®, Hamburg, Germany) and diluted with 1 mL of the extender (Tris-EY, Andromed® or BioxCell®) according to the case. The sperm samples were collected from all males and pooled to avoid any potential male effect. The sperm samples were kept at ~37 °C until transfer to the laboratory.

2.4. *Sperm Quality Assessment*

2.4.1. Sperm Concentration Determination

Raw sperm samples were collected from llama (*Lama glama*) vas deferens. After sperm collection, the pooled samples were kept at 37 °C in a water bath. The sperm concentration in each pooled sample was performed using the Neubauer hemacytometer (Marienfeld, Lauda-Königshofen, Germany). First, the sperm sample was aspirated and suspended in bi-distilled water (1: 200 v/v). Subsequently, a sample was placed in the Neubauer hemacytometer and evaluated under the inverted microscope (Leica, Leica Microsystems CMS GmbH, Mannheim, Germany) using the 200× magnification objective. The sperm concentration was determined by counting 5 squares randomly.

2.4.2. Sperm Morphology Assessment

Sperm morphological abnormalities were determined using an aliquot of the pooled sperm sample using phase-contrast inverted microscopy (Leica Microsystems CMS GmbH, Mannheim, Germany) under a 400× magnification objective. Briefly, 10 µL of sperm samples and 20 µL of the solution (1% formaldehyde solution in phosphate-buffered saline (PBS)) were suspended until a sperm suspension of 10×10^6 spz/mL was obtained. Then, 10 µL of the sperm suspension was placed on a tempered slide. Duplicate smears were performed, and at least 200 spermatozoa per smear were analyzed.

2.4.3. Acrosome Integrity Analysis

Similarly to sperm morphology evaluation, the sperm acrosome integrity was determined using an aliquot of the pooled sperm sample using phase-contrast inverted microscopy (Leica Microsystems CMS GmbH, Mannheim, Germany) under a 1000× magnification objective. Briefly, 10 µL of the sperm sample and 20 µL of the solution (3% glutaraldehyde solution in phosphate-buffered saline (PBS)) were suspended until a sperm suspension of 10×10^6 spz/mL was obtained. Then, 10 µL of the sperm suspension was placed on a tempered slide. Duplicate smears were performed, and at least 200 spermatozoa per smear were analyzed.

2.4.4. Sperm Membrane Integrity Evaluation

Sperm vitality was evaluated using phase-contrast optic microscopy by assessing the sperm membrane integrity using eosin and nigrosin staining [31]. Briefly, 10 µL of the sample and 20 µL of dye (sperm suspensions of 10×10^6 spz/mL) were placed on a tempered slide. Then, duplicate smears were performed. Spermatozoa were then examined under a Leica phase-contrast microscope (Leica Microsystems CMS GmbH, Mannheim, Germany) under a 40× magnification objective. Red-stained (eosin-positive) spermatozoa were considered viable cells, whereas dark-stained (nigrosin-positive) spermatozoa were considered non-viable cells. At least 200 spermatozoa per slide were analyzed, as described by Gomez-Quispe et al. [32].

2.4.5. Sperm Membrane Permeability Analysis

The Hypo-Osmotic Swelling test (HOST) was used for sperm membrane permeability assessment. A hypo-osmolar solution composed of citrate dihydrate (0.735 g) and fructose (1.351 g) diluted in 100 mL of bi-distilled water was used following the recommendations by Flores Huarco et al. [33]. Sperm evaluation was carried out considering those spermatozoa that showed swelling at the tail level (functional membrane permeability) as positive. At least 200 spermatozoa per sample were analyzed, as recommended by Gomez-Quispe et al. [32].

2.4.6. Sperm Motility Assessment

Sperm motility evaluation was performed before and after the cryopreservation process. Progressive individual motility scores were assessed under a coverslip (18 × 18 mm) on a warm stage (37 °C) by phase-contrast-inverted microscopy (10× objective) examined under a Leica microscope (Leica Microsystems CMS GmbH, Mannheim, Germany) after a dilution (1:100) in each extender (Tris-EY, Andromed® or BioxCell®, pH = 7.4, maintained at 37 °C). Motility was evaluated by measuring the following parameters: progressive motility (PM), oscillatory motility (OM), circular motility (CM), and total motility (TM), which is the sum of the previous parameters [6,15,21].

The objective motility assessment was performed using a commercially available computer-assisted semen analysis system (CASA; AndroVision®, Minitube, Tiefenbach, Germany) equipped with a phase-contrast microscope using a 10× negative contrast. The video signal was acquired using a digital camera attached to the microscope and its respective software (AndroVision® software, Minitube, Tiefenbach, Germany) for sperm motility analysis. In addition, digitized images were made up of 1,920,000 pixels (picture elements) and 256 gray levels for sperm functionality analysis, where the recorded videos were analyzed at 25 frames per second. At least 125 spermatozoa per field were randomly captured in duplicate in the acquisition mode of the software. Each sperm kinetic was measured for twelve primary kinematic parameters as follows: curvilinear velocity (VCL; μm/s), as the average path velocity of the sperm head along its true trajectory per unit time; straight-line velocity (VSL; μm/s), as the average path velocity of the sperm head along a straight line from its first to its last position per unit time; average path velocity (VAP; μm/s), as the average velocity of the sperm head along its average trajectory per unit of time; distance curved line (DCL; μm), as the distance that the sperm head moves per time unit on the curve path; distance straight line (DSL; μm), as the distance that the sperm head moves per time unit on the straight path; distance on average path (DAP; μm), as the distance that the sperm head moves per time unit on the average path; amplitude of lateral head displacement (ALH; μm), as the average value of the extreme side-to-side movement of the sperm head in each beat cycle; head cross frequency (BCF; Hz), as the frequency at which the actual sperm trajectory crosses the average path trajectory; head activity (HAC; rad), as the average of all calculated differences of the mean axis angle of two consecutive frames; wobble coefficient index (WOB; %), as the ratio between VAP and VCL (×100); linearity index (LIN; %), as the ratio between VSL and VCL (×100); and the straightness index (STR; %), as the ratio between VSL and VAP (×100). The measurements of individual sperm kinetics from each sample were saved in an Excel v. 2019 (Microsoft Corporation, Redmond, WA, USA) compatible database by the software for further analysis.

2.5. *Extender Preparation: Tris-EY, Andromed® and BioxCell®*

Tris-EY extender was prepared every week using Tris (tris-hydroxymethyl-aminomethane; 2.42 g), citric acid (1.34 g), fructose (1.25 g), egg yolk (20%), and glycerol (5%) in 100 mL of bi-distilled water. After the addition of egg yolk, the solution was centrifuged at 4000 r.p.m. for 30 min and filtered with a paper filter of 30 μm (Grade 113, Whatman®, Dassel, Germany). The first sperm dilution at 37 °C (collection phase) was performed with fraction A (Tris, citric acid, fructose, and egg yolk), and the second dilution was performed at 5 °C (equilibration phase) with fraction B (Tris, citric acid, fructose, egg yolk, and glycerol). The Andromed® (Minitube GmbH, Tiefenbach, Germany) extender [composition: Tris, citric acid, phospholipids, fructose, glucose, antioxidants, pH buffers, glycine, soy lecithin: 1%—1 g/L; glycerol: 6.7%—38.47 g/L; antibiotics (Tylosin: 0.57 g/L, gentamicin: 0.286 g/L, Spectinomycin: 0.343 g/L, Lincomycin: 0.172 g/L); bi-distilled ultrapure water] was prepared on the same day of sperm collection using a 1:4 ratio (extender: bi-distilled water). Briefly, 4 mL of bi-distilled water was placed in a graduated tube, and 1 mL of Andromed® was added. Then, the dilution was kept at 37 °C until use during the sperm collection phase. The second dilution at 5 °C was carried out using the Andromed® extender at the same temperature (equilibration phase). The BioxCell® (IMV Technologies, L'Aigle, France) extender (composition; Tris: 2.3 g/L; Sodium citrate: 6.2 g/L, Potassium chloride: 0.8 g/L; fructose: 1.2 g/L; Monohydrate lactose: 0.8 g/L; glycine: 0.2 g/L; anhydrous glucose: 0.5 g/L; Taurine: 0.005 g/L; gentamicin sulfate: 0.24 g/L; Tylosin tartrate: 0.33 g/L; Linco-Spectin 100: 0.383 g/L; glycerol: 7%—40.2 g/L; Hydrate of calcium lactate: 0.7 g/L; soy lecithin: 1.5%—1.5 g/L; Monohydrate citric acid: 2.5 g/L; bi-distilled ultrapure water) was prepared on the same day of the sperm collection in a similar to the Andromed® extender using a 1: 4 ratio (extender: bi-distilled water). Then, the dilution was mixed and kept at 37 °C until use during the sperm collection phase. The second dilution at 5 °C was carried out using the BioxCell® extender at the same temperature (equilibration phase).

2.6. Sperm Cryopreservation Process

For the sperm cryopreservation process, the sperm sample pools were diluted in each extender to ~120 × 10^6 spz/mL and subjected to the cooling phase (5 °C) for 150 min. In the case of the Tris-EY extender, fraction B was added at 5 °C. Regarding Andromed® and BioxCell®, the extender volume was adjusted and added until a proper concentration at 5 °C was reached. All sperm samples were kept at 5 °C for 30 min. (equilibration phase) irrespective of the extender used. The total duration of the cryopreservation process (cooling phase + equilibration phase) was 180 min. Subsequently, the sperm samples were stored in 0.25 mL straw (IMV Technologies®, L'Aigle, France) and cryopreserved in a semen bio-freezer device (TK3000 CSE, TK Technologies®, São Bernardo Do Campo, Brazil) at a freezing rate of −20 °C per min. (total 6 min) until the temperature of −120 °C was reached, as recommended by Perez et al. [20]. Then, the sperm samples were stored in liquid nitrogen at −196 °C. Finally, for the evaluation of the cryopreserved sperm samples, the straws were thawed at 37 °C for 45 s in a water bath before the quality and the fertility potential evaluation.

2.7. In Vitro Fertilization Process

Llama ovaries were collected from a slaughterhouse and transported to the laboratory in a thermos within 4 h after collection. The ovaries were kept in a physiological solution (NaCl 0.9%; Medifarma®, Lima, Peru) and 50 mg/mL of gentamicin (Gentamicina, Genfar®, Ate, Peru) at 32 to 35 °C. The oocyte collection and the washing process were performed, as recommended by [27], as well as the oocyte maturation process, which was performed in an incubator at 38.5 °C with 5% CO_2 and >90% humidity for 36 h. The oocyte maturation was assessed by the presence of the first polar body. Then, groups of 10 oocytes were washed twice in FERT TALP medium [$CaCl_2·2H_2O$ (29.4 mg); KCl (23.9 mg); $MgCl_2·6H_2O$ (10.1 mg); $NaH_2PO_4·H_2O$ (5.5 mg); Lactic Acid (Sodium salt; 60% *w/w* syrup; 186 µL); NaCl (666 mg); $NaHCO_3$ (210 mg); Na Pyruvate (1.0 mL); Penicillamine; Hypotaurine; Epinephrine; BSA (Fraction V; 600 mg); Gentamycin] and transferred to a drop of ~50 µL of clean FERT TALP medium [26,27,34]. Then, the in vitro fertilization process was performed using 2 µL of the sperm sample (final concentration: 1 × 10^6 spz/mL). Gamete interaction was performed under incubation conditions for 24 h at 38.5 °C in a humid atmosphere (>90%) and 5% CO_2. Fertilization was assessed by the identification of the second polar corpuscle at the level of the perivitelline space.

2.8. Statistical Analysis

All statistical analyses were performed using the R 4.3.0 software (R Core Team, 2020). All data derived from sperm concentration, morphological abnormalities, acrosome integrity, sperm viability, membrane permeability, and sperm motility traits were subjected to descriptive statistics (mean, standard error, homoscedasticity, and normality tests). The evaluated variables that fulfilled the assumptions were subjected to a completely randomized design (one-way ANOVA) and Tukey's test, and in the case of not fulfilling the assumptions, the variables were subjected to a Kruskal–Wallis test and the subsequent post hoc test. In addition, all data from all the sperm analyzed by the computer-assisted analysis were imported into a single data set or data matrix that represented more than 5000 observations, each one defined by the kinetic descriptors specified before. The objective motility analyses were evaluated by means of a multi-analysis of variance (MANOVA) to evaluate the influence of two independent variables on the mean kinetic parameters. The level of significance was set at $p < 0.05$.

3. Results

3.1. *Sperm Quality Traits in Samples Obtained from Llama (Lama glama) Vas Deferens*

3.1.1. Sperm Concentration

The sperm concentration in samples obtained in llama males ($n = 30$) was $41.1 \pm 5.3 \times 10^6$ spz/mL. No differences were observed regarding sperm concentration values after dilution using different cryopreservation extenders ($p > 0.05$).

3.1.2. Sperm Morphological Defects in Cryopreserved Sperm Samples Obtained from Llama (*Lama glama*) Vas Deferens Using Different Extenders

No significant differences were observed among cryopreservation extenders regarding the total sperm defects ($p < 0.05$). Moreover, no significant differences were observed among cryopreservation extenders with regard to cytoplasmic droplets, bent/coiled tails, midpiece defects, and microcephalia percentage ($p > 0.05$); however, statistical differences were observed between Andromed® and BioxCell® compared to Tris-EY regarding double-tailed sperm cells with the latter showing the highest percentage ($p < 0.05$). The detached head and detached tail percentage were significantly higher in Tris-EY compared to Andromed® extender samples; however, no differences were observed regarding the same parameters between Tris-EY and BioxCell® sperm samples ($p > 0.05$; Table 1).

Table 1. Sperm morphological defects (%) in fresh (F) and cryopreserved (C) sperm samples obtained from llama (*Lama glama*) vas deferens using different extenders.

Morphological Parameter	Cryopreservation Extender		
	Tris-EY	Andromed®	BioxCell®
Cytoplasmic droplet $_F$ (%)	6.27 ± 2.69^{aA}	10.58 ± 2.49^{aA}	8.53 ± 1.91^{aA}
Cytoplasmic droplet $_C$ (%)	6.37 ± 3.49^{aA}	15.58 ± 2.76^{aA}	8.37 ± 3.91^{bA}
Macrocephalia $_F$ (%)	0.21 ± 0.06^{abA}	0.08 ± 0.02^{aA}	0.41 ± 0.20^{bA}
Macrocephalia $_C$ (%)	0.31 ± 0.16^{aA}	0.18 ± 0.04^{bB}	0.51 ± 0.19^{aA}
Fusiform head $_F$ (%)	0.10 ± 0.01^{aA}	0.09 ± 0.02^{aA}	0.11 ± 0.03^{aA}
Fusiform head $_C$ (%)	7.46 ± 1.45^{aB}	6.69 ± 1.59^{bB}	7.71 ± 2.09^{aB}
Microcephalia $_F$ (%)	0.41 ± 0.33^{aA}	0.39 ± 0.19^{aA}	0.41 ± 0.33^{aA}
Microcephalia $_C$ (%)	2.41 ± 0.23^{aB}	2.39 ± 0.34^{aB}	2.17 ± 0.54^{aB}
Detached head $_F$ (%)	1.41 ± 0.52^{aA}	0.37 ± 0.14^{bA}	0.76 ± 0.25^{abA}
Detached head $_C$ (%)	11.12 ± 2.52^{aB}	9.37 ± 1.42^{aB}	12.56 ± 2.51^{aB}
Midpiece defects $_F$ (%)	30.21 ± 6.01^{aA}	27.84 ± 5.52^{aA}	28.97 ± 6.32^{aA}
Midpiece defects $_C$ (%)	29.10 ± 5.81^{aA}	28.98 ± 6.77^{aA}	30.55 ± 7.64^{aA}
Bent/Coiled tail $_F$ (%)	12.74 ± 1.99^{aA}	11.37 ± 1.71^{aA}	12.72 ± 1.58^{aA}
Bent/Coiled tail $_C$ (%)	16.32 ± 2.03^{aA}	14.25 ± 2.93^{aA}	17.01 ± 2.71^{aB}
Double-tailed $_F$ (%)	1.36 ± 0.29^{aA}	0.62 ± 0.12^{bA}	0.51 ± 0.21^{bA}
Double-tailed $_C$ (%)	1.86 ± 0.25^{aA}	1.26 ± 0.21^{aB}	1.51 ± 0.61^{aB}
Detached tail $_F$ (%)	5.43 ± 1.93^{aA}	2.35 ± 0.37^{bA}	3.05 ± 0.56^{abA}
Detached tail $_C$ (%)	9.09 ± 2.34^{aB}	7.87 ± 1.73^{bB}	9.95 ± 2.98^{aB}
Total defects $_F$ (%)	59.14 ± 2.00^{aA}	53.69 ± 3.40^{aA}	55.47 ± 3.30^{aA}
Total defects $_C$ (%)	84.00 ± 3.08^{aB}	86.50 ± 3.00^{aB}	90.34 ± 3.22^{aB}

Different superscript letters (^{a,b}) in a row represent significant differences among cryopreservation extenders ($p < 0.05$). Different superscript letters (^{A,B}) in a column represent significant differences before (F) and after the cryopreservation (C) process within the same morphological parameter ($p < 0.05$). F: fresh; C: cryopreserved.

3.1.3. Acrosome Membrane Integrity during the Cryopreservation Process Using Different Extenders in Spermatozoa Obtained from Llama (*Lama glama*) Vas Deferens

Significant differences were observed among cryopreservation extenders regarding acrosome membrane integrity at 37 °C and at 5 °C ($p < 0.05$). No significant differences were observed between Andromed® and BioxCell® at 5 °C ($p > 0.05$); however, statistical differences were observed between both and Tris-EY, with the latter being the highest at 5 °C ($p < 0.05$). Moreover, significant differences were observed regarding acrosome membrane integrity among cryopreservation steps irrespective of the cryopreservation extender evaluated ($p < 0.05$; Table 2).

Table 2. Acrosome membrane integrity (%) in fresh and cryopreserved sperm samples obtained from llama (*Lama glama*) vas deferens using different extenders.

Parameter	Cryopreservation Extender		
	Tris-EY	**Andromed®**	**BioxCell®**
Acrosome membrane integrity (AI) at 37 °C (%)	70.58 ± 1.02 ^{aA}	64.88 ± 0.48 ^{bA}	59.20 ± 0.49 ^{cA}
Acrosome membrane integrity (AI) at 5 °C (%)	60.83 ± 0.61 ^{aB}	55.35 ± 0.71 ^{bB}	53.64 ± 0.42 ^{bB}
Acrosome membrane integrity (AI) post-thawing (%)	23.63 ± 0.58 ^{aC}	31.52 ± 0.31 ^{bC}	9.97 ± 0.31 ^{cC}

Different superscript letters ($^{a–c}$) in a row represent significant differences among cryopreservation extenders ($p < 0.05$). Different superscript letters ($^{A–C}$) in a column represent significant differences among cryopreservation steps within the same extender ($p < 0.05$).

3.1.4. Plasma Membrane Integrity and Permeability during the Cryopreservation Process Using Different Extenders in Spermatozoa Obtained from Llama (*Lama glama*) Vas Deferens

Table 3 shows the plasma membrane integrity and plasma membrane permeability using different extenders during the cryopreservation process in sperm samples obtained from llama (*Lama glama*) vas deferens. Significant differences were observed among cryopreservation extenders regarding sperm plasma membrane integrity at 37 °C ($p < 0.05$). No significant differences were observed between BioxCell® and the other extenders, ($p > 0.05$); however, statistical differences were observed between Tris-EY and Andromed® with the first having the lowest and the second the highest percentage obtained when both extenders were compared ($p < 0.05$). Moreover, significant differences were observed regarding membrane permeability among cryopreservation extenders, with the Tris-EY extender being the highest and significantly different from the other extenders ($p < 0.05$). No differences were observed between the Andromed® and BioxCell® extenders regarding membrane permeability at 37 °C ($p > 0.05$) (Table 3).

3.1.5. Sperm Motility during the Cryopreservation Process Using Different Extenders in Spermatozoa Obtained from Llama (*Lama glama*) Vas Deferens

Table 4 shows the sperm motility parameters using different extenders during the cryopreservation process in sperm samples obtained from llama (*Lama glama*) vas deferens. Significant differences were observed among sperm cryopreservation extenders regarding total motility, progressive motility, and oscillatory motility post-thawing ($p < 0.05$). No significant differences were observed among the different sperm cryopreservation extenders with regard to circular motility post-thawing ($p > 0.05$). Moreover, statistical differences were observed in Tris-EY and Andromed® compared to BioxCell® in all motility parameters evaluated at 37 °C and 5 °C except for progressive motility ($p < 0.05$). In general, Tris-EY and Andromed® extenders showed better results in all motility parameters compared to BioxCell®, with the latter showing the lowest percentage in all motility parameters irrespective of the cryopreservation step evaluated. Overall, significant differences were observed between the pre-freezing (37 °C and 5 °C) steps and the post-thawing results in all motility parameters irrespective of the sperm cryopreservation extender studied ($p > 0.05$; Table 4).

Table 3. Plasma membrane integrity (viability; %) and plasma membrane permeability (HOST) using different extenders during the cryopreservation process in sperm samples obtained from llama (*Lama glama*) vas deferens.

Parameter	Cryopreservation Extender		
	Tris-EY	Andromed®	BioxCell®
Plasma membrane integrity (Viability) at 37 °C (%)	48.91 ± 5.55 ^{aA}	65.78 ± 3.38 ^{bA}	55.89 ± 1.27 ^{abA}
Plasma membrane integrity (Viability) at 5 °C (%)	45.23 ± 3.77 ^{aAB}	61.57 ± 2.37 ^{bA}	50.64 ± 0.86 ^{abA}
Plasma membrane integrity (Viability) post-thawing (%)	38.57 ± 1.52 ^{aB}	29.05 ± 0.48 ^{bB}	10.98 ± 0.32 ^{cB}
Hypo-Osmotic Swelling test (HOST) at 37 °C (%)	69.79 ± 2.83 ^{aA}	62.68 ± 1.19 ^{bA}	57.01 ± 2.27 ^{bA}
Hypo-Osmotic Swelling test (HOST) at 5 °C (%)	43.88 ± 6.54 ^{aB}	56.63 ± 1.89 ^{aB}	46.19 ± 1.25 ^{aB}
Hypo-Osmotic Swelling test (HOST) post-thawing (%)	24.78 ± 4.11 ^{aC}	30.73 ± 2.35 ^{bC}	20.07 ± 4.65 ^{cC}

Different superscript letters ($^{a–c}$) in a row represent significant differences among cryopreservation extenders ($p < 0.05$). Different superscript letters ($^{A–C}$) in a column represent significant differences among cryopreservation steps within the same extender ($p < 0.05$).

Table 4. Sperm motility parameters using different extenders during the cryopreservation process in sperm samples obtained from llama (*Lama glama*) vas deferens.

Motility Parameter	Cryopreservation Extender		
	Tris-EY	Andromed®	BioxCell®
Total Motility at 37 °C (%)	61.77 ± 3.00 ^{aA}	65.56 ± 2.11 ^{aA}	38.33 ± 0.63 ^{bA}
Total Motility at 5 °C (%)	54.71 ± 3.61 ^{aA}	55.90 ± 0.85 ^{aB}	33.21 ± 0.88 ^{bB}
Total Motility Post-Thawing (%)	26.54 ± 1.52 ^{aB}	31.37 ± 0.57 ^{bC}	9.17 ± 0.51 ^{cC}
Progressive Motility at 37 °C (%)	29.00 ± 3.23 ^{aA}	22.44 ± 1.67 ^{aA}	18.59 ± 1.38 ^{aA}
Progressive Motility at 5 °C (%)	23.12 ± 4.48 ^{aA}	15.81 ± 0.82 ^{aB}	15.01 ± 0.66 ^{aA}
Progressive Motility Post-Thawing (%)	11.10 ± 1.23 ^{aB}	7.24 ± 0.57 ^{bC}	2.05 ± 0.09 ^{cB}
Oscillatory Motility at 37 °C (%)	34.96 ± 1.17 ^{aA}	37.20 ± 2.16 ^{aA}	17.06 ± 1.00 ^{bA}
Oscillatory Motility at 5 °C (%)	27.65 ± 2.57 ^{aB}	32.52 ± 0.85 ^{aB}	16.30 ± 0.99 ^{bA}
Oscillatory Motility Post-Thawing (%)	14.01 ± 0.89 ^{aB}	23.01 ± 0.80 ^{bB}	6.24 ± 0.34 ^{cB}
Circular Motility at 37 °C (%)	7.82 ± 1.30 ^{aA}	5.96 ± 0.69 ^{aA}	2.69 ± 0.12 ^{aA}
Circular Motility at 5 °C (%)	6.26 ± 2.21 ^{aA}	7.58 ± 1.23 ^{aA}	1.89 ± 0.16 ^{bB}
Circular Motility Post-Thawing (%)	1.43 ± 0.62 ^{aB}	1.12 ± 0.20 ^{aB}	1.19 ± 0.05 ^{aC}

Different superscript letters ($^{a–c}$) in a row represent significant differences among cryopreservation extenders. Different superscript letters ($^{A–C}$) in a column represent significant differences among cryopreservation steps within the same extender ($p < 0.05$).

Table 5 shows the sperm kinematic parameters using different extenders before and after cryopreservation in sperm samples obtained from llama (*Lama glama*) vas deferens. Significant differences were observed among sperm cryopreservation extenders regarding sperm kinematic parameters before and after cryopreservation ($p < 0.05$). No significant differences were observed among the extenders regarding VCL, DSL, ALH, HAC, WOB, and LIN parameters before cryopreservation ($p > 0.05$). Moreover, no differences were observed between Tris-EY and Andromed® with regard to VSL, VAP, DCL, DAP, BCF, and STR parameters before cryopreservation ($p > 0.05$). On the contrary, significant differences were observed between Tris-EY and Andromed® compared to BioxCell® regarding the same parameters, with the latter presenting as the lowest for all the kinetic variables shown before ($p < 0.05$).

Table 5. Sperm kinetic parameters using different extenders before (F) and after cryopreservation (C) in sperm samples obtained from llama (*Lama glama*) vas deferens.

Kinetic Parameter	Cryopreservation Extender		
	Tris-EY	**Andromed®**	**Bioxcell®**
VCL $_F$ (μm/s)	36.16 ± 1.41 ^{aA}	34.76 ± 1.37 ^{aA}	33.06 ± 2.67 ^{aA}
VCL $_C$ (μm/s)	14.17 ± 1.04 ^{aB}	25.33 ± 1.94 ^{bB}	15.33 ± 1.56 ^{aB}
VSL $_F$ (μm/s)	20.50 ± 0.87 ^{aA}	20.63 ± 1.58 ^{aA}	10.91 ± 1.94 ^{bA}
VSL $_C$ (μm/s)	7.57 ± 0.67 ^{aB}	12.90 ± 1.08 ^{bB}	9.03 ± 0.90 ^{aA}
VAP $_F$ (μm/s)	24.71 ± 1.46 ^{aA}	24.52 ± 1.71 ^{aA}	16.64 ± 1.82 ^{bA}
VAP $_C$ (μm/s)	10.00 ± 0.86 ^{aB}	17.16 ± 1.38 ^{bB}	10.50 ± 1.09 ^{aB}
DCL $_F$ (μm)	9.73 ± 0.64 ^{aA}	8.55 ± 1.23 ^{aA}	14.14 ± 1.78 ^{bA}
DCL $_C$ (μm)	4.73 ± 0.34 ^{aB}	8.40 ± 0.88 ^{bA}	4.67 ± 0.51 ^{aB}
DSL $_F$ (μm)	2.82 ± 2.14 ^{aA}	2.56 ± 1.31 ^{aA}	4.04 ± 0.52 ^{aA}
DSL $_C$ (μm)	1.55 ± 0.14 ^{aA}	2.33 ± 0.16 ^{bA}	1.60 ± 0.19 ^{aB}
DAP $_F$ (μm)	4.67 ± 0.81 ^{aA}	4.24 ± 0.66 ^{aA}	6.76 ± 0.86 ^{bA}
DAP $_C$ (μm)	2.77 ± 0.21 ^{aB}	4.45 ± 0.46 ^{bA}	2.30 ± 0.28 ^{aB}
ALH $_F$ (μm)	0.36 ± 0.07 ^{aA}	0.33 ± 0.08 ^{aA}	0.49 ± 0.09 ^{aA}
ALH $_C$ (μm)	0.22 ± 0.02 ^{aB}	0.32 ± 0.04 ^{aA}	0.18 ± 0.03 ^{bB}
BCF $_F$ (Hz)	2.78 ± 0.81 ^{aA}	2.42 ± 0.24 ^{aA}	4.91 ± 0.27 ^{bA}
BCF $_C$ (Hz)	0.82 ± 0.11 ^{aB}	2.02 ± 0.24 ^{bB}	0.57 ± 0.13 ^{aB}
HAC $_F$ (rad)	0.09 ± 0.00 ^{aA}	0.08 ± 0.00 ^{aA}	0.10 ± 0.04 ^{aA}
HAC $_C$ (rad)	0.04 ± 0.00 ^{aB}	0.05 ± 0.00 ^{aB}	0.03 ± 0.00 ^{bB}
WOB $_F$ (VAP/VCL)	0.68 ± 0.01 ^{aA}	0.71 ± 0.07 ^{aA}	0.50 ± 0.09 ^{aA}
WOB $_C$ (VAP/VCL)	0.70 ± 0.01 ^{aA}	0.68 ± 0.02 ^{aA}	0.68 ± 0.02 ^{aB}
LIN $_F$ (VSL/VCL)	0.57 ± 0.02 ^{aA}	0.59 ± 0.04 ^{aA}	0.33 ± 0.03 ^{aA}
LIN $_C$ (VSL/VCL)	0.50 ± 0.02 ^{aB}	0.51 ± 0.04 ^{aA}	0.60 ± 0.02 ^{bB}
STR $_F$ (VCL/VAP)	0.83 ± 0.02 ^{aA}	0.84 ± 0.09 ^{aA}	0.66 ± 0.00 ^{bA}
STR $_C$ (VCL/VAP)	0.77 ± 0.02 ^{aB}	0.75 ± 0.03 ^{aA}	0.86 ± 0.00 ^{bB}

Different superscript letters (^{a,b}) in a row represent significant differences among cryopreservation extenders ($p < 0.05$). Different superscript letters (^{A,B}) in a column within the same parameter represent significant differences within the same extender ($p < 0.05$). Curved line velocity (VCL, μm s^{-1}), straight-line velocity (VSL, μm s^{-1}), average pathway velocity (VAP, μm s^{-1}), curved line distance (DCL, μm), straight line distance (DSL, μm), average path distance (DAP, μm), amplitude lateral head displacement (ALH, μm), beat cross frequency (BCF, Hz), head activity (HAC, rad), wobble (WOB, ratio), linearity (LIN, VSL/VCL ratio), and straightness (STR, VSL/VAP ratio). F: fresh; C: cryopreserved.

The Andromed® extender showed the highest VCL, VSL, VAP, DCL, DSL, DAP, and BCF values compared to Tris-EY and BioxCell® extenders after the cryopreservation process ($p < 0.05$); however, no differences were observed between Tris-EY and BioxCell® regarding the same kinetic parameters ($p > 0.05$). No significant differences were observed between Tris-EY and Andromed® with regard to ALH and HAC after cryopreservation ($p > 0.05$); however, statistical differences in the same parameters were observed when Tris-EY and Andromed® were compared to BioxCell®, with the latter presenting as the lowest compared to the others after cryopreservation ($p < 0.05$).

Overall, significant differences were observed when the sperm samples were compared before and after cryopreservation within the same extender and within the same kinetic parameter, obtaining lower kinetic values after cryopreservation irrespective of the parameter and extender analyzed ($p < 0.05$). The exception was observed in BioxCell®-derived samples where WOB, LIN, and STR were significantly higher after the cryopreservation process ($p < 0.05$; Table 5).

3.2. *In Vitro Fertilization Rates before and after Sperm Cryopreservation Process Using Different Extenders in Samples Obtained from Llama (Lama glama) Vas Deferens*

Figure 1 and Table 6 show the in vitro fertilization rates of fresh and cryopreserved sperm samples obtained from llama (*Lama glama*) vas deferens using different extenders. In all cases, within extenders, significant differences were observed with regard to fertilization rates before and after the cryopreservation process ($p < 0.05$; Figure 1A).

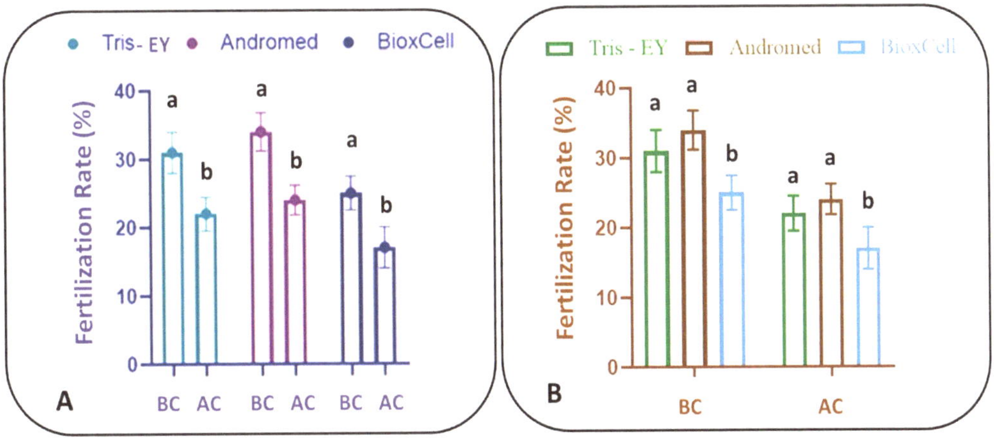

Figure 1. In vitro fertilization rates (%) using different cryopreservation extenders in sperm samples obtained from llama (*Lama glama*) vas deferens. (**A**) In vitro fertilization rates (%) before (BC) and after (AC) cryopreservation within each extender. Different letters (a,b) represent significant differences within each extender ($p < 0.05$). (**B**) In vitro fertilization rates (%) before (BC) and after (AC) cryopreservation among extenders. Different letters (a,b) represent significant differences among extenders before and after the cryopreservation process ($p < 0.05$).

Table 6. In vitro fertilization rates (%) using different extenders from fresh and cryopreserved sperm samples obtained from llama (*Lama glama*) vas deferens.

Type of Sperm	Sperm Extender		
	Tris-EY Fertilized Oocytes/ Total Oocytes Submitted to IVF (%)	Andromed® Fertilized Oocytes/ Total Oocytes Submitted to IVF (%)	BioxCell® Fertilized Oocytes/ Total Oocytes Submitted to IVF (%)
Fresh	47/149 (31.54) ^{aA}	51/149 (34.23) ^{aA}	36/150 (24.00) ^{bA}
Frozen-thawed	40/180 (22.22) ^{aB}	44/179 (24.58) ^{aB}	30/180 (16.66) ^{bB}

Different superscript letters (a,b) in a row represent significant differences among sperm extenders within the same type of sperm ($p < 0.05$). Different superscript letters (A,B) in a column within the same extender represent significant differences between fresh and cryopreserved sperm ($p < 0.05$). Total No. oocytes (fresh): 448; Total No. oocytes (frozen-thawed): 539.

Moreover, significant differences were observed among the different sperm extenders before and after cryopreservation regarding fertilization rates ($p < 0.05$; Figure 1B). No significant differences were observed between Tris-EY and Andromed® regarding fertilization rates ($p = 0.987$; Figure 1B); however, there were slightly better results when using the Andromed® compared with the Tris-EY extender. The lowest in vitro fertilization rate was obtained when the BioxCell® extender was used before and after the cryopreservation process (Figure 1B).

4. Discussion

The present study investigates potential differences in sperm quality in samples obtained from the vas deferens on llamas (*Lama glama*) during the cryopreservation process and the impact of using different extenders on fertility results tested by in vitro fertilization techniques. In general, better sperm quality values were observed during the pre-freezing phase in spermatozoa, cryopreserved using Tris-EY and Andromed® extenders compared to BioxCell® in most of the parameters evaluated, except for plasma membrane integrity and permeability where similar results were observed between Andromed® and BioxCell®.

In general, it can be concluded that the Andromed® extender could be the extender of choice to obtain the best results when applied to spermatozoa samples collected from the vas deferens in llamas. The present results in terms of sperm quality were slightly better than those observed by other authors who reported motility percentages between 56.8% and 56.2% using the commercial extender Triladyl®, probably due to composition differences between the different extenders [13,30]. However, all percentages observed in terms of sperm quality could be considered similar and acceptable, probably due to the fact that the buffer composition was similar, avoiding sudden changes in pH and also providing different energy sources, such as glucose or fructose [17].

Regarding sperm morpho-abnormalities, there is no information in the literature on the influence of the extenders used in the present study and the impact of cryopreservation on samples obtained from the vas deferens in llamas. Compared to other camelid species, the percentage of cytoplasmic droplets observed in alpaca studies was significantly higher compared to the present study in llamas using the same collection technique [30]. In addition, lower percentages of sperm head morpho-abnormalities were also observed in llamas compared to alpacas [30].

The sperm quality obtained in the present study was better than that reported by other authors using epididymal sample collection for sperm cryopreservation in alpacas [35,36]. The sperm motility of 14.0, 8.6 and 17.0% was observed in samples obtained from alpaca epididymis, which were maintained in a physiological buffer solution (Phosphate-Buffered Saline, PBS) and diluted using Tris, Tes, and skim milk, respectively. On the other hand, mean viability was 32.6%, and plasma membrane integrity was 34.5%. Moreover, using the artificial vagina collection method and evaluating three different extenders, Raymundo et al. [37] achieved a motility of 54.0 ± 8.0%. On the other hand, Giuliano et al. [38] reported sperm motility of 32.3 ± 20.4%, plasma membrane permeability (HOST) of 36.1 ± 13.1%, and viability of 54.1 ± 17.0% using electroejaculation in llamas, while the comparisons of total motility and oscillatory motility parameters were better in the present study except when BioxCell® was used. Progressive motility was higher (26.9%) in the study by Giuliano et al. [38] compared to the results of the present study. Furthermore, considering that, in general, most of the results in the study by Giuliano et al. [38] were lower than those of the present study, it could be due to the type of semen sample collection method that can directly or indirectly influence the sperm characteristics as the ejaculates present secretions from the adnexal glands increasing the semen viscosity, frequently seen in camelids [39]. A study with the same collection method using similar extenders and under similar environmental conditions reported slightly lower percentages of total motility, oscillatory motility, and circular motility and higher percentages of progressive motility in relation to the results of the present study considering that the use of extenders, such as Andromed® and Tris-EY, allowed acceptable sperm parameters to be obtained after the sperm collection from the vas deferens and the cryopreservation process [15].

In the present study, significant differences were observed when comparing the three cryopreservation extenders regarding total motility, with Andromed® being the best, followed by Tris-EY and finally BioxCell®. Membrane permeability was shown to be better in samples, using Andromed®, followed by samples using Tris-EY and, finally, BioxCell® samples. As for sperm viability analysis, a higher percentage of spermatozoa with an intact plasma membrane was observed when the Tris-EY extender was used, followed by Andromed® samples and finally BioxCell® samples. Lower motility percentages were reported by Torres Hualla [40] (12.04% for Tris-EY and 11.32% for the skimmed milk extender) compared to those obtained in the present study. Also, the sperm viability using Tris-EY was 18.96%, and using skimmed milk was 14.2%. Banda et al. [36] using different extenders such as Tris, Tea, and skimmed milk for alpaca sperm cryopreservation in samples collected by an artificial vagina, obtained motility percentages of 14.0, 8.6, and 17.0, respectively. In the same study, the results obtained for plasma membrane permeability (HOST) were 17.0, 19.1, and 17.9%, and viability was 32.6, 26.3, and 27.2%, respectively. These results were similar to those obtained in the present study, probably due to the similarity of the extender composition, such as Tris and Tes. Ordoñez et al. [41] carried out sperm collection using electroejaculation methods for alpaca sperm pellet cryopreservation using Andromed®- and Tris-obtaining motilities post-thawing of 0.00 and 11.08%, respectively. This was probably due to the type of sperm collection method (electroejaculation), which has the disadvantage of having different contaminants, such as urine debris, which can be harmful to sperm cells [39]. Overall, in the present study, a low population of physiologically functional spermatozoa was observed for BioxCell® and an average for Andromed® and Tris-EY extenders. This was maybe due to changes in sperm morphology and mitochondrial mass damage after the sperm freezing-thawing process. This may also be influenced by the sperm cryopreservation method since, in the present study, sperm cryopreservation was carried out using a curve of −20 °C/min and reaching a freezing temperature of −120 °C, which is apparently the appropriate methodology for camelids [20]. Some authors reported that sperm plasma membranes with a higher polyunsaturated fatty acids content are more sensitive to damage induced by lipid peroxidation [38]. Moreover, this kind of plasma membrane is more sensitive to damage by reactive oxygen species (ROS), which results in decreased sperm motility, probably due to the rapid loss of intracellular ATP, leading to axoneme damage and decreased sperm viability, among other consequences [42].

In the present study, the observed results indicate that there was similarity between the Tris-EY and Andromed® extenders in most of the sperm quality parameters except for progressive motility during the equilibration phase at 5 °C, where similarities were observed among the three extenders for llama sperm collected from the vas deferens. Using electroejaculation methods, Choez et al. [43] evaluated the sperm motility and plasma membrane permeability (HOST) during the first hour of the cooling process and observed a motility of 50.93%, which produced similar results to the present study using the Tris-EY extender. Other motility parameters, such as total motility, were higher than those reported by other authors [44]. However, the results were similar when using the BioxCell® extender [45]. On the other hand, progressive motility was lower in the present study and slightly higher when the Tris-EY extender was used, probably due to the fact that egg yolk improves progressive motility, as mentioned by Fumuso et al. [46]. Moreover, the results were slightly better than those obtained in the present study, for example, those reported for viability percentages using egg yolk-based extenders [31]. However, in the present study, in all extender groups, a decrease in viability of just 8 to 15% was observed compared to the collection timepoint, mainly due to the action of egg yolk (Tris-EY) and lecithins (Andromed® and BioxCell®), since these components have the main function of maintaining membrane stability during the cooling process. In addition, these components have a high percentage of cholesterol that maintains the structure of the sperm plasma membrane, facilitating the adhesion of several molecules that protect the plasma membrane at low temperatures (cold shock), although in camelids, this phenomenon has not yet been well studied. It has been described that this process can alter the sperm membrane cholesterol/phospholipid ratio; however, the use of egg yolk-based extenders could be variable due to their composition of fatty acids, phospholipids, cholesterol and lipoproteins [11,15,17,47,48]. The slight loss of motility and other parameters indicated that spermatozoa obtained from the vas deferens were resistant to cold shock when in contact with extenders containing Tris-EY as well as those that did not contain egg yolks such as Andromed® and BioxCell®. Furthermore, it has been reported that the process of a spontaneous or induced acrosome reaction occurs after the equilibrium phase in camelids [35,38,49]. Other recent studies in llamas have shown that the cooling process affects the sperm organelles, resulting in acrosome damage, mitochondrial loss, and the disorganization of axoneme and periaxoneme structures, increasing the level of reactive oxygen species (ROS) [45]. Finally, less binding at the oviduct level has been observed using refrigerated and frozen-thawed spermatozoa compared to fresh spermatozoa [31].

Regarding the objective motility parameters, most of the evaluated traits coincided with those determined by the subjective motility assessment (e.g., total motility), with higher values in samples using Andromed®, followed by Tris-EY and finally BioxCell® before and after cryopreservation process. The kinetic parameters related to different types of velocities were variable and corresponded to the motility obtained before sperm pre-freezing and post-thawing. On the other hand, parameters such as WOB, which is related to the ratio of the average trajectory velocity over curvilinear velocity expressed as a percentage, and LIN, which shows the ratio between the straight trajectory and the average sperm trajectory, were similar among extender samples. Similarly, the same was observed for STR, which was quite high in all samples analyzed regardless of the extender used. In summary, all the kinetic variables related to indices were similar, while those with values directly related to the type of movement had differences among different extender samples used in sperm obtained from the vas deferens in llamas. VCL, VSL, and VAP parameters were shown to be higher when using Andromed® compared to the Tris-EY extender, especially in thawed spermatozoa, which had higher velocity results (Andromed®) probably due to the soy lecithin content as opposed to egg yolk containing the Tris extender. This difference could be related to the higher proportion of low-density lipoproteins (LDLs) that provide greater sperm plasma membrane stability [17]. However, in egg yolk-containing extenders, such as Triladyl, interesting results have been obtained in other mammalian species with regard to STR, LIN, ALH, and BCF parameters, which were characterized by an efficient flagellar structure, ATP production, and consequently, the higher frequency of the sperm tail beating movement (BCF) [50].

The health concerns related to the use of animal proteins in sperm extenders have led to the development of alternative extenders for in vitro and in vivo use. The sperm samples diluted in the plant-based (soy lecithin-based) extender Andromed® resulted in better sperm quality and in vitro fertility compared to Tris-EY and BioXcell. Overall, all sperm quality parameters improved using the Andromed® extender, including objective and subjective motility, which is one of the most important parameters associated with semen fertilizing capacity. In other words, Andromed® was more effective at preserving sperm head and flagellar structures, stimulating ATP production, and improving fertility capacity. Regarding the results obtained in the present study using cryopreserved spermatozoa and the in vitro fertilization rates using oocytes obtained from the slaughterhouse, the Andromed®-derived results were similar to those obtained using epididymal cryopreserved seminal samples from alpacas [51]. Therefore, plant-based extenders are as effective at protecting spermatozoa during the cryopreservation process as egg yolk extenders. Another alpaca study showed that the in vitro fertilization rates were similar to those obtained in the present study, with the difference that non-cryopreserved spermatozoa were used [52]. However, there were studies in alpacas where higher fertilization rates have been observed compared to those reported in the present study [23]. The main findings of the present study in relation to fertility were that (1) vas deferens-derived sperm diluted in Tris-EY and Andromed® resulted in similar fertilization rates in both fresh and cryopreserved samples; (2) vas deferens-derived sperm diluted in BioXcell resulted in reduced sperm quality and in vitro fertilization rates compared to Tris-EY- and Andromed®-diluted samples in both fresh and cryopreserved samples; (3) the temperature fluctuation during the cryopreservation process was detrimental for sperm quality and in vitro fertilization rates in all samples irrespective of the extender used, with BioXcell®-derived samples being the most affected. In South American camelids, in vitro fertilization studies are very scarce, and the results are still quite low compared to other species, such as bovines, where much higher fertilization rates were reported [25,26,53]. However, the classical sperm quality parameters studied, such as total motility, progressive motility, oscillatory motility, plasma membrane integrity, and permeability, among others, and other studies performed with computerized equipment (CASA) showed better IVF results with the use of the commercial extender Andromed® and also the Tris-EY extender [25,26]. The reduction in fertility associated with BioXcell® could indicate that this extender would be less tolerant to temperature fluctuations, and thus, it would not be a suitable substitute for an egg yolk extender such as Tris-EY. When the semen temperature fluctuates, it triggers several morphological membrane changes, compacting or relaxing the packing of the phospholipid bilayer and causing membrane destabilization, resulting in a deleterious effect on sperm function and fertility. Finally, it should be noted that due to the differences obtained in the in vitro fertilization results, more studies are needed in order to increase reproductive efficiency through the use of Assisted Reproductive Technologies (ARTs) in South American camelids in order to obtain in vitro-produced embryos to be transferred [25]. For example, sperm metabolic activity is inversely related to extended sperm survival and during the sperm cryopreservation process, there is an increase in oxidative stress due to the production of reactive oxygen species (ROS). The addition of components, such as citric acid and glycerol, plays a role in reducing the levels of peroxide generated in the storage media. Moreover, modifying the percentage of egg yolk or lecithin could also be effective in reducing sperm metabolic activity and the detrimental effects derived from ROS, which are linked to a cell-aging effect and lead to an apoptotic stage losing the fertility capacity.

5. Conclusions

The sperm quality and in vitro fertilization rates of fresh and cryopreserved spermatozoa collected from llama (*Lama glama*) vas deferens were influenced by the extender used during the cryopreservation process. In general, Andromed® showed slightly better results post-thawing regarding acrosome membrane integrity, plasma membrane permeability, sperm kinetic parameters, and in vitro fertilization rates when using sperm samples obtained from llama (*Lama glama*) vas deferens. Although in vitro fertilization rates were similar to those obtained in other camelid species, further research is needed to understand the effects of different preservation extenders in order to improve the in vitro fertilization rates from sperm samples obtained from llama (*Lama glama*) vas deferens.

Author Contributions: All the authors listed in this paper have contributed to the preparation and execution of this research. C.W.B., P.P.M. and W.G.: Methodology and Writing—Review and Editing, and Funding Acquisition; E.A.C., R.M., E.F., Y.P.M. and M.A.G.-R.: Methodology and Critical Reading and Funding Acquisition; U.H.P.-G. and M.G.P.-D.: Conceptualization, Methodology, Data Curation, Writing—Review and Editing, Critical Reading, and Funding Acquisition; M.G.-H.: Conceptualization, Methodology, Writing—Original Draft, Critical Reading, and Writing—Review and Editing, and Funding Acquisition. All authors have read and agreed to the published version of the manuscript.

Funding: The present research was funded by the Peruvian institutional funds of the Universidad Nacional del Altiplano (UNAP) (Puno, Perú) and the Fondo Nacional de Desarrollo Científico, Tecnológico y de Innovación Tecnológica (FONDECYT-CONCYTEC; Agreement: 000399-2019-FONDECYT-DE; Peru) as part of the project entitled "Monitoring and control of follicular activity and its application in the in vitro embryo production in South American camelids in the Peruvian Altiplano".

Institutional Review Board Statement: This research was performed in strict accordance with the recommendations in the legal framework (Animal Welfare Law) for all Peruvian Public and Private Laboratories and Higher Education Institutions. Moreover, this study was conducted according to the guidelines of the Declaration of Helsinki and following the Code of Ethics for animal experiments as reflected in the ARRIVE guidelines available at http://www.nc3rs.org.uk/ARRIVEchecklist (accessed on 1 December 2023). This study was approved by the Bioethics Committee for the use of experimental animals at the Universidad Nacional del Altiplano, Puno, Peru (approval date: 17 January 2020, Code Number: DE-003866-2019).

Informed Consent Statement: Not applicable.

Data Availability Statement: All data generated or analyzed during this study are included in this article.

Acknowledgments: The authors would like to thank the institutional staff for the excellent technical assistance.

Conflicts of Interest: The authors declare no conflicts of interest.

References

1. Mamani-Mango, G.; Moina Gonzales, M.; Ramos Hidalgo, M.; Mendoza Mallma, J.; Ruiz Béjar, J.; Rivas Palma, V.; Mellisho Salas, E. Effect of Extender and Freezing Rate on Quality Parameters and In Vitro Fertilization Capacity of Alpaca Spermatozoa Recovered from Cauda Epididymis. *Biopreserv. Biobank.* **2019**, *17*, 39–45. [CrossRef] [PubMed]
2. Kershaw-Young, C.M.; Maxwell, W.M. The effect of seminal plasma on alpaca sperm function. *Theriogenology* **2011**, *76*, 1197–1206 [CrossRef] [PubMed]
3. Al-Essawe, E.M.; Abraham, C.; Kunkitti, P.; Axnér, E.; de Verdier, K.; Båge, R.; Morrell, J.M. Extenders for alpaca epididymal spermatozoa: Comparison of INRA96 and andromed. *Anim. Reprod. Sci.* **2020**, *223*, 106629. [CrossRef] [PubMed]
4. Santiani, A.; Ugarelli, A.; Evangelista-Vargas, S. Characterization of functional variables in epididymal alpaca (*Vicugna pacos*) sperm using imaging flow cytometry. *Anim. Reprod. Sci.* **2016**, *173*, 49–55. [CrossRef] [PubMed]
5. Bravo, P.W.; Moscoso, R.; Alarcon, V.; Ordonez, C. Ejaculatory Process and Related Semen Characteristics. *Arch. Androl.* **2002**, *48*, 65–72. [CrossRef] [PubMed]
6. Bertuzzi, M.L.; Fumuso, F.G.; Neild, D.M.; Udaquiola, M.; Carretero, M.I. Use of Commercial Extenders, with and without the Addition of Egg Yolk, for Cooling Llama Semen. *Anim. Reprod. Sci.* **2022**, *247*, 107073. [CrossRef] [PubMed]
7. Perez-Guerra, U.H.; Quispe, Y.M.; Gonzáles, H.I.; Luque, N.; Ruelas, D.A.; Carretero, M.I.; Gutiérrez-Reinoso, M.A.; Pérez-Durand, M.G.; García-Herreros, M. Ovarian Follicular Dynamics and Its Functional Significance in Relation with Follicle Deviation, Vaginal Cytology, and Hormone Profiles in Llamas (*Lama glama*). *Animals* **2022**, *12*, 3299. [CrossRef] [PubMed]
8. Abraham, M.C.; De Verdier, K.; Båge, R.; Morrell, J.M. Semen Collection Methods in Alpacas. *Vet. Rec.* **2017**, *180*, 613–614. [CrossRef] [PubMed]
9. Skidmore, J.A.; Morton, K.M.; Billah, M. Artificial Insemination in Dromedary Camels. *Anim. Reprod. Sci.* **2013**, *136*, 178–186. [CrossRef] [PubMed]
10. Skidmore, J.A. The Use of Some Assisted Reproductive Technologies in Old World Camelids. *Anim. Reprod. Sci.* **2019**, *207*, 138–145. [CrossRef]
11. Swelum, A.A.A.; Saadeldin, I.M.; Ba-Awadh, H.; Al-Mutary, M.G.; Moumen, A.F.; Alowaimer, A.N.; Abdalla, H. Efficiency of Commercial Egg Yolk-Free and Egg Yolk-Supplemented Tris-Based Extenders for Dromedary Camel Semen Cryopreservation. *Animals* **2019**, *9*, 999. [CrossRef] [PubMed]
12. Pérez-Durand, M.G.; Massa-Guzmán, A.; Luque-Mamani, N.; Ruelas-Calloapaza, D.A.; Urviola-Sánchez, J.M.; Condori-Chuchi, E.A.; Gutiérrez-Reinoso, M.A.; Perez-Guerra, U.H.; García-Herreros, M. Age-Related Differences in Testosterone Concentration and Its Relation to Testicular Biometrics, Hemodynamics, and Fertility in Alpacas (*Vicugna pacos*). *Vet. Sci.* **2023**, *10*, 429. [CrossRef] [PubMed]
13. Morton, K.M.; Thomson, P.C.; Bailey, K.; Evans, G.; Maxwell, W.M.C. Quality Parameters for Alpaca (*Vicugna pacos*) Semen Are Affected by Semen Collection Procedure. *Reprod. Domest. Anim.* **2010**, *45*, 637–643. [CrossRef] [PubMed]
14. Bravo, P.W.; Skidmore, J.A.; Zhao, X.X. Reproductive Aspects and Storage of Semen in Camelidae. *Anim. Reprod. Sci.* **2000**, *62*, 173–193. [CrossRef] [PubMed]
15. Bertuzzi, M.L.; Torres, E.Y.; Huanca, T.; Neild, D.; Carretero, M.I. Comparison of Extenders With the Addition of Egg Yolk for Cooling Alpaca Sperm Obtained from Deferent Ducts. *Front. Vet. Sci.* **2020**, *7*, 597954. [CrossRef] [PubMed]
16. Watson, P.F. Recent Developments and Concepts in the Cryopreservation of Spermatozoa and the Assessment of Their Post-Thawing Function. *Reprod. Fertil. Dev.* **1995**, *7*, 871–891. [CrossRef] [PubMed]
17. Yeste, M. Sperm Cryopreservation Update: Cryodamage, Markers, and Factors Affecting the Sperm Freezability in Pigs. *Theriogenology* **2016**, *85*, 47–64. [CrossRef] [PubMed]
18. Carretero, M.I.; Giuliano, S.M.; Arraztoa, C.C.; Santa Cruz, R.C.; Fumuso, F.G.; Neild, D.M. Comparison of Two Cooling Protocols for Llama Semen: With and without Collagenase and Seminal Plasma in the Medium. *Andrologia* **2017**, *49*, e12691. [CrossRef] [PubMed]
19. Guillén Palomino, C.Y.; Fumuso, F.G.; Bertuzzi, M.L.; Giuliano, S.M.; Velásquez González, N.; Bariani, M.V.; Carretero, M.I. Use of Androcoll-ETM to Separate Frozen-Thawed Llama Sperm from Seminal Plasma and Diluent. *Front. Vet. Sci.* **2021**, *7*, 594926. [CrossRef] [PubMed]
20. Pérez, M.D.; Bustamante, C.Q.; Coronel, I.U.; Carretero, M.I.; Manrique, Y.Q.; Condori, E.C.; Perez, U.G. Criopreservación de Espermatozoides de Llama Obtenidos Del Conducto Deferente Utilizando Tres Curvas de Congelamiento. *Rev. Investig. Vet. Perú* **2022**, *33*, e24099. [CrossRef]
21. Meza, A.; Caldeira, C.; Valverde, A.; Ordóñez, C.; Ampuero, E.; Cucho, H.; Soler, C. Sperm Kinematic Characterization of Alpaca (*Vicugna pacos* L.) during the Reproductive Season. *Reprod. Domest. Anim.* **2018**, *53*, 1415–1423. [CrossRef] [PubMed]
22. Trasorras, V.; Giuliano, S.; Miragaya, M. In Vitro Production of Embryos in South American Camelids. *Anim. Reprod. Sci.* **2013**, *136*, 187–193. [CrossRef] [PubMed]
23. Ruiz, J.; Paulo Santayana, R.; José Mendoza, M.; Leandra Landeo, J.; Huamán, E.; Ticllacuri, F.; Fidel Mujica, L.; Silva, M.; Ratto, M.H. Effect of Oocyte Maturation Time, Sperm Selection Method and Oxygen Tension on in Vitro Embryo Development in Alpacas. *Theriogenology* **2017**, *95*, 127–132. [CrossRef]
24. Aponte, P.M.; Gutierrez-Reinoso, M.A.; Garcia-Herreros, M. Bridging the Gap: Animal Models in Next-Generation Reproductive Technologies for Male Fertility Preservation. *Life* **2023**, *14*, 17. [CrossRef] [PubMed]

25. Trasorras, V.L.; Carretero, M.I.; Neild, D.M.; Chaves, M.G.; Giuliano, S.M.; Miragaya, M.H. Production, Preservation, and Transfer of South American Camelid Embryos. *Front. Vet. Sci.* **2017**, *4*, 190. [CrossRef] [PubMed]
26. Conde, P.A.; Herrera, C.; Trasorras, V.L.; Giuliano, S.M.; Director, A.; Miragaya, M.H.; Chaves, M.G.; Sarchi, M.I.; Stivale, D.; Quintans, C.; et al. In Vitro Production of Llama (*Lama glama*) Embryos by IVF and ICSI with Fresh Semen. *Anim. Reprod. Sci.* **2008**, *109*, 298–308. [CrossRef]
27. Durand, M.G.P.; Zevallos Aragon, J.P.; Perez Guerra, U.H. In Vivo Culture of Alpaca Embryos in the Oviduct versus Embryo Culture in Vitro. *Spermova* **2018**, *8*, 138–143. [CrossRef]
28. Perez Guerra, U.H.; Quispe Barriga, Y.M.; Luque Mamani, N.; Ruelas Calloapaza, D.A.; Palomino Cano, J.M.; Pérez Durand, M.G. Comparison of Superstimulatory Protocols with Different Doses of ECG on Ovarian Response and Oocyte Recovery by Follicular Aspiration in Llamas (*Lama glama*). *Anim. Reprod.* **2024**, *21*, e20220134. [CrossRef] [PubMed]
29. Cebra, C.; Van Saun, R.J.; Anderson, D.E.; Tibary, A.; Johnson, L.R.W. *Llama and Alpaca Care: Medicine, Surgery, Reproduction, Nutrition, and Herd Health*, 1st ed.; Elsevier Health Sciences: Amsterdam, The Netherlands, 2013; pp. 1–789. [CrossRef]
30. Perez, M.G.; Zevallos, J.; Perez, U.H. Recovery of Alpaca Sperm from Vas Deferens during the Breeding Season. *Spermova* **2014**, *4*, 139–144.
31. Zampini, R.; Castro-González, X.A.; Sari, L.M.; Martin, A.; Diaz, A.V.; Argañaraz, M.E.; Apichela, S.A. Effect of Cooling and Freezing on Llama (*Lama glama*) Sperm Ultrastructure. *Front. Vet. Sci.* **2020**, *7*, 587596. [CrossRef] [PubMed]
32. Gómez-Quispe, O.E.; Pérez, M.G.; Ojeda-Gutierrez, Y.G. Survival to Freezing of Sperm Collected by Vas Deferens from Vicugna Pacos with Tempol Antioxidant Addition. *Asian J. Anim. Vet. Adv.* **2016**, *11*, 377–382. [CrossRef]
33. Flores Huarco, N.H.; Giuliano, S.M.; Fumuso, F.G.; Miragaya, M.H.; Cucho Dolmos, H.C.; Carretero, M.I. Use of Dimethylformamide to Cryopreserve Alpaca Semen Previously Incubated with Collagenase. *Reprod. Domest. Anim.* **2021**, *56*, 1387–1397. [CrossRef] [PubMed]
34. Parrish, J.J.; Susko-Parrish, J.; Winer, M.A.; First, N.L. Capacitation of Bovine Sperm by Heparin. *Biol. Reprod.* **1988**, *38*, 1171–1180. [CrossRef]
35. Morton, K.M.; Bathgate, R.; Evans, G.; Maxwell, W.M.C. Cryopreservation of Epididymal Alpaca (*Vicugna pacos*) Sperm: A Comparison of Citrate-, Tris- and Lactose-Based Diluents and Pellets and Straws. *Reprod. Fertil. Dev.* **2007**, *19*, 792–796. [CrossRef] [PubMed]
36. Banda, J.R.; Evangelista, S.V.; Ruiz, L.G.; Sandoval, R.M.; Rodríguez, L.C.; Valdivia, M.C.; Santiani, A.A. Effect of extenders based on tris, tes and skim milk on cryopreservation of epididymal alpaca sperm. *Rev. Inv. Vet. Perú* **2010**, *21*, 145–153.
37. Raymundo, F.; Huanca, W.; Huanca, T.; Huerta, S.; Cordero, A. Efecto de tres dilutores en la conservación del semen de alpacas. *Rev. Inv. Vet. Perú* **2006**, *17*, 125–130.
38. Giuliano, S.; Santa Cruz, R.; Arraztoa, C.; Fumuso, F.G.; Bertuzzi, M.; Carretero, M.I.; Susana, D.; Giuliano, M.; María, D.; Carretero, I. Selección espermatica de semen refrigerado de llama con diluyente a base de yema de huevo Sperm Selection in Cooled Llama Semen with Egg Yolk Extender. *Spermova* **2019**, *9*, 35–41. [CrossRef]
39. Aisen, E.G.; Huanca López, W.; Pérez Durand, M.G.; Torres Mamani, E.; Villanueva Mori, J.C.; Ousset, M.J.; Medina, V.H.; Pérez Guerra, U.H.; Huanca Mamani, T. Spermatozoa Obtained From Alpaca Vas Deferens. Effects of Seminal Plasma Added at Post-Thawing. *Front. Vet. Sci.* **2021**, *8*, 611301. [CrossRef] [PubMed]
40. Torres Hualla, E.A. Comparación de 2 Tipos de Dilutores (Tris-Yema de Huevo y Leche Descremada) En La Sobrevivencia Espermática Post—Descongelamiento de Semen de Alpaca (*Vicugna pacos*) de La Raza Huacaya En La Zona Costa de Tacna. Bachelor´s Thesis, Universidad Nacional Jorge Basadre Grohmann, Tacna, Peru, 2018.
41. Ordoñez, C.; Ampuero, E.; Alarcón, V.; Franco, E.; Hanzen, C.; Cucho, H. Three Extenders Evaluation in Cryopreservation of Alpaca (*Vicugna pacos*) Semen in Pellets. *Spermova* **2015**, *5*, 119–123. [CrossRef]
42. Kumar, S.; Millar, J.D.; Watson, P.F. The Effect of Cooling Rate on the Survival of Cryopreserved Bull, Ram, and Boar Spermatozoa: A Comparison of Two Controlled-Rate Cooling Machines. *Cryobiology* **2003**, *46*, 246–253. [CrossRef] [PubMed]
43. Choez, K.; Arriaga, I.; Terreros, M.; Condori, R.; Arroyo, G.; Huanca, W. Características Del Semen de Alpacas Obtenido Por Electroeyaculación y Su Motilidad Durante La Refrigeración. *Spermova* **2015**, *5*, 42–46. [CrossRef]
44. Juyena, N.S.; Vencato, J.; Pasini, G.; Vazzana, I.; Stelletta, C. Alpaca Semen Quality in Relation to Different Diets. *Reprod. Fertil. Dev.* **2013**, *25*, 683–690. [CrossRef] [PubMed]
45. Fumuso, F.G.; Bertuzzi, M.L.; Velásquez González, N.; Miragaya, M.H.; Carretero, M.I. Cryopreservation of Llama Semen Using a Combination of Permeable Cryoprotectants. *Reprod. Domest. Anim.* **2021**, *56*, 958–964. [CrossRef]
46. Fumuso, F.G.; Giuliano, S.M.; Chaves, G.; Neild, D.M.; Miragaya, M.H.; Bertuzzi, M.L.; Carretero, M.I. Incubation of Frozen-Thawed Llama Sperm with Seminal Plasma. *Andrologia* **2020**, *52*, e13597. [CrossRef] [PubMed]
47. Hammerstedt, R.H.; Graham, J.K.; Nolan, J.P. Cryopreservation of Mammalian Sperm: What We Ask Them to Survive. *J. Androl.* **1990**, *11*, 73–88. [CrossRef] [PubMed]
48. García, W.; Maxi, E.; Macedo, V.; Ampuero, E.; Malaga, J. Cryopreservation of Alpaca Spermatozoa obtained via Post Copula in a Tris Extender with EGG Yolk from three Avian Species. *Spermova* **2021**, *11*, 11–16. [CrossRef]
49. Carretero, M.I.; Chaves, M.G.; Arraztoa, C.C.; Fumuso, F.G.; Gambarotta, M.C.; Neild, D.M. Air-Drying Llama Sperm Affects DNA Integrity. *Front. Vet. Sci.* **2020**, *7*, 597952. [CrossRef]
50. Celeghini, E.C.C.; de Arruda, R.P.; de Andrade, A.F.C.; Nascimento, J.; Raphael, C.F.; Rodrigues, P.H.M. Effects That Bovine Sperm Cryopreservation Using Two Different Extenders Has on Sperm Membranes and Chromatin. *Anim. Reprod. Sci.* **2008**, *104*, 119–131. [CrossRef] [PubMed]

51. Gamarra, G.; Huaman, E.; León, S.; Carpio, M.; Alvarado, E.; Asparrin, M.; Vivanco, H.W. First In Vitro Embryo Production in Alpacas (*Lama pacos*). *Reprod. Fertil. Dev.* **2008**, *21*, 177–178. [CrossRef]
52. Durand, M.G.P.; Aragón, J.P.Z.; Guerra, U.H.P. Comparación de Sistemas de Cultivo de Embriones de Alpacas. *Rev. Investig Altoandinas—J. High Andean Res.* **2017**, *19*, 157–164. [CrossRef]
53. Coleman, N.V.; Shagiakhmetova, G.A.; Lebedeva, I.Y.; Kuzmina, T.I.; Golubev, A.K. In Vitro Maturation and Early Developmental Capacity of Bovine Oocytes Cultured in Pure Follicular Fluid and Supplementation with Follicular Wall. *Theriogenology* **2007**, *67* 1053–1059. [CrossRef]

Disclaimer/Publisher's Note: The statements, opinions and data contained in all publications are solely those of the individual author(s) and contributor(s) and not of MDPI and/or the editor(s). MDPI and/or the editor(s) disclaim responsibility for any injury to people or property resulting from any ideas, methods, instructions or products referred to in the content.

animals

MDPI

Article

Sperm Incubation in Biggers–Whitten–Whittingham Medium Induces Capacitation-Related Changes in the Lizard *Sceloporus torquatus*

Uriel Ángel Sánchez-Rivera [1,2,3,*], Norma Berenice Cruz-Cano [1], Alfredo Medrano [2], Carmen Álvarez-Rodríguez [1] and Martín Martínez-Torres [1,*]

1 Laboratorio de Biología de la Reproducción, Facultad de Estudios Superiores Iztacala, Universidad Nacional Autónoma de México, Mexico City 54090, Mexico; bionormacruz.7@gmail.com (N.B.C.-C.)
2 Laboratorio de Reproducción, Facultad de Estudios Superiores Cuautitlán, Universidad Nacional Autónoma de México, Mexico City 54714, Mexico; amedrano@unam.mx
3 Posgrado en Ciencias de la Producción y de la Salud Animal, Universidad Nacional Autónoma de México, Mexico City 04510, Mexico
* Correspondence: uriel.sanchez@iztacala.unam.mx (U.Á.S.-R.); martor@unam.mx (M.M.-T.); Tel.: +52-555-623-1258 (M.M.-T.)

Simple Summary: Sperm acquire the ability to fertilize the egg during their transit through the female reproductive tract. This process, known as sperm capacitation, is well recognized in mammals and can be accomplished under laboratory conditions using specialized media. However, it remains unknown whether this process occurs in lizards. In this study, we investigated sperm incubation under conditions that promote capacitation to determine if similar changes occurred in their sperm. Our sperm assessment revealed functional changes, such as modifications in movement and staining patterns, commonly observed in mammals. This suggests that sperm capacitation may occur in this group of animals. Understanding sperm physiology is crucial for developing assisted reproduction technologies to aid conservation efforts for threatened species.

Abstract: Sperm capacitation involves biochemical and physiological changes that enable sperm to fertilize the oocyte. It can be induced in vitro under controlled conditions that simulate the environment of the oviduct. While extensively studied in mammals, its approach in lizards remains absent. Understanding the mechanisms that ensure reproduction is essential for advancing the implementation of assisted reproductive technologies in this group. We aimed to perform a sperm analysis to determine if capacitation-related changes were induced after incubation with capacitating media. Fifteen males of *Sceloporus torquatus* were collected during the early stage of the reproductive season. The sperm were isolated from the seminal plasma and then diluted up to a volume of 150 μL using BWW medium to incubate with 5% CO_2 at 30 °C for a maximum duration of 3 h. A fraction was retrieved hourly for ongoing sperm assessment. The sperm analysis included assessments of its motility, viability, the capacitation status using the chlortetracycline (CTC) assay, and the acrosome integrity with the lectin binding assay to detect changes during incubation. We found that total motility was maintained up to 2 h post incubation, after which it decreased. However, sperm viability remained constant. From that moment on, we observed a transition to a deeper and less symmetrical flagellar bending in many spermatozoa. The CTC assay indicated a reduction in the percentage of sperm showing the full (F) pattern and an increase in those exhibiting the capacitated (B) and reactive (RA) patterns, accompanied by an elevation in the percentage of damaged acrosomes as revealed by the lectin binding assay. In mammals, these changes are often associated with sperm capacitation. Our observations support the notion that this process may also occur in saurian. While sperm analysis is a valuable method for assessing certain functional changes, additional approaches are required to validate this process.

Keywords: sperm capacitation; BWW medium; lizard; incubation; assisted reproduction techniques

Citation: Sánchez-Rivera, U.Á.; Cruz-Cano, N.B.; Medrano, A.; Álvarez-Rodríguez, C.; Martínez-Torres, M. Sperm Incubation in Biggers–Whitten–Whittingham Medium Induces Capacitation-Related Changes in the Lizard *Sceloporus torquatus*. *Animals* **2024**, *14*, 1388. https://doi.org/10.3390/ani14091388

Academic Editor: Anna Wysokińska

Received: 21 March 2024
Revised: 19 April 2024
Accepted: 22 April 2024
Published: 6 May 2024

Copyright: © 2024 by the authors. Licensee MDPI, Basel, Switzerland. This article is an open access article distributed under the terms and conditions of the Creative Commons Attribution (CC BY) license (https://creativecommons.org/licenses/by/4.0/).

1. Introduction

Ejaculated mammalian sperm are morphologically mature but functionally unable to fertilize [1]. Sperm undergo biochemical and physiological changes to become capable of binding and interacting with the oocyte. These modifications include cholesterol efflux, increased membrane fluidity, changes in intracellular ion concentrations, pH elevation, alterations in protein kinase activity, and tyrosine phosphorylation, and occur during their passage through the female reproductive tract in a complex process known as sperm capacitation [2–4]. This process can be accomplished in vitro under controlled conditions by recreating the oviductal environment using defined media supplemented with essential ions such as bicarbonate, calcium, albumin, and energy substrates [5].

Although fully recognized in mammals, sperm capacitation remains uncertain in lizards, despite sharing characteristics such as internal fertilization and the possession of the epididymis, where sperm acquire motility [6]. Females also have the ability to store sperm in the oviducts, allowing an asynchrony between mating and ovulation [7,8]. These characteristics suggest that spermatozoa may require physiological changes after insemination to acquire fertilization ability. Thus far, only one study has conclusively shown sperm capacitation in crocodiles by noting increased intracellular levels of cyclic adenosine monophosphate (cAMP), which enhance motility and elevate protein phosphorylation levels [9]. Moreover, epididymal spermatozoa in *Lacerta vivipara* exhibit increased motility when incubated in a medium containing caffeine, a phosphodiesterase inhibitor known to elevate cAMP levels [10]. These observations suggest that this mechanism may indeed occur within lizards.

Given the ongoing global decline in herpetofauna [11,12], the comprehension of reproductive biology is crucial for the development of any assisted reproductive technologies (ARTs) [13]. We selected *Sceloporus torquatus* as a model for advancing ARTs due to our understanding of its reproductive biology [14,15]. We developed non-invasive semen collection methods by establishing the time to obtain greater volumes and generated sperm quality references [16,17]. However, our efforts in sperm cryopreservation have yielded a low success rate in post-thawing recovery [18]. The above highlights that it is crucial to grasp semen quality parameters, prevent spontaneous acrosome reactions [5], and enhance post-procedural recovery [19]. Moreover, artificial insemination has been unsuccessful (unpublished data), possibly due to inadequate manipulation and preparation of both gametes [20]. These challenges underscore the importance of studying sperm physiology for successful ART implementation. In order to fill the gaps in knowledge regarding sperm capacitation, we conducted an incubation study using Biggers–Whitten–Whittingham (BWW) medium. Our aim was to determine by means of sperm analysis if functional changes similar to those found in mammals were induced.

2. Materials and Methods

2.1. Animals

The capture of 15 adult males of *Sceloporus torquatus* (SVL > 70 mm) [21] was conducted in the Sierra de Guadalupe State Park, Coacalco, State of Mexico (19°61′ N, 99°11′ W, 2480 m altitude), under scientific collection licenses SPA/DGVD/086681/21 and SPARN/DGVD/12218/23 granted by the Secretaría del Medio Ambiente y Recursos Naturales. The collection occurred during the early stage of the reproductive season (October–November) in both 2021 and 2022. Morphometric data of the animals were recorded, including snout–vent length (using digital Vernier calipers to the nearest 0.01 mm) and body weight for each individual (measured using a digital scale with 0.1 g precision). The lizards were kept in outdoor enclosures measuring 3.0 × 5.0 × 2.0 m, with access to food and water, and then released into their natural habitat after completing experimental procedures.

2.2. Semen Collection and Incubation

Semen was collected by gently pressing the genital papillae, following the method described by Martínez-Torres et al. [16]. We registered the number of ejaculates, the total semen volume, and sperm concentration for each male. The ejaculates were washed with PBS and isolated from seminal plasma via double centrifugation at 978× *g*-force for 10 min at room temperature. The sperm samples were diluted to a final volume of 150 µL using BWW medium, with the following composition: 120 mM NaCl, 4.6 mM KCl, 1.7 mM $CaCl_2 \cdot 2H_2O$, 1.2 mM KH_2PO_4, 1.2 mM $MgSO_4 \cdot 7H_2O$, 5.6 mM D-glucose, 0.27 mM sodium pyruvate, 44 mM sodium lactate, 5 U/mL penicillin, 5 mg/mL streptomycin, 20 mM HEPES, 3 mg/mL BSA, and 25 mM $NaHCO_3$, at a pH of 7.4 and an osmolarity of 300 mOsm [9]. All chemicals were purchased from Sigma. A portion of 30 µL was retrieved for assessment at 0 h, while the remaining sample was incubated with 5% CO_2 at 30 °C for up to 3 h. Additionally, a fraction was retrieved each hour for ongoing semen assessment.

2.3. *Sperm Assessment*

2.3.1. Sperm Motility

To assess sperm motility, 5 µL of diluted sperm was placed on a slide and viewed under an optical microscope (Leica DM100) at 40× magnification. The percentage of active sperm was determined based on the presence of symmetrical flagellar movement, which propels them in nearly linear progressive trajectories, leading them out of the visual field [16]. It is crucial to differentiate between actively moving sperm and those passively carried away by the medium flow.

2.3.2. Sperm Viability

To perform a sperm viability test, 5 µL of diluted sperm was taken and mixed with an equal volume of eosin-nigrosin dye on a slide [18]. The mixture was allowed to dry and then observed under an optical microscope at 100× magnification. Any unstained sperm was considered live, while sperm stained in dark pink was considered dead (Figure 1A).

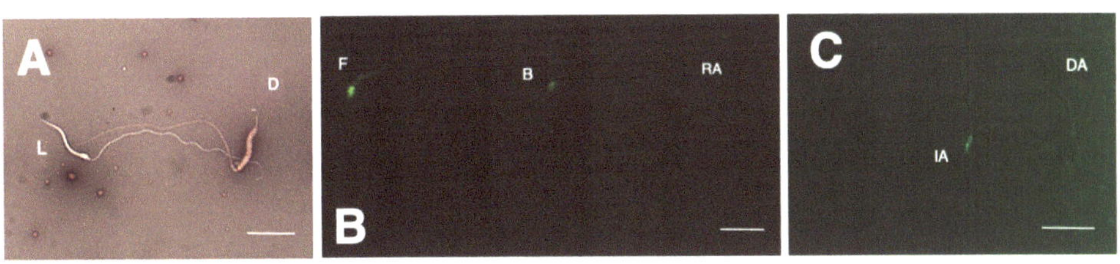

Figure 1. Representative images of *Sceloporus torquatus* sperm evaluation. (**A**) shows live (L) and dead (D) spermatozoa stained with eosin nigrosin, (**B**) shows full (F), band (B), and acrosome-reacted (RA) patterns indicative of capacitance state, as revealed by the CTC assay, and (**C**) shows sperm with intact (IA) and damaged acrosome (DA) spermatozoa, as revealed by the lectin binding assay. Linear bars correspond to 10 µm.

2.3.3. Capacitation Status

We prepared a CTC solution (805 µmol clortetracycline, 20 mM Tris, 130 mM NaCl, and 5 mM L-cysteine at a pH of 7.8) using the method described by Naijian et al. [22]. We mixed this solution in a 1:1 ratio with the same volume of 10 µL of diluted sperm. We stopped the reaction by adding 5 µL of 0.2% glutaraldehyde. Then, we prepared smears and examined them under an epifluorescent microscope at 100× magnification. We determined the sperm capacitation state based on the proportion of spermatozoa exhibiting CTC assay patterns (Figure 1B): full/F (uniform fluorescence head), band/B (post-acrosomal region without fluorescence), and acrosome-reacted/AR (fluorescent-free head or a thin fluorescent band on the equatorial segment).

2.3.4. Acrosome Integrity

The sperm were placed on a slide and left to dry. The slides were permeabilized by immersion in 96% ethanol. We spread 10 µL of fluorescein-conjugated *Pisum sativum* agglutinin (PSA-FITC) lectin and then incubated for 7 min in the dark. After gently washing, we mounted slides with a drop of an antifade solution (220 mM DABCO diluted in glycerol) [23]. The sample was examined under epifluorescence microscopy at 100× magnification to determine the percentage of cells with well-defined acrosomes (Figure 1C).

2.4. Statistical Analysis

The data are presented as mean ± standard error of the mean. Prior to data analysis, we assessed normality and homogeneity using the Shapiro–Wilk and Bartlett tests, respectively. As our data did not meet the assumptions for parametric statistics, we utilized the Kruskal–Wallis test to identify significant differences among the incubation time periods. Subsequently, we conducted Dunn's post hoc test to determine if there are differences between incubation times for each sperm assessment. We assessed the effect of sperm incubation using the Wilcoxon test, with T0 as the control for each pair. A p-value of less than 0.05 was considered statistically significant. We carried out all the analyses and plots using the R (version 4.3.2) software on iOS.

3. Results

According to morphometric values, all males (n = 15) were considered adults. We obtained semen showing consistent characteristics typical of an ejaculate [16], including volume and sperm concentration (Table 1).

Table 1. Morphometric and semen characteristics of male lizards (*Sceloporus torquatus*).

Body Weight (g)	Snout–Vent Length (cm)	Vent–Tail Length (cm)	Number of Ejaculates	Total Semen Volume (μL)	Sperm Concentration ($\times 10^6$/mL)
28.66 ± 4.2 (12.7–52.1)	8.73 ± 0.5 (6.0–11.5)	8.84 ± 0.6 (4.7–11.0)	2.00 ± 0.4 (1.0–3.0)	3.21 ± 1.31 (2.0–6.0)	94.23 ± 19.2 (18.3–220.0)

The data are the mean ± standard error of the mean; the range is shown in parentheses.

Following dilution, all samples exhibited high motility, ranging from 79% to 98%. We found a significant decrease in the second (70.9 ± 4.3%) and third hour (61.8 ± 6.8%, $p < 0.05$) post incubation (H = 21.05, $p < 0.05$, Figure 2a). A statistically significant difference was observed in the second and third hour post treatment ($p < 0.05$), but without difference between the hours. Sperm viability remained above 80% throughout the entire 240-min observation period ($p < 0.05$, Figure 2b). Of note, the spermatozoa displayed active linear movement, but a transition to deeper and less symmetrical flagellar bending, with non-linear movement in many spermatozoa, was observed starting at 2 h post incubation.

In the case of CTC patterns, a high percentage of sperm showed the full pattern (98.6 ± 0.4%) at time 0, with a significant decrease starting from the second hour (35.2 ± 4.6%) and third hour (16.6 ± 4.1%, $p < 0.05$) post incubation. The band pattern showed low levels immediately after dilution with BWW medium (1.2 ± 0.4%), which gradually increased from the first hour and reached 46.8 ± 3.5% in the third hour. Regarding the acrosome-reacted pattern, it was initially absent but appeared from the second hour of incubation, reaching an average value of 36.4 ± 5.6% ($p < 0.05$) at 3 h (Figure 3).

Significant changes in acrosome integrity were observed. At 2 h post incubation, 84.6 ± 3.1% of spermatozoa maintained integrity ($p < 0.05$). This percentage decreased to 74.2 ± 3.8% at 3 h post incubation ($p < 0.05$) (Figure 4).

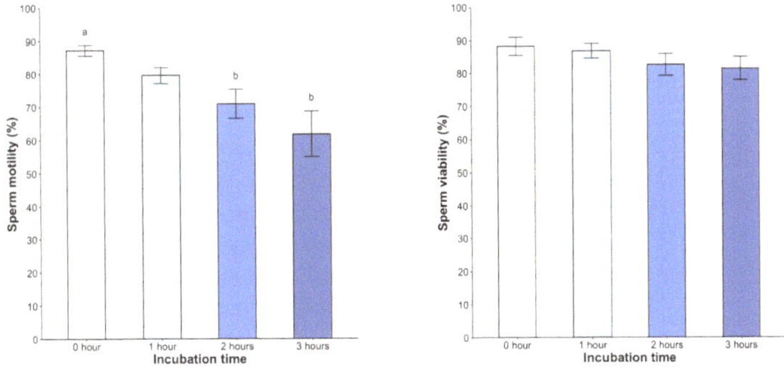

(a) (b)

Figure 2. (**a**) Sperm motility and (**b**) viability of *Sceloporus torquatus* incubated in BWW medium. The lines represent the standard error of the mean. Different letters indicate significant differences between incubation times (Dunn, $p < 0.05$).

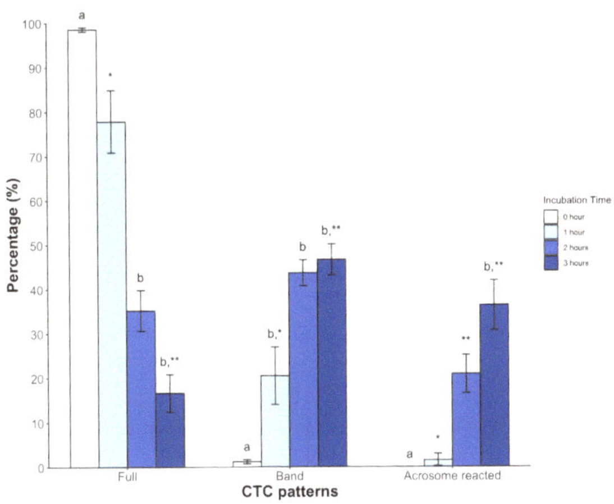

Figure 3. The clorthetracyclin (CTC) sperm patterns of *Sceloporus torquatus* incubated in BWW medium. The lines represent the standard error of the mean. Different letters and asterisks indicate significant differences between incubation times in each pattern (Dunn, $p < 0.05$).

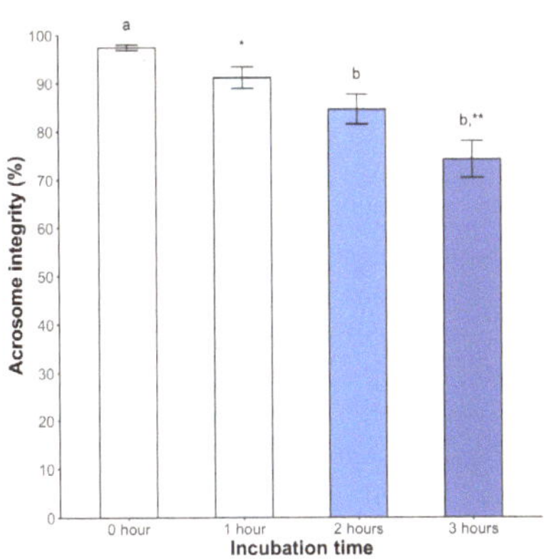

Figure 4. Sperm acrosome integrity of *Sceloporus torquatus* incubated in BWW medium. The lines represent the standard error of the mean. Different letters and asterisks indicate significant differences between incubation times (Dunn, $p < 0.05$).

4. Discussion

Mammalian sperm incubation in defined media reveals biochemical and physiological changes during capacitation [3,24]. Considering the lack of studies about sperm physiology in lizards, it is essential to determine if their sperm undergoes capacitation (as reported in crocodiles) [9] to advance the successful implementation of ARTs in this group. To address this, we incubated *Sceloporus torquatus* spermatozoa in BWW medium at 30 °C with 5% CO_2 for up to 3 h to assess functional capacitation-related changes in sperm quality.

We observed a consistently high percentage of motility (above 79%), which decreased over time starting at 2 h. This trend suggests that the medium may favor the metabolic processes of sperm [25], potentially because its composition improves cell longevity [26]. Considering the specific variations of each species, the choice of medium is crucial for an adequate manipulation of gametes [27]. BWW medium promotes sperm motility and consistently induces increased cAMP levels and protein tyrosine phosphorylation in mammals [28]. Additionally, it enhanced motility in *Crocodylus porosus* in a 120-min incubation [9]. In light of the observed effect of phosphodiesterase inhibitor (which increases cAMP levels) on *Lacerta vivipara* sperm, resulting in increased motility [10], we hypothesize that this mechanism is conserved in this group and *S. torquatus* would react similarly under capacitation conditions.

We also noted modifications in motility patterns in many sperm, with increased and deeper flagellar beat amplitude, less symmetrical bending, and nonlinear movement. These observations suggest the hyperactive movement, which may facilitate the zona pellucida penetration during fertilization, as observed in mammals [29]. However, we did not quantify the proportion of spermatozoa undergoing these changes. A comprehensive assessment of motility using computer-assisted sperm analysis (CASA) is essential to detect movement types and evaluate the proportion of hyperactivated sperm [30].

Evaluating the response of sperm metabolism can be challenging due to the complexity of the involved molecules. However, CTC binds to the membrane in a calcium-dependent manner, enabling the monitoring of sperm labeling and the detection of changes in fluidity [31,32]. The above corresponds with increased tyrosine phosphorylation, indicative of hyperactivated motility [33]. This assay is being applied for the first time in any non-avian reptile and represents a valuable reference for further studies on seminal parameters in other species. Incubation induced significant changes, with a decrease in the F pattern (non-capacitated sperm) over time and a higher percentage of the B pattern (capacitated sperm) after 2 h. Similar observations in dogs [34], mice [27], and boars [35] have been reported. The AR pattern increased from the second hour. These findings were confirmed by means of the lectin-binding assay, which specifically binds to spermatozoa with complete acrosome content [36]. Similar results were found in chickens under mammalian capacitating conditions [37]. Based on these findings, we inferred that sperm incubated under the described conditions underwent molecular changes consistent with capacitation, attributable to medium composition.

The constituents of the BWW medium induce diverse changes in the sperm. Albumin, for instance, modifies lipid composition and membrane fluidity by reducing plasma membrane cholesterol content [26,28]. Moreover, the presence of bicarbonate (above 15mM) in the medium initiates early changes that promote sperm capacitation by activating adenylate cyclase and elevating intracellular cAMP levels, resulting in the hyperpolarization of the plasma membrane and increased intracellular pH [2,38]. Also, calcium ions play a crucial role in this process by activating protein kinase A (PKA), which phosphorylates proteins involved in sperm functions like hyperactivation and the acrosome reaction [39,40]. Although the effects of glucose are unknown, it is essential for capacitation in mice, while pyruvate and lactate may inhibit it [27].

The incubation time has recently been recognized as a significant factor affecting sperm quality. In vitro studies on human sperm have shown a wide range from 1 to 24 h to capacitation induction, leading to sperm subpopulations with varying degrees of functionality [24]. We found the changes in sperm assessments starting at two hours, which accentuated at the third hour, accompanied by a significant decrease in total motility. However, the limited sperm volumes in lizards hamper our ability to incubate for longer periods or devise protocols to select the sperm with capacitation-associated effects.

Further research on changes in oviductal content and functional assays in both mated and unmated females would offer valuable insights [41] for optimizing the formulation of a more suitable medium. Incubation with calcium or progesterone ionophores can also be investigated, as both activate PKA-mediated signaling pathways, potentially improving efficiency in capacitation induction [39].

5. Conclusions

Our study revealed some suggestive changes associated with sperm capacitation, such as a change in the type of movement characterized by increased and deeper flagellar beat amplitude, an increased occurrence of capacitation patterns, and damaged sperm in the acrosome after two hours of incubation. These observations support the idea that this process also occurs in saurian. While sperm analysis is a valuable method for assessing certain functional changes, there are additional approaches required to validate this process. Establishing if sperm capacitation is a prerequisite for acquiring fertilization competence is crucial to improving the success of the implementation of any ART in this group of animals.

Author Contributions: Conceptualization, M.M.-T. and A.M.; investigation, U.Á.S.-R., N.B.C.-C. and C.Á.-R.; data curation, formal analysis, and visualization, U.Á.S.-R. and N.B.C.-C.; writing—original draft preparation, M.M.-T. and U.Á.S.-R.; writing—review and editing, A.M., N.B.C.-C. and C.Á.-R. All authors have read and agreed to the published version of the manuscript.

Funding: This research was funded by the Universidad Nacional Autónoma de México through the PAPIIT projects awarded to MMT and AM: IN217722 (FESI) and IN205421 (FESC), respectively. Additionally, support was provided by the Consejo Nacional de Ciencia y Tecnología scholarship (CVU 893879) granted to UÁSR.

Institutional Review Board Statement: All procedures were carried out with the approval of the Institutional Subcommittee for the Care and Use of Experimental Animals at the Faculty of Veterinary Medicine and Zootechnics (UNAM) under the protocol DC-2018/2-16.

Informed Consent Statement: Not applicable.

Data Availability Statement: Dataset available on request from the authors.

Acknowledgments: The authors would like to extend their gratitude to the General Coordination of Ecological Conservation of the State of Mexico for providing facilities for lizard collection, and special thanks to Enrique González-Hernández for his invaluable comments on the project and to Alicia Alcántar for laboratory support. This paper is part of the requirements for obtaining a Doctoral degree at the Posgrado en Ciencias de la Producción y de la Salud Animal, UNAM, of U.Á. Sánchez-Rivera.

Conflicts of Interest: All authors declare that they have no conflicts of interest.

References

1. Fujihara, Y.; Miyata, H.; Ikawa, M. Factors controlling sperm migration through the oviduct revealed by gene-modified mouse models. *Exp. Anim.* **2018**, *67*, 91–104. [CrossRef] [PubMed]
2. Ickowicz, D.; Finkelstein, M.; Breitbart, H. Mechanism of sperm capacitation and the acrosome reaction: Role of protein kinases. *Asian J. Androl.* **2012**, *14*, 816–821. [CrossRef] [PubMed]
3. Puga-Molina, L.C.; Luque, G.M.; Balestrini, P.A.; Marín-Briggiler, C.I.; Romarowski, A.; Buffone, M.G. Molecular basis of human sperm capacitation. *Front. Cell Dev. Biol.* **2018**, *6*, 72. [CrossRef] [PubMed]
4. Stival, C.; Puga-Molina, L.C.; Paudel, B.; Buffone, M.G.; Visconti, P.E.; Krapf, D. Sperm capacitation and acrosome reaction in mammalian sperm. In *Sperm Acrosome Biogenesis and Function During Fertilization*, 1st ed.; Buffone, M.G., Ed.; Springer: Cham, Switzerland, 2016; Volume 220, pp. 93–106. [CrossRef]
5. Lemoine, M.; Grasseau, I.; Brillard, J.P.; Blesbois, E. A reappraisal of the factors involved in in vitro initiation of the acrosome reaction in chicken spermatozoa. *Reproduction* **2008**, *136*, 391–399. [CrossRef] [PubMed]
6. Dosemane, D.; Bhagya, M. In vitro study of the spermatozoa motility in the lizard *Eutropis carinata*. *Int. J. Zool. Res.* **2015**, *11*, 89–95. [CrossRef]

7. Martínez-Torres, M. Almacenamiento de espermatozoides en la vagina de la lagartija vivípara *Sceloporus torquatus* (Sauria Prhynosomatidae). *Acta Zool. Mex.* **2009**, *25*, 497–506. [CrossRef]
8. Ortega-León, A.M.; Cruz, M.V.; Zúñiga-Vega, J.J.; Castillo, R.C.; Méndez-de la Cruz, F.R. Sperm viability in the reproductive tract of females in a population of *Sceloporus mucronatus* exhibiting asynchronous reproduction. *West N. Am. Nat.* **2009**, *69*, 96–104 [CrossRef]
9. Nixon, B.; Anderson, A.L.; Smith, N.D.; McLeod, R.; Johnston, S.D. The Australian saltwater crocodile (*Crocodylus porosus*) provides evidence that the capacitation of spermatozoa may extend beyond the mammalian lineage. *Proc. R Soc. B* **2016**, *283*, 20160495. [CrossRef]
10. Depeiges, A.; Dacheux, J.L. Acquisition of sperm motility and its maintenance during storage in the lizard, *Lacerta vivipara Reproduction* **1985**, *74*, 23–27. [CrossRef]
11. Böhm, M.; Collen, B.; Baillie, J.E.M.; Bowles, P.; Chanson, J.; Cox, N.; Hammerson, G.; Hoffmann, M.; Livingstone, S.R.; Ram, M.; et al. The conservation status of the world's reptiles. *Biol. Conserv.* **2013**, *157*, 372–385. [CrossRef]
12. Sinervo, B.; Méndez-de la Cruz, F.R.; Miles, D.B.; Heulin, B.; Bastiaans, E.; Villagrán-Santa Cruz, M.; Lara-Resendiz, R.; Martínez-Méndez, N.; Calderón-Espinosa, M.L.; Meza-Lázaro, R.N.; et al. Erosion of Lizard Diversity by Climate Change and Altered Thermal Niches. *Science* **2010**, *328*, 894–899. [CrossRef] [PubMed]
13. Comizzoli, P.; Holt, W.V. Recent Progress in Spermatology Contributing to the Knowledge and Conservation of Rare and Endangered Species. *Ann. Rev. Anim. Biosci.* **2022**, *10*, 469–490. [CrossRef] [PubMed]
14. Cruz-Cano, N.B.; Sánchez-Rivera, U.Á.; Álvarez-Rodríguez, C.; Dávila-Govantes, R.; Cárdenas-León, M.; Martínez-Torres, M. Sex steroids are correlated with environmental factors and body condition during the reproductive cycle in females of the lizard *Sceloporus torquatus*. *Gen. Comp. Endocrinol.* **2021**, *314*, 113921. [CrossRef] [PubMed]
15. Sánchez-Rivera, U.Á. Análisis hormonal y citológico de la espermatogénesis mediante biopsia testicular en *Sceloporus torquatus* (Sauria: Phrynosomatidae). Bachelor's Thesis, National Autonomous University of Mexico, Mexico City, Mexico, 2017.
16. Martínez-Torres, M.; Sánchez-Rivera, U.Á.; Cruz-Cano, N.B.; Castro-Camacho, Y.J.; Luis, J.; Medrano, A. A non-invasive method for semen collection and evaluation in small and median size lizards. *Reprod. Domest. Anim.* **2019**, *54* (Suppl. 4), 54–58. [CrossRef] [PubMed]
17. Martínez-Torres, M.; Álvarez-Rodríguez, C.; Luis, J.; Sánchez-Rivera, U.Á. Electroejaculation and semen evaluation of the viviparous lizard *Sceloporus torquatus* (Squamata: Phrynosomatidae). *Zoo Biol.* **2019**, *38*, 393–396. [CrossRef] [PubMed]
18. Sánchez-Rivera, U.Á.; Medrano, A.; Cruz-Cano, N.B.; Alcántar-Rodríguez, A.; Dávila-Govantes, R.; Castro-Camacho, Y.J.; Martínez-Torres, M. Implementation of a method for sperm cryopreservation in sceloporine lizards. *Conserv. Physiol.* **2022**, *10*, coac068. [CrossRef] [PubMed]
19. Campbell, L.; Cafe, S.I.; Upton, R.; Sean Doody, J.; Nixon, B.; Clulow, J.; Clulow, S. A model protocol for the cryopreservation and recovery of motile lizard sperm using the phosphodiesterase inhibitor caffeine. *Conserv. Physiol.* **2020**, *8*, coaa044. [CrossRef] [PubMed]
20. Mattioli, M.; Barboni, B.; Lucidi, P.; Seren, E. Identification of capacitation in boar spermatozoa by chlortetracycline staining. *Theriogenology* **1996**, *45*, 373–381. [CrossRef] [PubMed]
21. Feria-Ortiz, M.; Nieto-Montes de Oca, A.; Ugarte, I.H. Diet and Reproductive Biology of the Viviparous Lizard *Sceloporus torquatus torquatus* (Squamata: Phrynosomatidae). *J. Herpetol.* **2001**, *35*, 104–112. [CrossRef]
22. Naijian, H.R.; Kohram, H.; Shahneh, A.Z.; Sharafi, M.; Bucak, M.N. Effects of different concentrations of BHT on microscopic and oxidative parameters of Mahabadi goat semen following the freeze–thaw process. *Cryobiology* **2013**, *66*, 151–155. [CrossRef]
23. Ortega-Morales, L.D.; Alcantar-Rodriguez, A.; Espejel, M.C.; Medrano, A. The effect of non-traditional cooling on dog sperm cryosurvival and ability to perform the acrosome reaction. *Aust. J. Vet. Sci.* **2019**, *51*, 73–82. [CrossRef]
24. Sáez-Espinosa, P.; Huerta-Retamal, N.; Robles-Gómez, L.; Avilés, M.; Aizpurua, J.; Velasco, I.; Romero, A.; Gómez-Torres, M.J. Influence of in vitro capacitation time on structural and functional human sperm parameters. *Asian J. Androl.* **2020**, *22*, 447–453. [CrossRef]
25. Bustani, G.S.; Baiee, F.H. Semen extenders: An evaluative overview of preservative mechanisms of semen and semen extenders. *Vet. World.* **2021**, *14*, 1220–1233. [CrossRef] [PubMed]
26. Witte, T.S.; Schäfer-Somi, S. Involvement of cholesterol, calcium and progesterone in the induction of capacitation and acrosome reaction of mammalian spermatozoa. *Anim. Reprod. Sci.* **2007**, *102*, 181–193. [CrossRef]
27. Kito, S.; Ohta, Y. Medium effects on capacitation and sperm penetration through the zona pellucida in inbred BALB/c spermatozoa. *Zygote* **2005**, *13*, 145–153. [CrossRef]
28. McPartlin, L.A.; Littell, J.; Mark, E.; Nelson, J.L.; Travis, A.J.; Bedford-Guaus, S.J. A defined medium supports changes consistent with capacitation in stallion sperm, as evidenced by increases in protein tyrosine phosphorylation and high rates of acrosomal exocytosis. *Theriogenology* **2008**, *69*, 639–650. [CrossRef] [PubMed]
29. Ho, H.; Suarez, S.S. Hyperactivation of mammalian spermatozoa: Function and regulation. *Reproduction* **2001**, *122*, 519–526. [CrossRef] [PubMed]
30. Suarez, S.S. Control of hyperactivation in sperm. *Hum. Reprod. Update* **2008**, *14*, 647–657. [CrossRef]
31. Mendes Cunha, A.T.; Faria, O.; Guimarães, A.L.S. Bovine Sperm Capacitation: Physiological Changes and Evaluations. *JSM Invit. Fertil.* **2017**, *2*, 1011. [CrossRef]

32. Rathi, R.; Colenbrander, B.; Bevers, M.M.; Gadella, B.M. Evaluation of In Vitro Capacitation of Stallion Spermatozoa. *Biol. Reprod.* **2001**, *65*, 462–470. [CrossRef]
33. Gadella, B.M.; Boerke, A. An update on post-ejaculatory remodeling of the sperm surface before mammalian fertilization. *Theriogenology* **2016**, *85*, 113–124. [CrossRef] [PubMed]
34. Rota, A.; Peña, A.I.; Linde-Forsberg, C.; Rodríguez-Martínez, H. In vitro capacitation of fresh, chilled and frozen–thawed dog spermatozoa assessed by the chlortetracycline assay and changes in motility patterns. *Anim. Reprod. Sci.* **1999**, *57*, 199–215. [CrossRef] [PubMed]
35. Ded, L.; Dostalova, P.; Zatecka, E.; Dorosh, A.; Komrskova, K.; Peknicova, J. Fluorescent analysis of boar sperm capacitation process in vitro. *Reprod. Biol. Endocrinol.* **2019**, *17*, 109. [CrossRef]
36. Kekäläinen, J.; Larma, I.; Linden, M.; Evans, J.P. Lectin staining and flow cytometry reveals female-induced sperm acrosome reaction and surface carbohydrate reorganization. *Sci. Rep.* **2015**, *5*, 15321. [CrossRef]
37. Priyadarshana, C.; Setiawan, R.; Tajima, A.; Asano, A. Src family kinases-mediated negative regulation of sperm acrosome reaction in chickens (*Gallus gallus domesticus*). *PLoS ONE* **2020**, *15*, e0241181. [CrossRef] [PubMed]
38. Soriano-Úbeda, C.; Romero-Aguirregomezcorta, J.; Matás, C.; Visconti, P.E.; García-Vázquez, F.A. Manipulation of bicarbonate concentration in sperm capacitation media improves in vitro fertilisation output in porcine species. *J. Anim. Sci. Biotechnol.* **2019**, *10*, 19. [CrossRef]
39. Nixon, B.; Cafe, S.L.; Eamens, A.L.; De Iuliis, G.N.; Bromfield, E.G.; Martin, J.H.; Skerrett-Byrne, D.A.; Dun, M.D. Molecular insights into the divergence and diversity of post-testicular maturation strategies. *Mol. Cell. Endocrinol.* **2020**, *517*, 110955. [CrossRef] [PubMed]
40. Vyklicka, L.; Lishko, P.V. Dissecting the signaling pathways involved in the function of sperm flagellum. *Curr. Opin. Cell Biol.* **2020**, *63*, 154–161. [CrossRef]
41. Friesen, C.R.; Kahrl, A.F.; Olsson, M. Sperm competition in squamate reptiles. *Phil. Trans. R. Soc. B.* **2020**, *375*, 20200079. [CrossRef]

Disclaimer/Publisher's Note: The statements, opinions and data contained in all publications are solely those of the individual author(s) and contributor(s) and not of MDPI and/or the editor(s). MDPI and/or the editor(s) disclaim responsibility for any injury to people or property resulting from any ideas, methods, instructions or products referred to in the content.

animals

MDPI

Article

The Impact of Microorganisms on Canine Semen Quality

Kinga Domrazek [1,*], Paweł Konieczny [2,3], Marcin Majka [2,3], Michał Czopowicz [4] and Piotr Jurka [1]

[1] Department of Small Animal Diseases and Clinic, Institute of Veterinary Medicine, Warsaw University of Life Sciences-SGGW, Nowoursynowska 159c, 02-776 Warsaw, Poland; piotr_jurka@sggw.edu.pl
[2] Vet Cell Tech Sp. z o.o., 30-348 Kraków, Poland; pawelk@vetcelltech.com (P.K.); mmajka@cm-uj.krakow.pl (M.M.)
[3] Institute of Pediatrics, Faculty of Medicine, Jagiellonian University Medical College, 30-663 Krakow, Poland
[4] Division of Veterinary Epidemiology and Economics, Institute of Veterinary Medicine, Warsaw University of Life Sciences-SGGW, Nowoursynowska 159c, 02-776 Warsaw, Poland; michal_czopowicz@sggw.edu.pl
* Correspondence: kinga_domrazek@sggw.edu.pl

Simple Summary: Various microorganisms, including *Mycoplasma* spp., have been reported in canine ejaculate. The impact of these microorganisms on semen quality remains unclear. The aim of this study was to evaluate the prevalence of bacteria and *Mycoplasma* spp. (and various species) in canine semen. Interestingly, 36.5% of the examined dogs tested negative for both aerobic bacteria and mycoplasmas, while 12.7% tested positive for bacterial presence. Additionally, 60.3% of the dogs tested positive for *Mycoplasma* spp. using PCR, with most carrying 1–2 *Mycoplasma* species. We found no significant difference in semen characteristics between *Mycoplasma*-positive and -negative dogs. The detection of *Mycoplasma* was not significantly linked to the presence of bacteria in semen. All the microorganisms identified were classified as saprophytic flora. Some canine ejaculate is sterile. Our findings suggest the existence of undescribed species of canine mycoplasmas, necessitating advanced diagnostic techniques like NGS for their identification.

Abstract: Various microorganisms, including *Mycoplasma* spp., have been reported in canine ejaculate. The impact of these microorganisms on semen quality remains unclear. This study included 63 male intact healthy dogs aged 1–8 years. One dog exhibited azoospermia, indicating a relatively low incidence of this condition. Interestingly, 36.5% of the examined dogs tested negative for both aerobic bacteria and mycoplasmas, while 12.7% tested positive for bacterial presence. Additionally, 60.3% of the dogs tested positive for *Mycoplasma* spp. using PCR, with most carrying 1–2 *Mycoplasma* species. We found no significant difference in semen characteristics between *Mycoplasma*-positive and -negative dogs. The detection of *Mycoplasma* was not significantly linked to the presence of bacteria in semen. All the microorganisms identified were classified as saprophytic flora. Our findings indicate that *Mycoplasma* spp. is common in canine ejaculate. Semen quality parameters were not correlated with the presence of *Mycoplasma* spp. in semen. *Mycoplasma* HRC689 was the most common species. Some dogs exhibited no presence of aerobic bacteria or mycoplasmas in their semen. Our study highlights the common presence of *Mycoplasma* spp. in canine ejaculate. Semen quality shows no correlation with *Mycoplasma* presence. Some canine ejaculate is sterile. Our findings suggest the existence of undescribed species of canine mycoplasmas, necessitating advanced diagnostic techniques like NGS for their identification.

Keywords: mycoplasma; semen quality; CASA-system

Citation: Domrazek, K.; Konieczny, P.; Majka, M.; Czopowicz, M.; Jurka, P. The Impact of Microorganisms on Canine Semen Quality. *Animals* **2024**, *14*, 1267. https://doi.org/10.3390/ani14091267

Academic Editor: Anna Wysokińska

Received: 4 April 2024
Revised: 18 April 2024
Accepted: 20 April 2024
Published: 23 April 2024

Copyright: © 2024 by the authors. Licensee MDPI, Basel, Switzerland. This article is an open access article distributed under the terms and conditions of the Creative Commons Attribution (CC BY) license (https://creativecommons.org/licenses/by/4.0/).

1. Introduction

The importance of semen quality in canine reproduction cannot be overstated, as it directly influences the success of breeding programs and the health of the offspring [1]. The evaluation of semen quality encompasses various parameters, including sperm count, motility, morphology, and viability, all of which directly influence the likelihood of successful conception [1]. The contribution of the stud dog constitutes half of the factors that are to be considered when assessing the potential causes of infertility in canine breeding [2]. Due to this fact, the semen quality of the stud dog should be routinely evaluated before mating. Being a carrier of various pathogens is another major factor that needs to be controlled in a dog used for reproduction, as some pathogens may be transmitted via the sexual route to the bitch and lead to reproductive failure [3]. The presence of bacteria in canine ejaculate is a problematic issue in veterinary medicine. Many studies have shown that canine ejaculate is not sterile [4]. It is difficult, however, to distinguish between contamination from the urethra or foreskin and a primary infection of the urogenital tract [4]. Organisms commonly cultured from the semen of healthy male dogs include *Escherichia coli*, *Pasteurella multocida*, beta-hemolytic *Streptococcus*, coagulase-negative *Staphylococcus*, *Staphylococcus pseudintermedius*, *Canicola haemoglobinophilus*, *Klebsiella* spp., and *Pseudomonas* spp. [5–7]. Bacterial infections of the urogenital tract can have detrimental effects on canine semen quality, potentially leading to reproductive failures. Some studies describe a negative influence of bacteria belonging to the natural urogenital microbiome, e.g., *E. coli* on fertility [8]. However, except for *Brucella canis*, bacteria appear to be an uncommon cause of compromised fertility in dogs [9]. A negative impact on the seminal quality parameters is likely associated with an increasing number of bacterial species in canine sperm [7]. On the other hand, the bacteria commonly found in semen may play a protective role by inhibiting the growth of pathogenic microorganisms.

The negative influence of bacteria on sperm results from various mechanisms, including direct contact, competition for nutrients, and detritus production [10,11]. Bacterial contamination of ejaculate can lead to decreased spermatozoa motility, increased percentage of dead spermatozoa, and changes in morphology [8,12]. Moreover, after artificial insemination or natural mating, bacteria from ejaculate may induce uterine infections, fertilization failure, embryonic and fetal resorption, abortions, or stillbirths, contributing to decreased litter size and even leading to septicemia in the bitch [13]. One group of bacteria with potential negative impact on semen quality is mycoplasmas.

The data about the occurrence and role of *Mycoplasma* spp. in canine semen are contradictory. Some authors suggest that they have a negative influence on canine fertility [14] and can cause orchitis, epididymitis, and prostatitis [15]. In vitro studies have shown that *Mycoplasma* spp. can be attached to the spermatozoa by interlacing fibrils of variable diameter, which may reduce its motility [16]. Furthermore, Laber and Holtzmann [14] reported a significant increase in the percentage of abnormal spermatozoa and decrease in their motility caused by *M. canis*. *M. maculosum*, and *M. spumans* were described as a cause of 100% of dead forms and 70% of abnormalities in the head, midpiece, and tail of spermatozoa in Bernese Mountain Dogs [12].

Our study aimed to determine the prevalence of aerobic bacteria and mycoplasmas in Polish male dogs and the impact of these microorganism on semen quality.

2. Materials and Methods

2.1. Study Population and Sampling

This study enrolled adult male intact dogs between 1 and 8 years of age to avoid the potential influence of extreme age on their fertility. These dogs were sourced from kennels affiliated with the Polish Kennel Club (ZKwP, Poland), as well as from shelters for homeless animals. Subsequently, each dog underwent a routine clinical examination to ensure they were free of systemic diseases, and serum testosterone, estradiol, and total thyroxin concentrations were measured to eliminate the potential influence of endocrine disorders on semen quality.

All medical procedures were performed as a part of routine veterinary examination on the owners' request and thus, according to the European directive EU/2010/63 and Polish legal regulations, the approval of Ethical Committee for the described procedures was not required, as they could be qualified as nonexperimental clinical veterinary practices excluded from the directive (Act of 15 January 2015 on the protection of animals used for scientific or educational purposes).

Eventually, 63 clinically healthy male dogs with the aforementioned hormones within the reference intervals were enrolled in study. Semen was collected in the sterile containers by digital manipulation, and the sperm-rich fraction of the ejaculate was analyzed according to standards [1]. From each semen sample, the swab was collected and sent to the commercial laboratory (Vetlab, Warsaw, Poland) for the routine bacteriological examination. Additionally, three cotton swabs were taken from each semen sample and air-dried. One of these swabs was sent to the same commercial veterinary laboratory (Vetlab, Poland) for PCR for *Mycoplasma* spp., canine herpesvirus type 1 (CHV-1), and *Chlamydia* spp., while the remaining two were kept at −80 °C until the results of PCR had been obtained. Based on the PCR results, dogs were categorized into the *Mycoplasma*-positive or *Mycoplasma*-negative group. The samples from *Mycoplasma* spp.-positive dogs were further analyzed to identify the exact *Mycoplasma* species. No samples were positive for CHV-1 or *Chlamydia* spp., as described elsewhere [17].

2.2. Hormone Measurements

After clotting, blood samples were centrifuged at 2057× g for 5 min, and serum was harvested. Hormones were quantified using the competitive enzyme immunoassay competition method with final fluorescent detection (ELFA) (MINI, VIDAS, bioMérieux, Marcy l'Etoile, France) in accordance with the manufacturers' manuals. Reference intervals were defined as follows: testosterone $\geq$ 1 ng/mL [18], estradiol < 115 pg/mL [19], and total thyroxin within the range of 10–50 nmol/L [20].

2.3. PCR Analyses

To increase the efficiency of the reactions, two swabs were used to carry out the PCR reaction. DNA isolation was performed using the Swab-Extract DNA Purification Kit (Eurx, Gdańsk, Poland), following the manufacturer's guidelines. PCR reactions were performed using recently published primers specific to various *Mycoplasma* species [21], along with Taq PCR Master Mix (2x) (Eurx, Gdańsk, Poland). Protocols for PCR were adapted from standard procedures described previously [22,23]. Subsequently, the PCR products were analyzed via electrophoresis in a 2% agarose gel, and the approximate lengths of the amplicons were determined using a molecular-weight size marker (100 bp DNA ladder) as a reference.

2.4. Bacteriological Examination

The semen samples were collected for bacteriological tests using transport agar medium swabs and promptly send to the commercial laboratory (Vetlab, Poland). The samples were cultured on the following microbiological media: Columbia agar with 5% ovine blood, MacConkey agar, Columbia CNA agar with 5% ovine blood, and chocolate agar. Incubation conditions included maintaining a temperature of 35–37 °C for 48 h in an oxygen atmosphere (Columbia agar with 5% ovine blood, MacConkey agar, and Columbia CNA agar with 5% ovine blood) or an atmosphere with an elevated concentration of CO_2 (Chocolate Agar), facilitated by a CO_2 atmosphere generator (Gen Compact, bioMérieux, Marcy l'Etoile, France). The bacterial growth was reviewed 24 and 48 h post-incubation. Subsequently, the obtained bacterial colonies underwent analysis in the MALDI TOF Biotyper Sirius IV (Billerica, MA, USA).

2.5. Semen Quality Evaluation

2.5.1. Macroscopic Evaluation

The volume of the sperm-rich fraction was measured by using calibrated pipettes, and the color of the semen was visually assessed. Cloudy or milky opacity were considered normal, following guidelines outlined by Root Kustritz [24]. The pH value was determined in each semen sample by dipping litmus strips.

2.5.2. Morphology Evaluation

The morphology of spermatozoa was evaluated by preparing smears from the second fraction, which were then air-dried and immersed in the sperm stain (Microptic, Barcelona, Spain) for 5 min. Then, the samples were examined in the light microscope (ECLIPSE E 200, Nikon, Tokyo, Japan) at 100-fold magnification. At least 200 spermatozoa were reviewed and categorized according to the criteria established by Freshman [1] into normal spermatozoa and spermatozoa exhibiting defects of the head, midpiece, or tail. Semen samples with more than 70% of spermatozoa of normal morphology were classified as physiological [1].

2.5.3. Viability Evaluation

The conventional microscopic assessment of the proportion of viable (with intact cell membrane) and dead (with compromised cell membrane) spermatozoa was conducted using nigrosine–eosin stain according to established protocols [25]. A warm mixture of the stain and semen (comprising 3 μL of eosin, 3 μL of nigrosine, and 3 μL of semen) was smeared on a heated glass slide [26] and air-dried. Then, the samples were examined under the light microscope (ECLIPSE E 200, Nikon, Tokyo, Japan) at 100-fold magnification. Each assessment involved the evaluation of at least 200 cells, with the results presented as the percentage of viable and dead spermatozoa.

2.6. Computer-Assisted Sperm Analysis (CASA)

The computer-assisted sperm analysis (CASA) was conducted using the sperm class analyzer (SCA version 6.5.0.67, Microptic, Barcelona, Spain) in conjunction with the light microscope (ECLIPSE E 200, Nikon, Tokyo, Japan) and camera (Basler, Ahrensburg, Germany). The thermostable table of the analyzer was heated to a temperature of 37 °C according to the established protocols [27]. The sperm-rich semen fraction was diluted in the proportion 1:1–1:5 with phosphate-buffered saline (PBS, Sigma Aldrich, Saint Louis, MI, USA) and incubated for 5 min at 37 °C prior to evaluation. Analysis was performed using a 20-micron GoldCyto 4-chamber slide (Goldcyto Biotech corp., Shanghai, China), using the manufacturer settings for dogs, as follows: VLC Rapid 165 μm/s, Lin Rapid 55%, and the average head area 20 μm^2.

In each analysis, a minimum of 500 spermatozoa were counted and examined for the following characteristics: concentration, motility, mucus penetration, and round cell count. Additionally, spermatozoa were categorized into subpopulations based on their movement characteristics, including velocity (fast (RAPID), moderate (MEDIUM), slow (SLOW)), direction (progressive, moderately progressive, nonprogressive), and the percentage of spherical tracks. A total spermatozoa count exceeding 200×10^6 and the percentage of motile spermatozoa exceeding 70% was considered normal [24]. To ensure the reliability of the results, all semen samples were microscopically evaluated by the same highly qualified staff member.

2.7. Statistical Methods

Categorical variables were presented as counts of groups and percentages from this study population and compared between groups using the likelihood ratio G test or Fisher exact test (if any expected cell count in the contingency table was <5). The 95% confidence interval (CI 95%) for proportions was calculated using the Wilson score method [28]. Numerical variables were tested for normality of distribution through the inspection of normal probability Q-Q plots and using the Shapiro–Wilk W test. As normality assumption was violated in most cases, the numerical variables were expressed as the median, interquartile range (IQR), and range and compared between groups using the Mann–Whitney U test. Their correlations were tested using Spearman's rank correlation coefficient (R_s). All statistical tests were 2-tailed, and the significance level (α) was set at 0.05. Statistical analysis was performed using TIBCO Statistica 13.3 (TIBCO Software Inc., Palo Alto, CA, USA).

3. Results

3.1. Study Population

This study included sixty-three male, intact, clinically healthy dogs aged from 1 to 8 years with a median (IQR) of 3.0 (1.5–4.5) years; twenty-three dogs (36.5%) were 1 year old, eight dogs (12.7%) were 2 years old, nine dogs (14.3%) were 3 years old, nine dogs (14.3%) were 4 years old, five dogs (7.9%) were 5 years old, two dogs (3.2%) were 6 years old, three dogs (4.8%) were 7 years old, and four dogs (6.3%) were 8 years old. Three dogs were crossbreeds (4.8%), and the remaining sixty dogs belonged to forty-seven breeds, among which Border Collie was represented by six dogs and springer spaniel and English Mastiff were represented by three dogs (Table S1). Body weight ranged from 3 to 120 kg, with a median (IQR) of 24 (15–31) kg.

Testosterone and estradiol concentrations were within the reference interval in all dogs. Total thyroxin concentration was lowered in twelve dogs (only in two dogs had <10 nmol/L) and slightly elevated in one dog (56.5 nmol/L).

3.2. Semen Characteristics

Only 1/63 dogs (1.6%; CI 95%: 0.3–8.5%) had azoospermia, and 5/63 dogs (7.9%; CI 95%: 3.4–17.3%) had oligospermia ($<200 \times 10^6$ sperms). Abnormal spermatozoa morphology ($\leq$70% of normal spermatozoa in semen) was found in 3/62 dogs (4.8%, CI 95%: 1.7–13.3%). Details of semen characteristics are presented in Table 1. Semen volume was significantly positively correlated with the body weight of dogs ($R_s = 0.37$, $p = 0.003$), while round cell count and the proportion of normal spermatozoa was significantly correlated with the age of dogs, the former positively ($R_s = 0.34$, $p = 0.006$) and the latter negatively ($R_s = -0.35$, $p = 0.005$)

Table 1. General characteristics of the semen of study dogs.

Semen Characteristics [a]	Median	Interquartile Range (Range)
	General semen characteristic	
Semen volume [mL]	2.5	1.5–3.5 (0.4–5.0)
pH	6.0	6.0–6.5 (3.0–7.5)
Sperm concentration [$\times 10^6/^{mL}$]	365.7	204.3–596.6 (42.2–1649.3)
Total sperm number [$\times 10^6$]	671.2	398.9–1401.7 (113.4–3298.7)
Oval cell count [$\times 10^6/^{mL}$]	1.0	0.4–3.0 (0–17.7)
	Spermatozoa morphology	
Normal spermatozoa [%]	92.5	88.0–95.0 (58.5–99.0)
Head abnormalities [%]	1.8	1.0–4.0 (0–18.0)
Midpiece abnormalities [%]	2.0	1.5–3.5 (0–20.0)
Tail abnormalities [%]	2.8	1.0–5.0 (0–29.5)
Abnormal spermatozoa [%]	7.5	5.0–12.0 (1.0–41.5)
	Spermatozoa motility	
Total motility [%]	93.4	87.7–96.4 (34.4–99.8)
Progressive motility [%]	31.7	21.2–39.4 (0.7–55.9)
Medium-progressive motility [%]	32.1	26.5–45.9 (4.6–78.7)
Non-progressive motility [%]	23.6	17.8–29.5 (5.1–49.5)
Spherical tracks [%]	38.3	31.4–49.7 (1.2–83.7)
Rapid motility [%]	57.4	45.6–68.4 (3.7–94.2)
Medium motility [%]	24.7	17.6–31.5 (5.1–53.4)
Slow motility [%]	7.0	4.4–10.9 (0.5–38.8)

Table 1. *Cont.*

Semen Characteristics [a]	Median	Interquartile Range (Range)
Mucus penetration [%]	31.2	20.7–41.1 (2.4–64.9)
Viability [%]	91.8	87.0–95.0 (40.0–98.5)

[a] spermatozoa characteristics for 62 dogs that had spermatozoa in semen.

3.3. Bacteriological and PCR Findings

In 8/63 dogs (12.7%, CI 95%: 6.6–23.1%), the following aerobic bacteria were cultured from the semen: *Staphylococcus pseudintermedius* in 3 dogs, *Streptococcus canis* in 2 dogs, followed by *Staphylococcus vitulinus*, *E. coli*, and *Pseudomonas* sp. in 1 dog each. *Mycoplasma* spp. was detected using PCR in 38/63 dogs (60.3%; CI 95%: 48.0–71.5%). In 10/38 *Mycoplasma*-positive dogs (26.3%), the *Mycoplasma* species could not be determined using routine PCR primers. In the remaining 28 dogs, 54 *Mycoplasma* strains belonging to twelve species were identified (Table 2)—one species in 11/28 dogs (39.3%), two species in 10 dogs (35.7%), three species in 5 dogs (17.9%), and four species in 2 dogs (7.1%). Except for three dogs with *M. canis*, two dogs with *M. haemocanis*, and two dogs with M. HRC689, all other *Mycoplasma*-positive dogs had unique combinations of various *Mycoplasma* species (Table 3). The detection of *Mycoplasma* was not significantly associated with the presence of bacteria in the semen ($p = 0.461$).

Table 2. Species of *Mycoplasma* spp. detected in canine semen.

Mycoplasma Species	Number of Dogs	Prevalence (CI 95%) [%]
M. HRC689	13/38	34.2 (21.2–50.1)
M. canis	7/38	18.4 (9.2–33.4)
M. haemocanis	6/38	15.8 (7.4–30.4)
M. arginini	5/38	13.2 (5.8–27.3)
M. VJC365	4/38	10.5 (4.2–24.1)
M. molare	3/38	7.9 (2.7–20.8)
M. maculosum	3/38	7.9 (2.7–20.8)
M. feliminutum	3/38	7.9 (2.7–20.8)
M. edwardii	3/38	7.9 (2.7–20.8)
M. opalescens	3/38	7.9 (2.7–20.8)
M. cynos	3/38	5.3 (1.5–17.3)
M. bovigenitalium	3/38	5.3 (1.5–17.3)
Unidentified	10/38	26.3 (15.0–42.0)

Table 3. The combinations of *Mycoplasma* species detected in canine semen.

Mycoplasma spp.	Number of Dogs
1 *Mycoplasma* species	
M. canis	3
M. haemocanis	2
M. HRC689	2
M. arginini	1
M. edwardii	1
M. molare	1
M. VJC 358	1

Table 3. *Cont.*

Mycoplasma spp.	Number of Dogs
10 Combinations of 2 *Mycoplasma* species	
M. HRC689 and *M. canis*	1
M. HRC689 and *M. cynos*	1
M. HRC689 and *M. edwardii*	1
M. HRC689 and *M. arginini*	1
M. HRC689 and *M. feliminutum*	1
M. HRC689 and *M. bovigenitalium*	1
M. HRC689 and *M. haemocanis*	1
M. VJC358 and *M. haemocanis*	1
M. VJC358 and *M. feliminutum*	1
M. arginini and *M. molare*	1
5 Combinations of 3 *Mycoplasma* species	
M. HRC689 and *M. cynos* and *M. arginini*	1
M. HRC689 and *M. maculosum*	1
M. HRC689 and *M. canis* and *M. molare*	1
M. VJC358 and *M. feliminutum* and *M. opalescens*	1
M. haemocanis and *M. maculosum* and *M. opalescens*	1
2 Combinations of 4 *Mycoplasma* species	
M. HRC689 and *M. canis* and *M. maculosum* and *M. bovigenitalium*	1
M. canis and *M. edwardii* and *M. opalescens* and *M. haemocanis*	1

3.4. Relationship between the Presence of Mycoplasma and Semen Characteristics

There was no significant difference in demographic and hormonal characteristics between *Mycoplasma*-positive and *Mycoplasma*-negative dogs (Table S2). The presence of *Mycoplasma* in the semen did not prove to be associated with any significant changes in the semen characteristics (Table S3).

4. Discussion

To the best of our knowledge, this is the first study in which PCR testing for all known canine mycoplasmas has been performed on canine semen material. Lechner et al. tried to detect only six of them [6]. Schafer-Somi et al. evaluated nine species of canine *Mycoplasma* in semen by culturing them [29]. Tamiozzo performed gene sequencing and detected only two species of these bacteria [12]. Currently, in routine veterinary practice, the gold standard for mycoplasma diagnosis is PCR testing, so our study focused on this method. Commercially available laboratories detect only *Mycoplasma* spp. without species identification of this bacteria. This leads to a lack of available statistics on the prevalence of exact species. Moreover, the knowledge regarding which species are pathogenic and which are not makes the obtained results difficult to interpret.

Studies suggest that *Mycoplasma* spp. may be present in the reproductive tract of dogs at varying rates, with estimates ranging from 30% to 89% [15,30]. In our study, the prevalence of *Mycoplasma* spp. in canine ejaculate was 60%. Schafer-Somi et al. detected these bacteria in 55% of samples, including 35% of samples of good-quality ones [29]. The prevalence seems to be similar, but the methodology is significantly different. Interestingly, the most frequently detected species of *Mycoplasma* was in our study—*Mycoplasma* HRC689. The presence of this *Mycoplasma* species in canine semen has not been investigated so far. *M. cynos* [6] or *M. canis* [29] appear to be the most common *Mycoplasma* species in ejaculate.

In our study, these two species were detected in 5.3% and 18.4% of tested dogs, respectively. Our data did not show any significant correlation between various species of this bacteria and semen quality, while Tamiozzo suggested that *M. spumans* and *M. maculosum* negatively affected male dogs' fertility [12]. Also, in another study, *Mycoplasma* was detected in a significantly higher percentage of poor-quality ejaculate samples compared to ejaculate samples of good quality [29].

Our results show that the detection of *Mycoplasma* spp. was not significantly associated with the presence of bacteria in the semen. This phenomenon could be caused by several factors. First, too small of a sample size could have undermined the statistical power required to detect meaningful differences. Consequently, even if a genuine association had existed, it may have remained undetected. Secondly, high variability in the methodologies employed for *Mycoplasma* and bacterial detection, encompassing diverse approaches, such as culture-based methods and molecular assays like polymerase chain reaction (PCR), could introduce disparities in their diagnostic sensitivity or specificity. In our opinion, using NGS technology could shed more light on these aspects. More research in this field is needed. The last explanation of this phenomenon could be the coincidental presence of those microorganisms. The co-occurrence of *Mycoplasma* and bacterial species in semen may be incidental rather than reflective of a direct causal relationship. Shared transmission routes, such as sexual activity, or similar ecological niches within the reproductive tract, could facilitate coincidental cohabitation without necessitating an intrinsic association.

Among the aerobic bacteria isolated from the semen samples in our study, various species were identified, including *Staphylococcus* spp., *Streptococcus* spp., and *E. coli*. These results are consistent with data reported in the available literature [6]. Our study also showed that not all ejaculate contained aerobic bacteria. In only 12.7% of samples, aerobic bacteria were cultured. This result is contradictory to other studies, which suggested that canine semen is not sterile [5–7]. On the other hand, data evaluated by another author suggested that the source of bacteria could be an environment, bacteria on the urethra [31], or a lack of proper hygiene of the person who collect the samples. Regardless of the quality of semen, bacterial growth is observed in various fractions of dog semen. However, higher concentrations are typically found in the first fraction, which is primarily attributable to the presence of bacteria originating from the urethra [4]. Dogs included in our study did not show any signs of urinary tract infection. The samples were collected with clean gloves in sterile containers. The samples for bacteriology were collected according to rules that are used, e.g., during urine collection, which means that the middle stream of semen was collected for bacteriology [32]. On the one hand, there are studies that describe the presence of bacteria as physiological [33], and on the other hand, some others consider bacteriospermia as pathology [7,8]. The number of bacteria and the immune status of the organism matter. Typically, the detection of over 10,000 colony-forming units of aerobic bacteria per milliliter of semen indicates an infection of the genital tract [34]. The infection is generally correlated with presence of inflammatory cells [35]. In our study there were no significant differences in round cell concentration in semen and bacterial or mycoplasmal contamination. Similar results have been obtained in the analysis of the cytology of seminal fluids performed by Kustritz et al. [36].

The predominant components of the physiological microflora in female dogs typically comprise β-hemolytic *Streptococcus* spp., *Staphylococcus* spp., *E. coli*, *Enterococcus faecalis*, *Pasteurella multocida*, *Proteus* spp., *Bacillus* spp., *Corynebacterium* spp., *Klebsiella pneumoniae*, *Actinomyces* spp., and *Neisseria* spp. Additionally, certain authors propose the presence of *Lactobacillus* spp., *Mycoplasma* spp., and *Ureaplasma* spp. [37,38]. Our bacteriology results obtained from fertile dogs indicate that the saprophytic flora of the male reproductive tract is similar. This suggests that prophylactic antibiotic therapy after positive bacteriology results in dogs with normal parameters describing semen is not justified because similar microorganisms inhabit the body of the bitch, and there is no risk of infecting her.

The findings of this study have practical implications for veterinary practice and breeding programs. By identifying the microbial flora present in canine ejaculate and its influence on fertility parameters, this study contributes to the development of targeted screening and management protocols to improve breeding success rates and reproductive outcomes in dogs. In the current veterinary practice, the carriers of *Mycoplasma* spp. are mainly treated using doxycycline [39]. Our findings show that not every carrier of this bacteria should be treated. This result appears to be extremely important, as it will help reduce the use of antibiotic therapy in veterinary medicine. The overuse and misuse of antibiotics contribute to the development of antibiotic-resistant bacteria [40]. When antibiotics are used too frequently or inappropriately, bacteria can evolve and become resistant, making infections more difficult to treat [40]. In addition, antibiotics not only target harmful bacteria but can also affect the beneficial bacteria in the body, disrupting the natural balance of the microbiome. This disruption can lead to various health issues, including digestive problems and increased susceptibility to infections [41]. Due to this fact, it is highly recommended to evaluate semen quality, including the presence of inflammatory cells after obtaining bacteriology or PCR results from ejaculate. The decision on treatment should be made after careful consideration of all the factors.

Our study also has some limitations. While this study provides valuable insights into the prevalence and impact of aerobic bacteria and mycoplasmas in Polish male dogs, the findings may not be directly generalizable to other canine populations in different geographic regions. Factors such as breed diversity, environmental conditions, and management practices could influence the microbial composition of semen. This study focuses specifically on male dogs from Poland, which may limit the applicability of the findings to dogs from other countries or regions with different environmental conditions and management practices. Including participants from multiple geographic locations could enhance the external validity of this study. This study provides a cross-sectional snapshot of semen quality and microbial presence in male dogs at a specific point in time. Longitudinal data tracking changes in semen quality and microbial composition over time could provide deeper insights into the dynamic nature of these factors. The final limitation of our study is that the diagnostic methods we used are qualitative, not quantitative. It is possible that the quantity of bacteria has a greater impact on semen quality than the species themselves. More research in this area is needed. Our methodology, which contains PCR reactions and bacteriology culturing, is not cutting-edge technology, but it is available to both scientists and veterinarians. In the future, we want to expand our research to include the use of technology next-generation sequencing (NGS) [42]. Another limitation correlated with methodology is using basic diagnostic tools like CASA-system and microscopy evaluation of the morphology and viability of sperms. Our methodology, which includes diagnostic tools such as the CASA system, microscopic evaluation of sperm morphology and viability, PCR reactions, and bacteriological culture, are widely available to both researchers and clinical veterinarians. The use of these testing methods provides reproductive veterinarians, including practitioners and clinicians, with accessible tools for diagnosing infertility in their routine practice. Therefore, we also chose to use a simple eosin–nigrosine test as a surrogate for assessing functional membrane integrity. This test distinguishes between damaged and intact cell membranes, with damaged membranes staining pink while intact membranes remain unstained. While more sophisticated techniques, such as the hypoosmotic edema test (HOS), are typically available primarily in research settings, we anticipate incorporating them into our future research, similarly to next-generation sequencing (NGS) technology. Using these basic but effective tests, we aim to improve the diagnostic process for reproductive veterinarians and facilitate the timely and accurate identification of semen quality problems in dogs.

This study encountered challenges in identifying specific *Mycoplasma* species in some cases, with 10 out of 38 *Mycoplasma*-positive dogs having unidentified species. This limitation could affect the accuracy of the associations between *Mycoplasma* species and semen quality parameters. The presence of a positive result for *Mycoplasma* spp. and the absence of a positive result in PCR reactions for known species may suggest that these dogs were carriers of another species. The ideal solution to this situation would be to sequence genes from samples obtained from these dogs. This will be the direction of our further research.

The significance of this study lies in its contribution to understanding the factors affecting reproductive health in dogs. This research addresses a critical gap in current knowledge by investigating the prevalence and impact of aerobic bacteria and mycoplasmas on semen quality, which is a crucial aspect of canine fertility and breeding success. By identifying and characterizing the microbial flora present in canine ejaculate, this study sheds light on potential sources of contamination and infection that may compromise semen quality. Understanding the microbial composition of semen and its influence on fertility parameters is essential for developing effective strategies to optimize reproductive outcomes in dogs. Furthermore, the study findings may have practical implications for veterinary practice and breeding programs. By elucidating the role of aerobic bacteria and mycoplasmas in semen quality, veterinarians and breeders can implement targeted screening and management protocols to minimize the risk of reproductive tract infections and improve breeding success rates.

5. Conclusions

In conclusion, in canine ejaculate, *Mycoplasma* spp. is common in dogs that have not been used for reproduction. The semen quality parameters are not related to the general presence of *Mycoplasma* spp. The most common species is *Mycoplasma* HRC689. There are dogs in whose semen neither aerobic bacteria nor mycoplasmas are present, which indicates that in some cases, the semen could be sterile. It is likely, however, that there are yet undescribed species of canine mycoplasmas that cannot be detected using conventional diagnostic tools. Therefore, further investigations employing advanced techniques, such as NGS, are imperative to unveil these elusive pathogens.

Supplementary Materials: The following supporting information can be downloaded at: https://www.mdpi.com/article/10.3390/ani14091267/s1, Table S1. Breeds of dogs of the study population; Table S2. Demographic and hormonal characteristics of dogs from Mycoplasma-positive and Mycoplasma-negative group; Table S3. Influence of the presence of *Mycoplasma* in the semen on semen characteristics.

Author Contributions: Conceptualization, K.D. and P.J.; methodology, P.K., M.M. and K.D.; formal analysis, K.D.; investigation, K.D.; writing—original draft preparation, K.D.; writing—review and editing, P.J., M.C. and P.K.; statistical analysis M.C. All authors have read and agreed to the published version of the manuscript.

Funding: This research received no external funding.

Institutional Review Board Statement: All medical procedures were performed as a part of routine veterinary examination, on the owners' request and thus, according to the European directive EU/2010/63 and Polish legal regulations the approval of Ethical Committee for the described procedures was not required as they could be qualified as non-experimental clinical veterinary practices, excluded from the directive (Act of 15 January 2015 on the protection of animals used for scientific or educational purposes).

Informed Consent Statement: Not applicable.

Data Availability Statement: No new data were created or analyzed in this study. Data sharing is not applicable to this article.

Conflicts of Interest: The authors have read the journal's guideline and have the following competing interests: the co-authors (Konieczny P., Majka M.) are employees of Vet Cell Tech Sp. z o.o. company, which partially financed the project and provided the PCR analysis. The other authors have no competing interests.

References

1. Freshman, J.L. Semen Collection and Evaluation. *Clin. Tech. Small Anim. Pract.* **2002**, *17*, 104–107. [CrossRef] [PubMed]
2. Kolster, K.A. Evaluation of Canine Sperm and Management of Semen Disorders. *Vet. Clin. N. Am. Small Anim. Pract.* **2018**, *48*, 533–545. [CrossRef] [PubMed]
3. England, G.C.W.; Burgess, C.M.; Freeman, S.L. Perturbed Sperm–Epithelial Interaction in Bitches with Mating-Induced Endometritis. *Vet. J.* **2012**, *194*, 314–318. [CrossRef] [PubMed]
4. Luño, V.; González, N.; Martínez, F.; Revert, A.; Morrell, J.M.; Gil, L. Colloid Centrifugation Reduces Bacterial Load in Chilled Dog Semen. *Anim. Reprod. Sci.* **2020**, *219*, 106539. [CrossRef] [PubMed]
5. Kustritz, M.V.R. Collection of Tissue and Culture Samples from the Canine Reproductive Tract. *Theriogenology* **2006**, *66*, 567–574. [CrossRef] [PubMed]
6. Lechner, D.; Aurich, J.; Spergser, J.; Aurich, C. Double Semen Collection at a 1-h Interval in Dogs Decreases the Bacterial Contamination of Canine Ejaculates. *Theriogenology* **2023**, *208*, 126–131. [CrossRef] [PubMed]
7. Agudelo-Yepes, S.; Puerta-Suárez, J.; Carrillo-Gonzalez, D.F.; Cardona Maya, W.D. Bacteriospermia Assessment and Its Relationship with Conventional Seminal Parameters in Stud Dogs Ejaculates (Canis Familiaris). *J. Hell. Vet. Med. Soc.* **2023**, *73*, 4785–4794. [CrossRef]
8. Oguejiofor, C. Sperm Defects and Infertility Caused by Bacterial Infection of the Reproductive Tract in an Adult Male Dog: A Case Report. *Asian Pac. J. Reprod.* **2018**, *7*, 236. [CrossRef]
9. Luciani, M.; Krasteva, I.; Di Febo, T.; Perletta, F.; D'Onofrio, F.; De Massis, F.; D'Alterio, N.; Sacchini, F.; Tittarelli, M. Proteomics and Bioinformatics Investigations to Improve Serological Diagnosis of Canine Brucellosis. *Proteom. Clin. Appl.* **2023**, *17*, 2200116. [CrossRef]
10. Razin1, S.; Tully, J.G. Cholesterol Requirement of Mycoplasmas. *J. Bacreriology* **1970**, *102*, 306–310. [CrossRef]
11. Schulz, M.; Sánchez, R.; Soto, L.; Risopatrón, J.; Villegas, J. Effect of *Escherichia coli* and Its Soluble Factors on Mitochondrial Membrane Potential, Phosphatidylserine Translocation, Viability, and Motility of Human Spermatozoa. *Fertil. Steril.* **2010**, *94*, 619–623. [CrossRef] [PubMed]
12. Tamiozzo, P.J. Mycoplasma Maculosum and Mycoplasma Spumans Associated with Fertility Disorders in Dogs from a Bernese Mountain Dog Kennel. *Rev. Argent. Microbiol.* **2022**, *54*, 39–42. [CrossRef] [PubMed]
13. Meyers-Wallen, V.N. Unusual and Abnormal Canine Estrous Cycles. *Theriogenology* **2007**, *68*, 1205–1210. [CrossRef] [PubMed]
14. Laber, G.; Holzmann, A. Experimentally Induced Mycoplasmal Infection in the Genital Tract of the Male Dog. *Theriogenology* **1977**, *7*, 177–188. [CrossRef]
15. Johnston, S.D.; Root Kustritz, M.V.; Olson, P.N.S. *Canine and Feline Theriogenology*; Evlsevier: Amsterdam, The Netherlands, 2001; ISBN 0-7216-5607-2.
16. Fowlkes, D.M.; Dooher, G.B.; O'leary, W.M. Evidence by Scanning Electron Microscopy for an Association between Spermatozoa and T-Mycoplasmas in Men of Infertile Marriage Sc. *Fertil Steril* **1975**, *26*, 1203–1211. [CrossRef] [PubMed]
17. Domrazek, K.; Jurka, P. Prevalence of *Chlamydophila* Spp. and Canid Herpesvirus-1 in Polish Dogs. *Vet. World* **2024**, *17*, 226–232. [CrossRef] [PubMed]
18. Martins, M.I.M.; Souza, F.F.d.; Oba, E.; Lopes, M.D. The Effect of Season on Serum Testosterone Concentrations in Dogs. *Theriogenology* **2006**, *66*, 1603–1605. [CrossRef]
19. Frank, L.A.; Mullins, R.; Rohrbach, B.W. Variability of Estradiol Concentration in Normal Dogs. *Vet. Dermatol.* **2010**, *21*, 490–493. [CrossRef] [PubMed]
20. Shiel, R.E.; Sist, M.; Nachreiner, R.E.; Ehrlich, C.P.; Mooney, C.T. Assessment of Criteria Used by Veterinary Practitioners to Diagnose Hypothyroidism in Sighthounds and Investigation of Serum Thyroid Hormone Concentrations in Healthy Salukis. *J. Am. Vet. Med. Assoc.* **2010**, *236*, 302–308. [CrossRef]
21. Domrazek, K.; Kaszak, I.; Kanafa, S.; Sacharczuk, M.; Jurka, P. The Influence of *Mycoplasma* Species on Human and Canine Semen Quality: A Review. *Asian J. Androl.* **2023**, *25*, 29. [CrossRef]
22. Chalker, V.J. Canine Mycoplasmas. *Res. Vet. Sci.* **2005**, *79*, 1–8. [CrossRef]
23. Chalker, V.J.; Brownlie, J. Taxonomy of the Canine Mollicutes by 16S RRNA Gene and 16S/23S RRNA Intergenic Spacer Region Sequence Comparison. *Int. J. Syst. Evol. Microbiol.* **2004**, *54*, 537–542. [CrossRef] [PubMed]
24. Root Kustritz, M.V. The Value of Canine Semen Evaluation for Practitioners. *Theriogenology* **2007**, *68*, 329–337. [CrossRef]
25. Partyka, A.; Nizanski, W.; Ochota, M. Methods of Assessment of Cryopreserved Semen. In *Current Frontiers in Cryobiology*; InTech: London, UK, 2012.
26. Björndahl, L.; Söderlund, I.; Kvist, U. Evaluation of the One-step Eosin-nigrosin Staining Technique for Human Sperm Vitality Assessment. *Human Reprod.* **2003**, *18*, 813–816. [CrossRef] [PubMed]
27. Takagi, H. Interactive Evolutionary Computation: Fusion of the Capabilities of EC Optimization and Human Evaluation. *Proc. IEEE* **2001**, *89*, 1275–1296. [CrossRef]
28. Altman, D.; Machin, D.; Bryant, T.; Gardner, M. *Statistics with Confidence: Confidence Intervals and Statistical Guidelines*, 2nd ed.; BMJ Books: Bristol, UK, 2000.
29. Schäfer-Somi, S.; Spergser, J.; Aurich, C. Bacteria and Mycoplasms in Canine Ejaculates a Retrospective Survey. *Wien. Tierarztl. Monatsschr.* **2009**, *96*, 240–245.

30. Maksimović, Z.; Maksimović, A.; Halilbašić, A.; Rifatbegović, M. Genital Mycoplasmas of Healthy Bitches. *J. Vet. Diagn. Investig.* **2018**, *30*, 651. [CrossRef] [PubMed]
31. Goericke-Pesch, S.; Weiss, R.; Wehrend, A. Bacteriological Findings in Different Fractions of Canine Ejaculates Showing Normospermia, Teratozoospermia or Azoospermia. *Aust. Vet. J.* **2011**, *89*, 318–322. [CrossRef] [PubMed]
32. Manoni, F.; Gessoni, G.; Alessio, M.G.; Caleffi, A.; Saccani, G.; Silvestri, M.G.; Poz, D.; Ercolin, M.; Tinello, A.; Valverde, S.; et al. Mid-Stream vs. First-Voided Urine Collection by Using Automated Analyzers for Particle Examination in Healthy Subjects: An Italian Multicenter Study. *Clin. Chem. Lab. Med.* **2011**, *50*, 679–684. [CrossRef]
33. Banchi, P.; Bertolotti, L.; Spanoghe, L.; Ali Hassan, H.; Lannoo, J.; Domain, G.; Henzel, K.S.; Gaillard, V.; Rota, A.; Van Soom, A. Characterization of the Semen Microbiota of Healthy Stud Dogs Using 16S RNA Sequencing. *Theriogenology* **2024**, *216*, 1–7. [CrossRef]
34. Johnston, S.D. Performing a Complete Canine Semen Evaluation in a Small Animal Hospital. *Vet. Clin. N. Am. Small Anim. Pract.* **1991**, *21*, 545–551. [CrossRef] [PubMed]
35. Chen, L.; Deng, H.; Cui, H.; Fang, J.; Zuo, Z.; Deng, J.; Li, Y.; Wang, X.; Zhao, L. Inflammatory Responses and Inflammation-Associated Diseases in Organs. *Oncotarget* **2018**, *9*, 7204–7218. [CrossRef] [PubMed]
36. Root Kustritz, M.V.; Johnston, S.D.; Olson, P.N.; Lindeman, C.J. Relationship between Inflammatory Cytology of Canine Seminal Fluid and Significant Aerobic Bacterial, Anaerobic Bacterial or Mycoplasma Cultures of Canine Seminal Fluid: 95 Cases (1987–2000). *Theriogenology* **2005**, *64*, 1333–1339. [CrossRef]
37. Jagódka, D.; Kaczorek-Łukowska, E.; Graczyk, R.; Socha, P. Vaginal Aerobic Bacteria of Healthy Bitches and Those with Fertility Problems. *Pol. J. Vet. Sci.* **2023**, *26*, 733–739. [CrossRef] [PubMed]
38. Zduńczyk, S.; Jurczak, A. Incidence of *Mycoplasma canis* in the Vagina in Three Groups of Bitches. *Bull. Vet. Inst. Pulawy* **2008**, *52*, 533–535.
39. Pitorri, F.; Dell'Orco, M.; Carmichael, N.; Barker, E.N.; Hollywood, M.; Tasker, S. Use of Real-time Quantitative PCR to Document Successful Treatment of *Mycoplasma haemocanis* Infection with Doxycycline in a Dog. *Vet. Clin. Pathol.* **2012**, *41*, 493–496. [CrossRef] [PubMed]
40. Bytesnikova, Z.; Richtera, L.; Smerkova, K.; Adam, V. Graphene Oxide as a Tool for Antibiotic-Resistant Gene Removal: A Review. *Environ. Sci. Pollut. Res.* **2019**, *26*, 20148–20163. [CrossRef] [PubMed]
41. Patangia, D.V.; Anthony Ryan, C.; Dempsey, E.; Paul Ross, R.; Stanton, C. Impact of Antibiotics on the Human Microbiome and Consequences for Host Health. *Microbiologyopen* **2022**, *11*, e1260. [CrossRef]
42. Melgarejo, T.; Oakley, B.B.; Krumbeck, J.A.; Tang, S.; Krantz, A.; Linde, A. Assessment of Bacterial and Fungal Populations in Urine from Clinically Healthy Dogs Using Next-Generation Sequencing. *J. Vet. Intern. Med.* **2021**, *35*, 1416–1426. [CrossRef]

Disclaimer/Publisher's Note: The statements, opinions and data contained in all publications are solely those of the individual author(s) and contributor(s) and not of MDPI and/or the editor(s). MDPI and/or the editor(s) disclaim responsibility for any injury to people or property resulting from any ideas, methods, instructions or products referred to in the content.

animals

MDPI

Article

Effect of Different Dilution Methods and Ratios of Ram Semen on Sperm Parameters after Cryopreservation

Liuming Zhang, Xuyang Wang, Caiyu Jiang, Tariq Sohail, Yuxuan Sun, Xiaomei Sun, Jian Wang and Yongjun Li *

Key Laboratory for Animal Genetics & Molecular Breeding of Jiangsu Province, College of Animal Science and Technology, Yangzhou University, Yangzhou 225009, China; 18352767281@163.com (L.Z.); 18705271593@163.com (X.W.); m19352674787@163.com (C.J.); drtariqsohail34@yahoo.com (T.S.); 19802622001@163.com (Y.S.); xiaomeisun_yz@163.com (X.S.); jianwang1223@163.com (J.W.)

* Correspondence: yzliyongjun@163.com

Simple Summary: The appropriate dilution rate is crucial for the spermatozoa's survival. Both excessively large and excessively small dilution rates will diminish the effectiveness of semen preservation. However, the dilution method and ratio for preserving *Hu* ram semen through cryopreservation are currently unclear. Therefore, the study aimed to analyze the effects of various dilution methods and ratios on the spermatozoa motility parameters and functional integrity of *Hu* ram semen after cryopreservation. The results showed that employing the correct dilution method and ratio (two-step dilution 1:3, 1:2) could improve the preservation effect of semen.

Abstract: The dilution method and ratio were tested to assess their effects on the *Hu* ram semen after cryopreservation. Experiment I aimed to explore the effect of various dilution ratios (1:1, 1:2, 1:3, 1:4) of diluent I (Tris-based and egg yolk) under the condition of 1:1 dilution of diluent II (diluent I and glycerol) on the *Hu* ram semen preserved in liquid nitrogen regarding spermatozoa motility and kinetic parameters. Experiment II aimed to investigate the effect of various dilution ratios (1:1, 1:2, 1:3, 1:4) of diluent I under the condition of 1:2 dilution of diluent II to the *Hu* ram semen for cryopreservation on spermatozoa motility and kinetic parameters. The purpose of experiment III is to assess the effect of various dilution methods and ratios on the cryopreservation of *Hu* ram semen by detecting spermatozoa motility, kinetic parameters, plasma membrane integrity, acrosome integrity and reactive oxygen species (ROS) level. Experiment III includes four groups: one-step dilution method and two-step dilution method. The two-step dilution method includes two groups: 1:2, 1:1 and 1:3, 1:2, and the one-step dilution method includes two groups: 1:5 and 1:11. The results indicated that the post-thawed spermatozoa total motility (TM), progressive motility (PM) and average motion degree (MAD) were highest in the 1:2 group and significantly higher ($p < 0.05$) than those in the 1:1 and 1:4 groups under the condition of 1:1 dilution of diluent II. The post-thawed spermatozoa TM and PM of the 1:3 group were significantly higher ($p < 0.05$) than those of the other groups under the condition of 1:2 dilution of diluent II. The post-thawed spermatozoa TM, PM, plasma membrane integrity and acrosome integrity of the two-step group (1:3, 1:2) were the highest and significantly higher ($p < 0.05$) than those in the other groups. Additionally, the post-thawed spermatozoa ROS level of the two-step group (1:3, 1:2) was significantly lower ($p < 0.05$) than that in the one-step groups (1:5 and 1:11). Therefore, a two-step dilution (1:3, 1:2) was found to be the most suitable method and ratio for diluting the *Hu* ram semen after cryopreservation.

Keywords: dilution rates; *Hu* ram; cryopreservation; spermatozoa motility; spermatozoa functional integrity; spermatozoa ROS

Citation: Zhang, L.; Wang, X.; Jiang, C.; Sohail, T.; Sun, Y.; Sun, X.; Wang, J.; Li, Y. Effect of Different Dilution Methods and Ratios of Ram Semen on Sperm Parameters after Cryopreservation. *Animals* **2024**, *14*, 907. https://doi.org/10.3390/ani14060907

Academic Editor: Anna Wysokińska

Received: 24 January 2024
Revised: 29 February 2024
Accepted: 2 March 2024
Published: 15 March 2024

Copyright: © 2024 by the authors. Licensee MDPI, Basel, Switzerland. This article is an open access article distributed under the terms and conditions of the Creative Commons Attribution (CC BY) license (https://creativecommons.org/licenses/by/4.0/).

1. Introduction

Artificial insemination (AI) refers to the method of collecting semen artificially and transferring the processed semen to the reproductive organs of female animals in estrus [1].

AI is an advanced method of mating. Compared to natural mating, AI has made significant advancements and is an important technical method for the advancement of modern animal husbandry [2]. AI technology offers several advantages, including improving the utilization rate of superior male animals, increasing the conception rate of female animals, preventing the spread of certain reproductive tract diseases, addressing challenges related to crossbreeding and mating, overcoming regional limitations in female breeding and facilitating embryo transfer, synchronous estrus and other reproductive techniques [3–5]. In addition to the influence of female body condition, estrus detection and insemination methods, the efficiency of AI is also closely related to semen quality. Several factors affect semen quality, including semen preservation methods, dilution components, antibiotics, semen collection frequency and dilution times [6,7].

The concentration of spermatozoa is crucial for the survival time of spermatozoa. Appropriate dilution rates can increase the volume of semen, enhance the utilization rate of spermatozoa, and improve the effectiveness of multiple AI procedures [8,9]. During the preservation of semen, the metabolism of spermatozoa continuously consumes nutrients and produces metabolic wastes and oxidative factors, which can affect the effectiveness of semen preservation. Therefore, when the spermatozoa concentration is too high, the spermatozoa density exceeds the capacity of the diluent to provide sufficient nutrients for sperm movement, and the buffer and antioxidant substances in the diluent are unable to maintain pH stability and redox balance [10]. When the concentration of spermatozoa is too low, the level of beneficial substances in seminal plasma decreases. This reduction diminishes the protective ability of spermatozoa, resulting in a dilution effect and a shortened effective preservation time of semen [11]. Excessive dilution rate can lead to a decrease in spermatozoa density and the number of effective spermatozoa while increasing the insemination volume, and due to the unique structure of the sheep's reproductive tract, the increase in semen volume can exacerbate the reflux of spermatozoa from the cervix, ultimately reducing the pregnancy rate [12,13]. Therefore, it is crucial to find the appropriate dilution method and ratio for semen preservation and AI.

The *Hu* sheep is a unique breed of sheep in China, known for its early sexual maturity, rapid growth and development, high fecundity, suitability for indoor rearing, strong adaptability to various environments, and excellent meat production performance [14]. China has a large population and limited land, leading to a growing demand for mutton [15]. As a result, indoor sheep farming has become the predominant method of breeding, and the *Hu* sheep breed aligns with this current development trend. Currently, the breeding scale of *Hu* sheep in China is expanding, but there is still a shortage, especially in meeting the market demand for meat products and lambskin processing. In order to promote the sustained and rapid development of the *Hu* sheep breeding industry, it is essential to utilize AI and other related reproduction technologies.

In the study of frozen semen in sheep, the freezing methods such as dilution method and ratios are different. Chen [16] adopted a one-step dilution method of 7.5 times, and the post-thawed Mongolian sheep's spermatozoa progressive motility (PM) reached 39%. He [17] used a two-step dilution method of 5 times, and the post-thawed Dorset rams' spermatozoa total motility (TM) reached 31%. Fang [18] used a one-step dilution method of 2.5 times, and the post-thawed small-tailed Han sheep's spermatozoa TM reached 13%. However, the dilution method and ratios of *Hu* ram semen in cryopreservation are not clear. Therefore, the purpose of this study is to provide a reference for improving the preservation technology of *Hu* ram semen in production practice by examining spermatozoa TM, PM, kinematic parameters, membrane integrity, acrosome integrity and reactive oxygen species (ROS) level.

2. Materials and Methods

2.1. Experimental Design

Experiment 1: The study investigated the impact of different dilution ratios of diluent I (1:1, 1:2, 1:3, 1:4) under the condition of 1:1 dilution of diluent II on the motility and kinetic parameters of *Hu* ram semen preserved in liquid nitrogen (−196 °C). TM, PM, straight-line velocity (VSL, μm/s), curvilinear velocity (VCL, μm/s), average path velocity (VAP, μm/s), amplitude of lateral head displacement (ALH, μm) and average motion degree (MAD, °/s) of the four groups were evaluated.

Experiment 2: The study aimed to investigate the impact of different dilution ratios of diluent I (1:1, 1:2, 1:3, 1:4) under the condition of 1:2 dilution of diluent II on spermatozoa motility and kinetic parameters after cryopreservation. TM, PM, VSL, VCL, VAP, ALH and MAD of the four groups were evaluated.

Experiment 3 aimed to investigate the one-step and two-step dilution methods. The two-step dilution method included two groups: 1:2, 1:1 and 1:3, 1:2, and the one-step dilution method included two groups: 1:5 and 1:11. The purpose of this experiment was to assess the effect of various dilution methods and ratios on the cryopreservation of *Hu* ram semen by measuring spermatozoa motility (TM and PM), kinetic parameters (VSL, VCL, VAP, ALH and MAD), plasma membrane integrity, acrosome integrity and ROS level.

2.2. Semen Collection and Diluent Preparation

The procedures for animals during the experiment complied with the regulations of the Animal Ethics Committee of Yangzhou University (Approval ID: 202206132). The study utilized five mature and healthy *Hu* rams, which were housed at the sheep facility within Yangzhou University (Wenhuilu campus, Yangzhou, China, 119.429731°, 32.395371°). The rams aged 2~3 years were fed 0.7 kg concentrate/day and straw–hay mixtures during the study period following fertility examination. Furthermore, water and mineral blocks were provided freely to the rams. *Hu* sheep were known to be in estrus all year round. From April to June 2023, ram semen was collected three times a week (a total number of 75 ejaculates) using an artificial vagina. After semen collection, all samples were maintained at 37 °C and transported to the laboratory within 30 min for semen sample quality evaluation, including assessment of volume (0.5 to 1.5 mL), concentration ($\geq 2.5 \times 10^9$ spermatozoa/mL) and total motility ($\geq$80%). The qualified semen samples were pooled together for subsequent experiments.

The basic diluent was Tris-based, composed of 3.64 g Tris, 1.82 g citric acid, 0.5 g glucose, 20,000 IU sodium penicillin and streptomycin sulfate for 100 mL of distilled water. The freezing diluent I contained 80% (v/v) basic diluent and 20% (v/v) egg yolk. The freezing diluent II contained 94% (v/v) freezing diluent I and 6% (v/v) glycerol.

2.3. Semen Cryopreservation and Post Thaw

The semen freezing protocol of *Hu* rams was based on the research results of Wu [14] and improved upon this foundation. In experiment 1, semen samples were diluted at the ratios of 1:1, 1:2, 1:3 and 1:4, respectively, using freezing diluent I. Afterward, the samples were wrapped in a towel and stored at 4 °C for 2.5 h. Subsequently, isothermal freezing diluent II was added in a 1:1 ratio, and the samples were kept at 4 °C for 2.5 h. In experiment 2, semen samples were diluted at ratios of 1:1, 1:2, 1:3 and 1:4, respectively, using freezing diluent I. The samples were then wrapped in a towel and kept at 4 °C for 2.5 h. Subsequently, isothermal freezing diluent II was added at a ratio of 1:2, and the samples were kept at 4 °C for 2.5 h. In experiment 3, for the two-step dilution method, the semen samples were initially diluted at ratios of 1:2 and 1:3, respectively, using freezing diluent I. Afterward, the samples were wrapped in a towel and kept at 4 °C for 2.5 h. Subsequently, isothermal freezing diluent II was added in ratios of 1:1 and 1:2, respectively, and the samples were kept at 4 °C for 2.5 h. For the one-step dilution method, semen samples were diluted at ratios of 1:5 and 1:11, respectively, using freezing diluent II. Afterward, the samples were wrapped in a towel and kept at 4 °C for 5 h. Subsequently, all semen samples were promptly transferred into 0.25 mL straws at 4 °C. All the straws were held 2 cm above the liquid nitrogen for 20 min and then immediately plunged into the liquid nitrogen.

After being stored in liquid nitrogen for at least 1 week, four straws from each group were randomly selected, thawed at 70 °C for 5 s, and then utilized for further investigation and analysis.

2.4. Spermatozoa Motility and Kinetic Parameters Assessment

The spermatozoa motility (TM and PM) and kinetic parameters (VSL, VCL, VAP, ALH and MAD) were analyzed using a computer-assisted spermatozoa analyzer (CASA, ML-608JZ II Mailang, Nanning, China) equipped with a warm stage. Samples were diluted to 3.85×10^7 spermatozoa/mL using a Tris-based diluent and then incubated in a water bath at 37 °C for 3 min. A total of 1.4 µL of semen was placed on a MACRO spermatozoa counting chamber (YA-1, Yucheng, Nanjing, China), and a phase-contrast microscope (ML-800, Mailang, Nanning, China) equipped with a CCD-camera (MD06200C, Mailang, Nanning, China) was used for the assessment. And the CASA software (ML-608JZ II) recorded data at 30 frames per second.

2.5. Spermatozoa Membrane Integrity Assessment

The spermatozoa membrane integrity was analyzed using the hypo-osmotic swelling test (HOST). Briefly, the samples were diluted to a concentration of 2×10^8 spermatozoa/mL using a Tris-based diluent and 20 µL of diluted semen sample was added to 200 µL of hypotonic solution (0.245 g of sodium citrate and 0.45 g of fructose dissolved in 50 mL of water). The samples were incubated in a 37 °C water bath for 30 min. After incubation, 1.8 µL of the mixture was spread onto a ruby spermatozoa counting plate and examined under a microscope (CX31, Olympus, Tokyo, Japan) at a magnification of 400×. A total of 200 spermatozoa per group were assessed. Spermatozoa with a curled tail were considered to have intact plasma membrane.

2.6. Spermatozoa Acrosome Integrity Assessment

The spermatozoa acrosome integrity was assessed using FITC-PNA combined with PI fluorescence staining. Briefly, the samples were diluted to 3.21×10^8 spermatozoa/mL using a Tris-based diluent, and 100 µL diluted samples were mixed with 2 µL of FITC-PNA (200 µg/mL) and 2 µL of PI (0.5 mg/mL). The samples were incubated at 37 °C for 10 min in the dark and then mixed by adding 700 µL of PBS. They were immediately assayed for spermatozoa acrosome integrity using flow cytometry (Beckman Coulter, Shanghai, China). Flow cytometry was configured to detect 10,000 spermatozoa, and those exhibiting $FITC^-/PI^+$ and $FITC^-/PI^-$ were counted as spermatozoa with intact acrosomes.

2.7. Spermatozoa ROS Level Assessment

The spermatozoa ROS level was detected using the DCFH-DC probe. Briefly, the samples were diluted to 3.21×10^8 spermatozoa/mL using a Tris-based diluent, and 50 µL of diluted semen was mixed with 2 µL of DCFH-DC (10 mM). We incubated the sample at 37 °C in the dark for 30 min, then added PBS to wash the sample. The sample was then resuspended in 400 µL of PBS, and the fluorescence intensity was measured using a multifunctional microplate reader with excitation at 488 nm and emission at 525 nm. The level of ROS was indicated by fluorescence intensity.

2.8. Statistical Analysis

All data were analyzed using Statistical Product and Service Solutions (SPSS 25.0 for windows; SPSS Inc., Chicago, IL, USA). The Shapiro–Wilk normality analysis was performed to detect whether the data conform to the normal distribution. The data showed normal distribution, and the Duncan test by one-way ANOVA tests was performed to assess the difference in these parameters. All results were expressed as the Mean ± SEM, and a p value of <0.05 ($p < 0.05$) was considered significant. All experiments were performed with five replicates.

3. Results

3.1. Effect of Different Dilution Ratios of Diluent I under the Condition of 1:1 Dilution of Diluent II on Spermatozoa Motility and Kinetic Parameters after Cryopreservation

As shown in Table 1, the post-thawed spermatozoa TM, PM and MAD of the 1:2 group were the highest and significantly greater ($p < 0.05$) than those of the 1:1 and 1:4 groups. However, they were not significantly higher ($p > 0.05$) than those of the 1:3 group. The post-thawed spermatozoa VSL of the 1:2 group was the highest and significantly higher ($p < 0.05$) than that of the 1:1 group. However, it was not significantly higher ($p > 0.05$) than the 1:3 and 1:4 groups. Additionally, these groups did not show significant ($p > 0.05$) differences from each other in terms of spermatozoa VCL, VAP and ALH.

Table 1. Effect of different dilution ratios of diluent I under the condition of 1:1 dilution of diluent II on spermatozoa motility and kinetic parameters after cryopreservation.

Different Dilution Ratios of Diluent I	TM (%)	PM (%)	VSL (μm/s)	VCL (μm/s)	VAP (μm/s)	ALH (μm)	MAD (°/s)
1:1	50.33 ± 1.44 ^{c}	31.11 ± 0.99 ^{c}	29.28 ± 1.96 ^{b}	48.07 ± 2.78	33.99 ± 1.97	14.08 ± 0.82	27.96 ± 0.80 ^{c}
1:2	76.86 ± 1.29 ^{a}	55.16 ± 0.87 ^{a}	34.11 ± 0.59 ^{a}	53.03 ± 0.78	37.50 ± 0.55	15.53 ± 0.23	55.71 ± 0.12 ^{a}
1:3	71.44 ± 1.57 ^{ab}	52.67 ± 0.43 ^{ab}	32.68 ± 1.62 ^{ab}	50.46 ± 1.50	35.68 ± 1.06	14.78 ± 0.44	51.67 ± 3.36 ^{ab}
1:4	70.13 ± 2.25 ^{b}	49.65 ± 1.89 ^{b}	31.97 ± 0.53 ^{ab}	47.88 ± 0.88	33.86 ± 0.62	14.02 ± 0.26	47.76 ± 1.02 ^{b}

TM, total motility; PM, progressive motility; VSL, straight-line velocity; VCL, curvilinear velocity; VAP, average path velocity; ALH, amplitude of lateral head displacement; MAD, average motion degree. Different lowercase letter superscripts in the same column show significant differences ($p < 0.05$).

3.2. Effect of Different Dilution Ratios of Diluent I under the Condition of 1:2 Dilution of Diluent II on Spermatozoa Motility and Kinetic Parameters after Cryopreservation

The post-thawed spermatozoa TM and PM of the 1:3 group were significantly higher ($p < 0.05$) than those of the other groups, as shown in Table 2. The post-thawed spermatozoa VCL, VAP and ALH of the 1:3 group were the highest and significantly higher ($p < 0.05$) than those of the 1:2 group, but they were not significantly higher ($p > 0.05$) than the 1:1 and 1:4 groups. The post-thawed spermatozoa MAD of the 1:3 group was significantly higher ($p < 0.05$) than that of the 1:1 group, but it was not significantly higher ($p > 0.05$) than that of the 1:2 and 1:4 groups. There was no significant difference ($p > 0.05$) in spermatozoa VSL between the groups.

Table 2. Effect of different dilution ratios of diluent I under the condition of 1:2 dilution of diluent II on spermatozoa motility and kinetic parameters after cryopreservation.

Different Dilution Ratios of Diluent I	TM (%)	PM (%)	VSL (μm/s)	VCL (μm/s)	VAP (μm/s)	ALH (μm)	MAD (°/s)
1:1	69.51 ± 0.74 ^{c}	49.10 ± 0.52 ^{c}	33.43 ± 2.14	55.04 ± 1.59 ^{ab}	38.92 ± 1.12 ^{ab}	16.12 ± 0.46 ^{ab}	40.91 ± 3.21 ^{b}
1:2	73.45 ± 0.39 ^{b}	56.08 ± 0.53 ^{b}	32.89 ± 0.53	52.50 ± 1.40 ^{b}	37.12 ± 0.99 ^{b}	15.38 ± 0.41 ^{b}	51.03 ± 2.48 ^{ab}
1:3	79.86 ± 0.92 ^{a}	63.44 ± 1.55 ^{a}	36.54 ± 1.38	58.06 ± 1.40 ^{a}	41.05 ± 0.99 ^{a}	17.00 ± 0.41 ^{a}	58.89 ± 5.59 ^{a}
1:4	71.74 ± 1.49 ^{bc}	56.50 ± 1.16 ^{b}	35.57 ± 0.44	55.77 ± 0.88 ^{ab}	39.44 ± 0.62 ^{ab}	16.34 ± 0.26 ^{ab}	52.97 ± 3.02 ^{ab}

TM, total motility; PM, progressive motility; VSL, straight-line velocity; VCL, curvilinear velocity; VAP, average path velocity; ALH, amplitude of lateral head displacement; MAD, average motion degree. Different lowercase letter superscripts in the same column show significant differences ($p < 0.05$).

3.3. Effect of Different Dilution Methods and Ratios of Diluent on Spermatozoa Motility and Kinetic Parameters after Cryopreservation

As shown in Table 3, the post-thawed spermatozoa TM and PM of the two-step group (1:3, 1:2) were the highest and significantly higher ($p < 0.05$) than those of the other groups. The post-thawed spermatozoa VCL, VAP and ALH of the two-step group (1:3, 1:2) were significantly higher ($p < 0.05$) than those of the one-step groups (1:5 and 1:11). However, they were not significantly higher ($p > 0.05$) than the two-step group (1:2, 1:1). The post-thawed spermatozoa VSL of the two-step group (1:3, 1:2) was the highest and significantly higher ($p < 0.05$) than that of the two-step group (1:2, 1:1). The post-thawed spermatozoa MAD of the two-step group (1:3, 1:2) was the highest and significantly higher ($p < 0.05$) than that of the one-step group 1:11.

Table 3. Effect of different dilution methods and ratios of diluent on spermatozoa motility and kinetic parameters after cryopreservation.

Different Dilution Methods and Ratios	TM (%)	PM (%)	VSL (μm/s)	VCL (μm/s)	VAP (μm/s)	ALH (μm)	MAD (°/s)
Two-step: 1:2, 1:1	77.01 ± 2.07^{b}	60.83 ± 1.89^{b}	38.08 ± 1.23^{b}	57.68 ± 1.84^{ab}	40.78 ± 1.30^{ab}	16.89 ± 0.54^{ab}	49.19 ± 4.98^{ab}
Two-step: 1:3, 1:2	85.44 ± 2.07^{a}	72.68 ± 0.59^{a}	42.28 ± 0.34^{a}	62.41 ± 0.66^{a}	44.13 ± 0.46^{a}	18.28 ± 0.19^{a}	53.90 ± 1.46^{a}
One-step: 1:5	78.83 ± 1.24^{b}	60.57 ± 1.40^{b}	40.14 ± 1.15^{ab}	56.34 ± 2.33^{b}	39.83 ± 1.65^{b}	16.50 ± 0.68^{b}	42.38 ± 1.73^{ab}
One-step: 1:11	76.83 ± 1.71^{b}	60.09 ± 1.01^{b}	40.79 ± 0.91^{ab}	55.47 ± 1.44^{b}	39.22 ± 1.02^{b}	16.24 ± 0.42^{b}	39.07 ± 5.86^{b}

TM, total motility; PM, progressive motility; VSL, straight-line velocity; VCL, curvilinear velocity; VAP, average path velocity; ALH, amplitude of lateral head displacement; MAD, average motion degree. Different lowercase letter superscripts in the same column show significant differences ($p < 0.05$).

3.4. Effect of Different Dilution Methods and Ratios of Diluent on Spermatozoa Membrane Integrity after Cryopreservation

As shown in Figure 1A, the post-thawed spermatozoa plasma membrane integrity of the two-step group (1:3, 1:2) was significantly higher ($p < 0.05$) than that of the other groups. As shown in Figure 1B, the blue arrow (a) represents an intact spermatozoa membrane, and the red arrow (b) represents a damaged spermatozoa membrane.

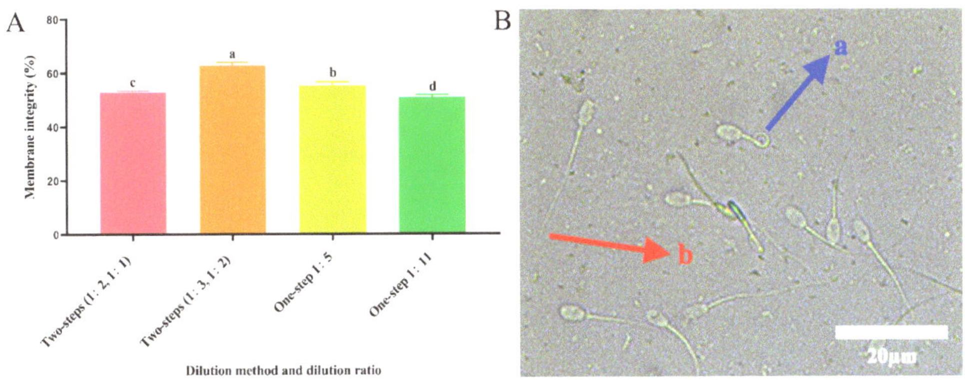

Figure 1. Assessment of plasma membrane integrity. (**A**) Effect of different dilution methods and ratios of diluent on spermatozoa membrane integrity after cryopreservation. Different lowercase letters show significant differences ($p < 0.05$). (**B**) Microscopic results of sperm in HOST experiment. (a) Tail curl represents intact membrane and (b) tail non-curl represents damaged membrane.

3.5. Effect of Different Dilution Methods and Ratios of Diluent on Spermatozoa Acrosome Integrity after Cryopreservation

As shown in Figure 2, the post-thawed spermatozoa acrosome integrity of the two-step group (1:3, 1:2) was significantly higher ($p < 0.05$) than that of the other groups. The post-thawed spermatozoa acrosome integrity of the two-step group (1:2, 1:1) was significantly higher ($p < 0.05$) than that of the one-step groups (1:5 and 1:11). The post-thawed spermatozoa acrosome integrity of the one-step group 1:5 was significantly higher ($p < 0.05$) than that of the one-step group 1:11.

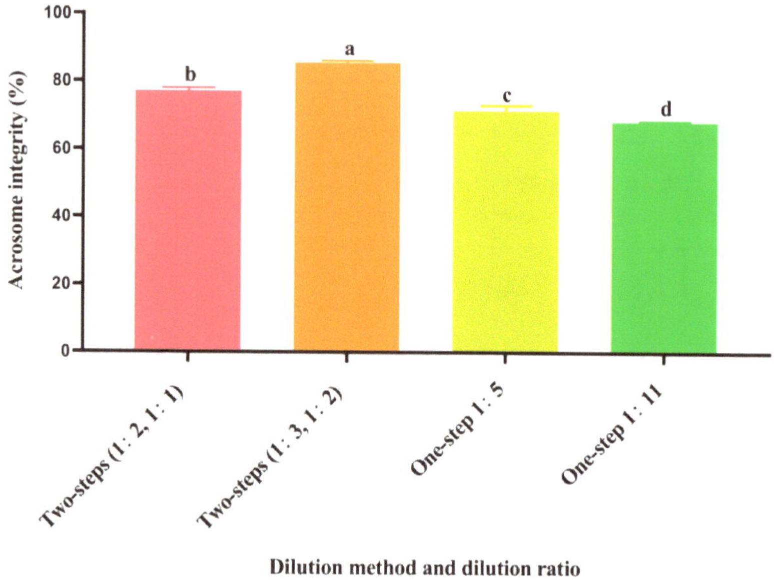

Figure 2. Effect of different dilution methods and ratios of diluent on spermatozoa acrosome integrity after cryopreservation. Different lowercase letters show significant differences ($p < 0.05$).

3.6. Effect of Different Dilution Methods and Ratios of Diluent on Spermatozoa ROS Level after Cryopreservation

As shown in Figure 3, the post-thawed spermatozoa ROS levels of the two-step groups (1:2, 1:1 and 1:3, 1:2) were significantly lower ($p < 0.05$) than those of the one-step groups (1:5 and 1:11), but there was no significant difference ($p > 0.05$) in the two-step groups.

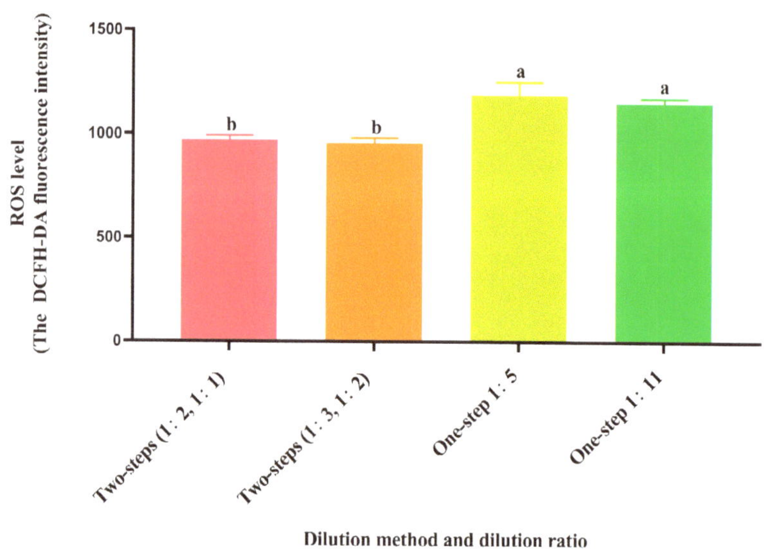

Figure 3. Effect of different dilution methods and ratios of diluent on spermatozoa ROS level after cryopreservation. Different lowercase letters show significant differences ($p < 0.05$).

4. Discussion

Semen dilution simulates the living environment of spermatozoa in vitro. By adding an appropriate diluent, it is possible to reduce the spermatozoa movement rate, inhibit the spermatozoa metabolism, prolong the spermatozoa survival time, increase the semen volume and improve the spermatozoa utilization, thereby increasing the number of breeding female animals [19–21]. In experiments I and II, the post-thawed spermatozoa TM and PM of the 1:1 group were significantly lower ($p < 0.05$) than those of the other groups. This may be due to the high concentration of spermatozoa diluted by a factor of two times, the excessive accumulation of metabolic waste and the inability of the buffer material in the diluent to maintain its balance [22]. On the other hand, it may be caused by the rapid consumption of nutrients such as sugar in the diluent [10]. In experiment I, the post-thawed spermatozoa TM and PM of the 1:4 group were significantly lower ($p < 0.05$) than those of the 1:2 group. In experiment II, the post-thawed spermatozoa TM and PM of the 1:4 group were significantly lower ($p < 0.05$) than those of the 1:3 group. This may be due to the fact that a high dilution of semen can cause the glutamic–oxaloacetic transaminase enzyme in spermatozoa to leak out, leading to the death of spermatozoa [23]. This is consistent with Wang's [24] findings on pig semen, which is highly diluted and causes the leakage of biological enzymes.

In experiment III, the post-thawed spermatozoa in the one-step group 1:11 exhibited significantly lower ($p < 0.05$) plasma membrane integrity and acrosome integrity compared to the other groups. This could be attributed to the reduced antifreeze performance of spermatozoa caused by the higher dilution rate in the one-step dilution method, or it could be due to the absence of essential seminal plasma components when the dilution rate is excessively high [25]. This is consistent with the findings of Wales [26] and Ashworth [27], indicating that high dilution impacts the efficacy of protective components in seminal plasma and the surface structure of the spermatozoa membrane, leading to a reduction in the freezing effect of semen. There is also a saying that the phenomenon is referred to as the dilution effect, which reduces the concentration of natural antioxidants, low-molecular-weight proteins and fatty acids in semen. These substances are beneficial for maintaining the structure and function of the spermatozoa membrane. Excessive dilution of semen can harm the structure of spermatozoa, cause spermatozoa to clump together, significantly shorten the effective preservation time of semen, and subsequently reduce the motility, metabolic activity and fertilization ability of spermatozoa [28,29]. Catt [30] reported that adding seminal plasma had a more positive effect on the cryopreservation of pig semen. Prathalingam [31] reported that the acrosome integrity of bovine semen decreased after thawing when a higher dilution rate was used in cryopreservation. Prochowska [32] reported that the dilution of semen can affect spermatozoa PM and functional integrity in cat semen. Castellini [33] reported that diluting rabbit semen by more than five times will lead to spermatozoa degeneration. These results are consistent with the findings that the integrity of spermatozoa plasma membrane and acrosome decreased as a result of high dilution in this study. The same results were obtained in the changes of spermatozoa membrane integrity and acrosome integrity in the one-step group 1:11. High dilution damages the spermatozoa structure.

In experiment III, the post-thawed spermatozoa TM and PM of the two-step group (1:3, 1:2) were significantly higher ($p < 0.05$) than those of the other groups. Due to the different dilution ratios, the glycerol concentration differs in each diluted group. Therefore, it is difficult to differentiate between the effects of dilution ratios and glycerol concentration in the experiment. On the one hand, this may be due to the fact that a glycerol concentration of 4% in this group is most appropriate for this dilution method and ratio. Proper concentration of glycerol not only enhances the frost resistance of spermatozoa but also prevents the toxic effects of high glycerol concentrations. Glycerol binds to metal ions, dehydrates the cells, reduces the total volume of ice during water solidification, and alleviates the growth of ice crystals, and ultimately reduces the damage caused by freezing [34,35]. On the other hand, this effect may be attributed to the two-step dilution method, which reduces the duration of contact with the spermatozoa and is less harmful when exposed to spermatozoa at low temperatures [36]. This is because the temperature of the diluent is higher when the one-step dilution method is used, which increases the toxic effect of glycerol. In experiment III, the post-thawed spermatozoa ROS level was significantly higher ($p < 0.05$) in the one-step groups (1:5 and 1:11) compared to the other groups. The increase in temperature at the beginning of the process, along with the addition of a glycerol-containing dilution, may have extended the interaction between spermatozoa and glycerol [37]. This prolonged exposure could have intensified the toxic effect of glycerol on spermatozoa, leading to the generation of oxidative stress and an increase in ROS levels.

5. Conclusions

In conclusion, the two-step dilution (1:3, 1:2) was found to be the most suitable method and ratio for diluting *Hu* ram semen after cryopreservation. Preserving semen using this dilution method and ratio can effectively enhance the motility parameters and functional integrity of spermatozoa. An excessively high or low spermatozoa concentration will not be conducive to the preservation of *Hu* ram semen.

Author Contributions: L.Z.: conceptualization, methodology, investigation, formal analysis, writing—original draft. X.W. and Y.S.: conceptualization, writing—original draft, writing—review and editing. C.J. and X.S.: methodology, investigation. J.W. and T.S.: formal analysis, writing—review and editing. Y.L.: writing—review and editing, visualization, supervision, project administration. All authors have read, revised, and approved the final manuscript. All authors have read and agreed to the published version of the manuscript.

Funding: This research was funded by [Jiangsu Modern Agricultural Industrial Technology System Construction Project] grant number [JATS [2023]448]. And The APC was funded by [Jiangsu Modern Agricultural Industrial Technology System Construction Project].

Institutional Review Board Statement: All experimental procedures conform to the Yangzhou University for the protection of experimental animals (ID: 202206132; 1 June 2022; Yangzhou, China).

Informed Consent Statement: Not applicable.

Data Availability Statement: All data sets collected and analyzed during the current study are available from the corresponding author (Y.L.) upon reasonable request.

Conflicts of Interest: The authors declare no conflicts of interest.

References

1. Mason, S.J. Current review of artificial insemination in dogs. *Vet. Clin. N. Am. Small Anim. Pract.* **2018**, *48*, 567–580. [CrossRef]
2. Yeste, M.; Rodríguez-Gil, J.E.; Bonet, S. Artificial insemination with frozen-thawed boar sperm. *Mol. Reprod. Dev.* **2017**, *84*, 802–813. [CrossRef]
3. Avanzi, B.R.; Farrás, M.C.; Melo, C.M.; Alvarenga, M.A.; Dell'Aqua, J.A.; Medeiros, A.S.L.; Araújo, G.H.M.; Papa, F.O. Efficiency of different cooling and storage systems for maintaining equine semen viability in a hot environment. *Anim. Reprod. Sci.* **2006**, *94*, 152–154.
4. Weiss, R.R.; Vianna, B.C.; Muradás, P.R. Inseminação eméguas com sêmen in natura e diluído. *Arch. Vet. Sci.* **2003**, *8*, 19–22. [CrossRef]
5. Macan, R.C.; Camargo, C.E.; Zielinski, B.L.; Cardoso, N.; Lara, N.; Bergstein-Galan, T.G.; Weiss, R.R.; Kozicki, L.E. Timed artificial insemination in crossbred mares: Reproductive efficiency and costs. *Reprod. Domest. Anim.* **2021**, *56*, 459–466. [CrossRef] [PubMed]
6. Yeste, M. Recent advances in boar sperm cryopreservation: State of the art and current perspectives. *Reprod. Domest. Anim.* **2015**, *50*, 71–79. [CrossRef]
7. García-Vázquez, F.A.; Mellagi, A.; Ulguim, R.R.; Hernández-Caravaca, I.; Llamas-López, P.J.; Bortolozzo, F.P. Post-cervical artificial insemination in porcine: The technique that came to stay. *Theriogenology* **2019**, *129*, 37–45. [CrossRef]
8. Ballester, J.; Johannisson, A.; Saravia, F.; Håård, M.; Gustafsson, H.; Bajramovic, D.; Rodriguez-Martinez, H. Post-thaw viability of bull AI-doses with low-sperm numbers. *Theriogenology* **2007**, *68*, 934–943. [CrossRef] [PubMed]

9. Hussaini, S.; Zhandi, M.; Shahneh, Z.A.; Sharafi, M. High dilution rate of bull semen affects cryopreservation outcomes: Kinematic and flow cytometric parameters. *Iran J. Vet. Med.* **2018**, *4*, 377–382.
10. Bonato, M.; Smith, M.A.M.J.; Malecki, I.A.; Cloete, S.W.P. The effect of dilution rate and successive semen collections on sperm quality and sexual motivation of sexually mature South African Merino rams. *Trop. Anim. Health Prod.* **2021**, *53*, 182. [CrossRef] [PubMed]
11. Mann, T. *The Biochemistry of Semen and of the Male Reproductive Tract*; Methuen and Co.: London, UK, 1964; p. 727.
12. Alvarez, M.; Anel-Lopez, L.; Boixo, J.C.; Chamorro, C.; Neila-Montero, M.; Montes-Garrido, R.; Paz, P.; Anel, L. Current challenges in sheep artificial insemination: A particular insight. *Reprod. Domest. Anim.* **2019**, *54*, 32–40. [CrossRef]
13. Hayden, S.S.; Blanchard, T.L.; Brinsko, S.P.; Varner, D.D.; Hinrichs, K.; Love, C.C. Pregnancy rates in mares inseminated with 0.5 or 1 million sperm using hysteroscopic or transrectally guided deep-horn insemination techniques. *Theriogenology* **2012**, *78*, 914–920. [CrossRef]
14. Wu, C.; Dai, J.; Zhang, S.; Sun, L.; Liu, Y.; Zhang, D. Effect of thawing rates and antioxidants on semen cryopreservation in Hu sheep. *Biopreserv. Biobank.* **2021**, *19*, 204–209. [CrossRef]
15. Yang, H.; Wan, Z.; Jin, Y.; Wang, F.; Zhang, Y. SMAD2 regulates testicular development and testosterone synthesis in Hu sheep. *Theriogenology* **2021**, *174*, 139–148. [CrossRef]
16. Chen, Y.; Meng, F.; Liu, Y.; Zhu, C.; Ling, Y.; Liu, C.; Li, L.; Liu, Y.; He, X.; Cao, J.; et al. Effects of resveratrol on DLD and NDUFB9 decrease in frozen semen of Mongolian sheep. *Cryobiology* **2023**, *114*, 104791. [CrossRef]
17. He, Y.; Wang, K.; Zhao, X.; Zhang, Y.; Ma, Y.; Hu, J. Differential proteome association study of freeze-thaw damage in ram sperm. *Cryobiology* **2016**, *72*, 60–68. [CrossRef]
18. Fang, Y.; Zhao, C.; Xiang, H.; Jia, G.; Zhong, R. Melatonin improves cryopreservation of ram sperm by inhibiting mitochondrial permeability transition pore opening. *Reprod. Domest. Anim.* **2020**, *55*, 1240–1249. [CrossRef] [PubMed]
19. Faigl, V.; Vass, N.; Jávor, A.; Kulcsár, M.; Solti, L.; Amiridis, G.; Cseh, S. Artificial insemination of small ruminants—A review. *Acta Vet. Hung.* **2012**, *60*, 115–129. [CrossRef]
20. Jimoh, O.A.; Akinola, M.O.; Oyeyemi, B.F.; Oyeyemi, W.A.; Ayodele, S.O.; Omoniyi, I.S.; Okin-Aminu, H.O. Potential of watermelon (*Citrullus lanatus*) to maintain oxidative stability of rooster semen for artificial insemination. *J. Anim. Sci. Technol.* **2021**, *63*, 46–57. [CrossRef] [PubMed]
21. Di-Iorio, M.; Rusco, G.; Iampietro, R.; Colonna, M.A.; Zaniboni, L.; Cerolini, S.; Iaffaldano, N. Finding an effective freezing protocol for turkey semen: Benefits of ficoll as non-permeant cryoprotectant and 1:4 as dilution rate. *Animals* **2020**, *10*, 421. [CrossRef] [PubMed]
22. Griggers, S.; Paccamonti, D.L.; Thompson, R.A.; Eilts, B.E. The effects of ph, osmolarity and urine contamination on equine spermatozoal motility. *Theriogenology* **2001**, *56*, 613–622. [CrossRef]
23. Pace, M.M.; Graham, E.F. The release of glutamic oxaloacetic transaminase from bovine spermatozoa as a test method of assessing semen quality and fertility. *Biol. Reprod.* **1970**, *3*, 140–146. [CrossRef]
24. Wang, G. *Study of the Effect of Puerarin on Pig Semen Preservation Atroom Temperature*; Anhui Agriculture University: Hefei, China, 2023.
25. Patil, S.; Kumar, P.; Singh, G.; Bala, R.; Jerome, A.; Patil, C.S.; Kumar, D.; Singh, S.; Sharma, R.K. 'Semen dilution effect' on sperm variables and conception rate in buffalo. *Anim. Reprod. Sci.* **2020**, *214*, 106304. [CrossRef]
26. Wales, R.G.; White, I.G. Viability of diluted dog spermatozoa in vitro. *J. Reprod. Fertil.* **1963**, *5*, 67–76. [CrossRef] [PubMed]
27. Ashworth, P.J.; Harrison, R.A.; Miller, N.G.; Plummer, J.M.; Watson, P.F. Survival of ram spermatozoa at high dilution: Protective effect of simple constituents of culture media as compared with seminal plasma. *Reprod. Fertil. Dev.* **1994**, *6*, 173–180. [CrossRef]
28. Maxwell, W.M.; Johnson, L.A. Chlortetracycline analysis of boar spermatozoa after incubation, flow cytometric sorting, cooling, or cryopreservation. *Mol. Reprod. Dev.* **1997**, *46*, 408–418. [CrossRef]
29. Hayden, S.S.; Blanchard, T.L.; Brinsko, S.P.; Varner, D.D.; Hinrichs, K.; Love, C.C. The "dilution effect" in stallion sperm. *Theriogenology* **2015**, *83*, 772–777. [CrossRef] [PubMed]
30. Catt, S.L.; O'Brien, J.K.; Maxwell, W.; Evans, G. Assessment of ram and boar spermatozoa during cell-sorting by flow cytometry. *Reprod. Domest. Anim.* **2010**, *32*, 251–258. [CrossRef]
31. Prathalingam, N.S.; Holt, W.V.; Revell, S.G.; Jones, S.; Watson, P.F. Dilution of spermatozoa results in improved viability following a 24 h storage period but decreased acrosome integrity following cryopreservation. *Anim. Reprod. Sci.* **2006**, *91*, 11–22. [CrossRef]
32. Prochowska, S.; Niżański, W.; Ochota, M.; Partyka, A. Effect of dilution rate on feline urethral sperm motility, viability, and DNA integrity. *Theriogenology* **2014**, *82*, 1273–1280. [CrossRef] [PubMed]
33. Castellini, C.; Lattaioli, P.; Moroni, M.; Minelli, A. Effect of seminal plasma on the characteristics and fertility of rabbit spermatozoa. *Anim. Reprod. Sci.* **2000**, *63*, 275–282. [CrossRef] [PubMed]
34. Lohmann, W.; Fowler, C.F.; Moss, A.J.J.R.; Perkins, W.H. Some remarks about the effect of glycerol on cells during freezing and thawing: Electron-spin resonance investigations concerning this effect. *Experientia* **1964**, *20*, 290. [CrossRef] [PubMed]
35. Bredderman, P.J.; Foote, R.H. Volume of stressed bull spermatozoa and protoplasmic droplets, and the relationship of cell size to motilityd fertility. *J. Anim. Sci.* **1969**, *28*, 496–501. [CrossRef] [PubMed]

36. Gil, J.; Lundeheim, N.; Söderquist, L.; Rodriíuez-Martínez, H. Influence of extender, temperature, and addition of glycerol on post-thaw sperm parameters in ram semen. *Theriogenology* **2003**, *59*, 1241–1255. [CrossRef]
37. Macedo, S.; Bliebernicht, M.; Rocha, A. Effect of two different non-permeable cryoprotectants (NPC) on equine sperm viability after vitrification. *Anim. Reprod. Sci.* **2018**, *194*, 18–19. [CrossRef]

Disclaimer/Publisher's Note: The statements, opinions and data contained in all publications are solely those of the individual author(s) and contributor(s) and not of MDPI and/or the editor(s). MDPI and/or the editor(s) disclaim responsibility for any injury to people or property resulting from any ideas, methods, instructions or products referred to in the content.

animals

MDPI

Article

The Distribution of Boars Spermatozoa in Morphometrically Distinct Subpopulations after In Vitro Exposure to Radiofrequency Electromagnetic Radiation at 2500 MHz and Their Motility

Ivona Žura Žaja [1], Silvijo Vince [2], Ivan Butković [2,*], Kim Senaši [3], Nina Poljičak Milas [4], Krešimir Malarić [5], Martina Lojkić [2], Ivan Folnožić [2], Suzana Milinković Tur [1], Mario Kreszinger [6,*], Marko Samardžija [2], Snježana Čipčić [7], Nikolino Žura [7], Mario Ostović [8] and Marinko Vilić [1]

1 Department of Physiology and Radiobiology, Faculty of Veterinary Medicine, University of Zagreb, 10 000 Zagreb, Croatia; izzaja@vef.unizg.hr (I.Ž.Ž.); tur@vef.unizg.hr (S.M.T.); mvilic@vef.unizg.hr (M.V.)
2 Clinic of Obstetrics and Reproduction, Faculty of Veterinary Medicine, University of Zagreb, 10 000 Zagreb, Croatia; svince@vef.hr (S.V.); mlojkic@vef.unizg.hr (M.L.); folnozic@vef.unizg.hr (I.F.); smarko@vef.unizg.hr (M.S.)
3 Faculty of Veterinary Medicine, University of Zagreb, 10 000 Zagreb, Croatia; kim.senasi996@gmail.com
4 Department of Pathological Physiology, Faculty of Veterinary Medicine, University of Zagreb, 10 000 Zagreb, Croatia; nmilas@vef.unizg.hr
5 Department of Communication and Space Technologies, Faculty of Electrical Engineering and Computing, University of Zagreb, 10 000 Zagreb, Croatia; kresimir.malaric@fer.hr
6 Clinic of Surgery, Orthopedics and Ophthalmology, Faculty of Veterinary Medicine, University of Zagreb, 10 000 Zagreb, Croatia
7 University of Applied Health Sciences, 10 000 Zagreb, Croatia; bielienoci@gmail.com (S.Č.); nikolino.zura@zvu.hr (N.Ž.)
8 Department of Animal Hygiene, Behaviour and Welfare, Faculty of Veterinary Medicine, University of Zagreb, 10 000 Zagreb, Croatia; mostovic@vef.unizg.hr
* Correspondence: ibutkovic@vef.unizg.hr (I.B.); kreszinger@vef.unizg.hr (M.K.); Tel.: +385-12390168 (I.B.); +385-12390128 (M.K.)

Citation: Žaja, I.Ž.; Vince, S.; Butković, I.; Senaši, K.; Milas, N.P.; Malarić, K.; Lojkić, M.; Folnožić, I.; Tur, S.M.; Kreszinger, M.; et al. The Distribution of Boars Spermatozoa in Morphometrically Distinct Subpopulations after In Vitro Exposure to Radiofrequency Electromagnetic Radiation at 2500 MHz and Their Motility. *Animals* **2024**, *14*, 828. https://doi.org/10.3390/ani14060828

Academic Editor: Anna Wysokińska

Received: 5 February 2024
Revised: 23 February 2024
Accepted: 25 February 2024
Published: 7 March 2024

Copyright: © 2024 by the authors. Licensee MDPI, Basel, Switzerland. This article is an open access article distributed under the terms and conditions of the Creative Commons Attribution (CC BY) license (https://creativecommons.org/licenses/by/4.0/).

Simple Summary: The global use of anthropogenic radiofrequency electromagnetic radiation (RF-EMR) in wireless technologies is increasing exponentially and presents a potential risk to animals, especially domestic animals and pets. Additionally, the semen of boar is, in the process of collection, manipulation and storage for the artificial insemination (AI) of sows, surrounded by and exposed to these different sources of wireless technologies devices. A frequency of 2.5 GHz (this frequency band is used in 5G technology) is of particular interest because many studies have used the frequency bands of 4G technology. For the efficiency of pig production and breeding, it is extremely important to determine the effects of such radiation on semen quality and sow fertilization success. Therefore, we aimed to investigate the effect of RF-EMR at 2500 MHz on in vitro exposed breeding boar semen spermatozoa motility and the proportions of spermatozoa subpopulations according to morphometric parameters. The progressive spermatozoa motility and the proportion of the spermatozoa subpopulation with a higher fertilizing potential were significantly reduced in the experimental group. These results indicate the importance of further research on the effects of RF-EMR on different animal species, especially in those undergoing AI procedures, which are important both in terms of the quality of semen and fertilization and production and breeding goals.

Abstract: Anthropogenic radiofrequency electromagnetic radiation (RF-EMR) from wireless technologies has increased dramatically. The boar semen used for artificial insemination is essential in sustaining the pig industry, and additionally it is also exposed to the effects of the RF-EMR of wireless technologies. Furthermore, there are no data on the effects of RF-EMR on semen quality, and this is the first analysis of sperm's morphometric parameters for assessing the effect of RF-EMR on the spermatozoa subpopulations of boars. This study investigated the effect of RF-EMR on in vitro exposed breeding boar semen spermatozoa motility and the proportions of spermatozoa subpopulations according to their morphometric head and tail parameters. The semen samples of 12 boars were divided into control and experimental groups. The samples in the experimental group were exposed in a gigahertz transverse electromagnetic chamber at a frequency of 2500 MHz (the frequency band used in 5G technology) and an electric field strength of 10 Vm^{-1} for two hours. After exposure, the spermatozoa motility was evaluated for both groups. A morphometric analysis of the semen smears was performed using SFORM software (Version 1.0; VAMS, Zagreb, Croatia). The progressive spermatozoa motility was significantly reduced in the experimental group (74.7% vs. 85.7%). PC analysis and cluster analysis revealed two spermatozoa subpopulations: S1, spermatozoa with a more regular head shape and a smaller midpiece outline, and S2, spermatozoa with a more elongated head shape and a larger midpiece outline. The experimental semen samples had a greater proportion of the S1 spermatozoa subpopulation (68.2% vs. 64.4%). The effect of RF-EMR at 2500 MHz on the in vitro exposed boar semen resulted in decreased progressive spermatozoa motility and a lower proportion of the spermatozoa subpopulation with a higher fertilizing potential.

Keywords: boars; radiofrequency electromagnetic radiation; exposure; morphometric analysis; spermatozoa subpopulations

1. Introduction

An enormous surge in wireless communication, with the consequent increase in human and animal exposure to radiofrequency electromagnetic radiation (RF-EMR), has been evident in recent decades. Based on many studies, there is sufficient evidence that RF-EMR of anthropogenic origin has increased many times over in nature and that this radiation affects the environment. For example, anthropogenic radiofrequency electromagnetic radiation from wireless technologies has increased the natural levels of the around 1 GHz frequency band in nature by about 10^{18} times [1]. Therefore, RF-EMR at today's intensity is known as "electro-pollution". Wi-Fi-based technology and receivers, such as laptops, tablets and mobile phones with their base stations, as well as Bluetooth devices, are now routinely used [2–4]. Although such technology has significantly improved our quality of life, it cannot be ruled out that it is also the cause of many ailments and diseases. The harmful effect of RF-EMR can be manifested in most organ systems; however, one of the most sensitive organ systems is the male reproductive system [3,4]. It is well known that RF-EMR reduces the quality of semen and has genotoxic effects on humans and animals both in vitro and in vivo [5–7]. The exposure of men to RF-EMR through various devices, such as mobile phones, wireless internet and laptops, causes abnormal spermatozoa morphology, a decrease in spermatozoa count due to apoptosis, reduced spermatozoa motility and viability, increased testosterone levels, decreased luteinizing hormone levels and increased spermatozoa DNA fragmentation [8–14]. These effects are correlated with the time of exposure [15]. A decrease in semen quality in humans (a decrease in progressive spermatozoa motility and an increase in spermatozoa DNA fragmentation) is also evident when the semen is exposed ex vivo to a laptop connected to a wireless network, i.e., a Wi-Fi frequency of 2.4 GHz for 4 h [16]. Moreover, an increased percentage of damaged epididymal spermatozoa heads was found in rats exposed to 24 h RF-EMR at a frequency of 2.4 GHz for a year [17]. The biological targets of RF-EMR are cell structures such as the plasma membrane (causing cell membrane permeability, including changes in calcium levels, ionic distribution and ion permeability), mitochondria and DNA [5,18]. Spermatozoa motility and morphology are important characteristics to assess in determining semen quality [19]. The introduction of computer-assisted sperm analysis (CASA) has advanced quality assessments of human and animal semen and the diagnosis of fertility, enabling the assessment of parameters such as motility and morphology [19]. The current computer-assisted sperm morphometric analysis (ASTMA) can be used to more accurately analyze individual spermatozoa morphometrics [20,21]. By applying ASTMA technology and multivariate procedures such as cluster analysis, it was observed that boar semen samples contained spermatozoa subpopulations of different morphometric characteristics that are not detectable by conventional subjective methods [22,23]. Morphometric results can vary depending on internal factors such as individual variability, species, breed, sexual maturity and age [23–25], and external factors including environmental factors, sample preparation and the morphometric analysis of semen [20,26].

To date, there are no data on the effects of RF-EMR on the reproductive system of domestic animals. Moreover, it cannot be ignored that the environment surrounding boars at farms is subject to constant increases in radiofrequency electromagnetic fields from different sources, including mobile phone base stations and wireless communication devices. Therefore, for the efficiency of pig production and breeding, it is extremely important to determine the effects of such radiation on the reproductive system of boars. The aim of this study was to investigate the effect of RF-EMR at 2500 MHz (the frequency band used in 5G technology), since it is very close to 2.45 GHz, the frequency of wireless communication devices (Wi-Fi and Bluetooth), on in vitro exposed breeding boar semen spermatozoa motility and proportions of spermatozoa subpopulations using principal component (PC) and cluster analyses of morphometric head and tail parameters.

2. Materials and Methods

2.1. Animals, Housing and Feeding

The study included a total of 12 boars of the Pietrain (8) and German Landrace (4) breeds, aged 1.5–3.5 years, from which semen was routinely taken twice a week, on a certain day of the week for each boar for the artificial insemination of sows. The boars are owned by the Centre for Artificial Insemination, Stočar d.o.o., Varaždin, Croatia. The boars were housed individually in 12 m^2 (4 × 3 m) pens, with straw bedded floors and natural lighting, and were fed twice a day, around 6:00 a.m. and 2:00 p.m., with a mixture produced by Stočar d.o.o., Varaždin, Croatia. The daily requirements of the boar are approximately 2 kg of mixture, with the following composition: crude protein (17.15%), crude fat (3.13%), crude cellulose (6.29%), ash (5.30%), metabolic energy (12.49 MJ/kg), lysine (1.24%), methionine (0.47%), tryptophan (0.18%), methionine + cystine (0.76%), zinc (126 mg/kg), magnesium (56 mg/kg), digestible protein (38.28%), copper (31.50 mg/kg), selenium (0.42 mg/kg), calcium (0.83%), total phosphorus (0.53%), usable phosphorus (0.22%), sodium (0.23%), vitamin D3 (1750.01 IU/kg), vitamin A (9100 IU/kg) and vitamin E (70 IU/kg).

2.2. Collection and Evaluation of Semen from the Boars

Collection of semen samples was performed in the morning hours (around 7:00 a.m.). The procedure for obtaining the ejaculate was carried out by the method of manual fixation of the penis. Standard evaluation of boar ejaculate was performed at the Centre for Artificial Insemination, Stočar d.o.o., Varaždin, Croatia. The semen was collected in wide-mouthed glass containers. The semen volume was determined using a measuring cup, and the concentration of sperm in the ejaculate was determined using an Accucell photometer type 60CI0394 (IMV technologies, Normandy, France). Mass motility was determined in native semen using an Olympus BX50F (Olympus, Tokyo, Japan) microscope with a built-in spermotherm. After determining sperm concentration and mass motility, a certain amount of Cronos diluent (Medi-Nova, Reggio Emilia, Italy) was added to the semen and it was transferred to 80 mL plastic bottles. The samples of semen were transported in specialized containers with thermometer (to protect against the harmful effects of light and temperature) for 45 min from the collection place to the laboratory.

2.3. Exposure of Samples to RF-EMR in Laboratory Conditions

Upon arrival at the clinic, samples of each boar (n = 12) were divided into two Petri dishes (control and experimental sample (12 each for a total of 24 samples). The experimental samples were exposed in a gigahertz transverse electromagnetic (GTEM) chamber to RF-EMR at a frequency of 2500 MHz and an electric field strength of 10 Vm^{-1} for two hours. GTEM was located at the Clinic for Reproduction and Obstetrics at the Faculty of Veterinary Medicine University of Zagreb (45°48′25.91″ North and 16°0′20.49″ East). A

GTEM-chamber was made at the Department of Communication and Space Technologies Department of Radiocommunications and High-Frequency Electronics, Faculty of Electrical Engineering and Computing, University of Zagreb, Zagreb, Croatia. The chamber contained a digital thermometer to measure temperature to determine if there was a thermal effect of the radiation (Figure 1). The average temperature inside the chamber during semen exposure was 19.1 °C (range 18.7–19.5 °C). Control samples (unexposed samples) were placed in a metal container (imitation for GTEM-chamber) and kept under the same conditions (temperature and time) as the experimental groups but without exposure to RF-EMR. Experimental samples (exposed) and control samples (unexposed) were assessed after exposure/sham-exposure of 2 h. In addition to the chamber, an HP 8657A signal generator and an RFGA0101-05 linear amplifier were used to achieve electromagnetic field strength. The amplifier was connected to a computer and the desired frequency was set using the SynthNV program (Windfreak Technologies, LLC., New Port Richey, FL, USA). The GTEM-chamber is a transmission line that is based on TEM-chambers where the letter G indicates that the GTEM chamber works in the GHz range up to 18 GHz. The GTEM-chamber is adjustable in the pyramidal part of the TEM-chamber with an impedance of 50 Ω. After exposure, the control and experimental semen samples were evaluated for spermatozoa motility.

Figure 1. Gigahertz transverse electromagnetic (GTEM) chamber. View of the GTEM chamber in which the boar semen samples were placed together with a digital thermometer.

2.4. Computer-Assisted Sperm Analysis

Spermatozoa motility was determined using a computer-assisted sperm analysis (CASA) device (Integrated Visual Optical System, Version 12; Hamilton Thorne Research, Beverly, MA, USA) located at the Clinic of Obstetrics and Reproduction, Faculty of Veterinary Medicine, University of Zagreb, Zagreb, Croatia. After the exposure procedure of the experimental samples, diluted experimental and control semen samples (5 μL each) were applied to a 20 μm deep Leja-chamber (Leja Products B.V., Nieuw Vennep, The Netherlands) and placed on a heated spermotherm (Minitub, Tiefenbach, Germany). After the cessation of passive spermatozoa movement, imaging was performed on all eight fields of the chamber. The program was set to analyze 45 frames obtained per field at a frame rate of 60 Hz. The analysis determined spermatozoa motility (%) and progressive motility (%).

2.5. Preparation and Staining of Semen Smears

The control and experimental semen samples were then used to make a smear on a glass slide. The semen smears were then stained with the Spermac set of reagents (Minitube, Tiefenbach, Germany) for diagnostic staining of spermatozoa, which is generally used to visualize the head, acrosome, equatorial region, central part and tail of spermatozoa. The Spermac method in brief: a thin smear was made on a clean slide, then placed in fixative for 5 min. After fixation, the slides were placed on a heating plate at 37 °C for 15 min. Then the slides were washed with distilled water and staining followed by first immersing the smear in 50 mL of red liquid (Spermac "A") for 1 min, followed by rinsing again with distilled water; afterwards, the smear was immersed in 50 mL of pale green liquid (Spermac "B") for 1 min, after which the smear was rinsed with distilled water and the last smear was immersed in 50 mL of dark green liquid (Spermac "C") for 1 min, with a final rinse. After staining, the final drying of the preparation followed on a heating plate at a temperature of 37 °C.

2.6. Morphometric Analysis of Spermatozoa

In total, 24 stained semen smears were analyzed, with approximately 70 spermatozoa measured for each smear (n = 1691). The spermatozoa morphometric analysis was performed using the SFORM program for image processing and analysis (VAMSTEC, Zagreb, Croatia). Only spermatozoa heads that did not overlap with those of other spermatozoa and non-banded tails were measured and analyzed. The borders of the head, midpiece and tail of the spermatozoa were marked automatically using the marking option of SFORM (first the head, then the midpiece and finally the tail) with manual correction using a computer mouse and then the calculated data were printed in the program table [27]. The area (μm^2), outline (μm), minimum radius (μm), maximum radius (μm), length (μm) and width (μm) were the primary morphometric parameters (size parameters) calculated for the spermatozoa head and midpiece, while for spermatozoa tail, only length was calculated. Different ratios of morphometric parameters were also calculated, such as the total length of the spermatozoa, which is the sum of the head length and the tail length, then head length/total length, head length/tail length, tail length/total length, head outline/total length, head area/total length and head length × head width/total length. Some primary morphometric parameters (head size morphometric parameters) were used to calculate head shape morphometric parameters using the following formulas: ellipticity (length/width), rugosity ($4\pi \times$ area/outline2), elongation (length − width)/(length + width) and regularity ($\pi \times$ length × width/4 × area).

2.7. Statistical Data Processing

A statistical data analysis was performed using the SAS 9.4 software package (Statistical Analysis Software 2002–2012 by the SAS Institute Inc., Cary, NC, USA). A descriptive data analysis was performed using MEANS and FREQ procedures. The dependent parameters between groups were analyzed via a multivariate analysis of variance (MANOVA) based on Wilks' lambda criterion using the GLM procedure. The results are expressed as least squares means (LSM) and 95% confidence intervals. To compare mean values, the TukeyKramer method of multiple comparisons was used at the level of statistical significance $p < 0.05$.

Multivariate clustering analyses (CLUSTERS) of data were performed through several steps to obtain spermatozoa subpopulations based on the data of the main morphometric parameters of the spermatozoa head and tail. The first step was the analysis of the main components (PROC FACTOR) to obtain the characteristic values (eigenvalues) of the morphometric parameters using the Kaiser criterion ($\lambda \geq 1$) to determine the number of main components. The number of clusters in the K-means cluster analysis was determined using the HPCLUS procedure, which selects the best k value (number of clusters or subpopulations) using the aligned box criterion value (Figure 2). The third step was to group the data using non-hierarchical analysis (K-means method and Euclidean distance) of the most important parameters for each component from the previous analysis using the FASTCLUS procedure. In order to better interpret the data of the obtained spermatozoa subpopulations, stepwise discrimination analysis (PROC STEPDISC) and testing of atypical values (PROC FASTCLUS) were performed. Testing for differences in the distribution of spermatozoa subpopulations between the control and experimental semen sample groups was done using Chi-square and Mantel-Haenszel Chi-square tests (PROC FREQ).

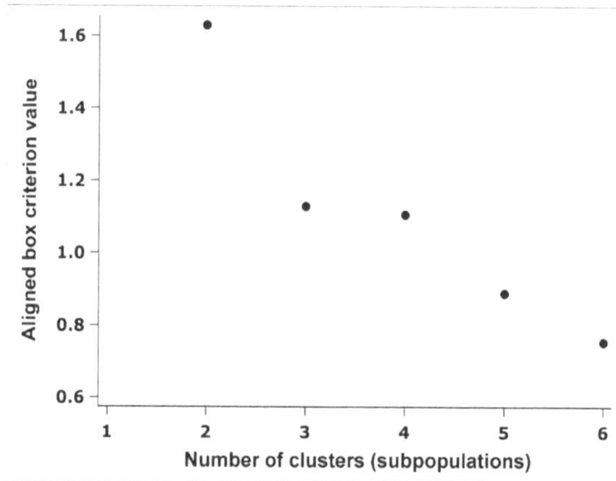

Figure 2. Two subpopulations of boar spermatozoa. The figure shows that two optimal subpopulations were obtained using the values of the equalized box criterion.

3. Results

3.1. Overall Semen Variables

Standard evaluation of ejaculate revealed that the semen samples correspond to the criteria valid for boars (minimum motility > 60%) and in this respect the conditions were met since average spermatozoa motility was about 80%. The mean volume (±SD) of the boar semen sample was 354.5 ± 126.3 mL. Sperm concentration in mL of semen (±SD) was 358.7 ± 169.4 million. The total sperm concentration (±SD) was 113.5 ± 34.6 billion.

3.2. Individual Morphometric Parameters of the Spermatozoa Head and Tail

Morphometric analysis was performed on a total of 1691 spermatozoa, of which 839 spermatozoa (49.6%) in the control sample and 852 (50.4%) in the experimental group. Statistical analysis of individual spermatozoa morphometric parameters between the experimental group exposed to RF-EMR at a frequency of 2500 MHz and the control group (Tables 1–3) revealed no statistically significant differences. The only morphometric parameter whose value was close to statistical significance was the midpiece convex ($p = 0.06$). Sperm motility determined by CASA between the experimental and control groups was not statistically significantly different, although motility was reduced in the experimental group (74.7%) as compared to the control group (85.7%). Progressive sperm motility was statistically significantly reduced ($p < 0.001$) in the experimental group (35.0%) compared to the control group (60.1%).

Table 1. Boar spermatozoa morphometric head parameters in the control and experimental semen sample groups.

Spermatozoa Morphometric Parameters		Control Group		Experimental Group		
		Mean	95% Confidence Interval	Mean	95% Confidence Interval	*p* Value
Morphometric parameters of the head	Area (μm^2)	46.79	46.47–47.11	46.75	46.43–47.07	0.84
	Outline (μm)	27.31	27.21–27.41	27.26	27.16–27.36	0.49
	Minimal radius (μm)	2.48	2.47–2.49	2.48	2.47–2.49	0.81
	Maximal radius (μm)	5.34	5.32–5.37	5.34	5.32–5.37	0.93
	Convex (μm)	47.39	47.06–47.72	47.43	47.11–47.76	0.85
	Length (μm)	10.31	10.26–10.36	10.3	10.26–10.35	0.86
	Bredth (μm)	5.45	5.42–5.47	5.45	5.42–5.47	0.89
	Ellipticity	1.9	1.89–1.91	1.89	1.89–1.91	0.78
	Rugosity	0.79	0.785–0.789	0.79	0.787–0.791	0.27
	Elongation	0.31	0.305–0.311	0.31	0.30–0.31	0.8
	Regularity	0.94	0.941–0.944	0.94	0.94–0.95	0.26

Ellipticity = length/bredth; Rugosity = (4π × area/outline2); Elongation = [(lenght − bredth)/(length + bredth)]; Regularity = (π × length × bredth/4 × area).

Table 2. Boar spermatozoa morphometric midpiece and tail parameters in the control and experimental semen sample groups.

Spermatozoa Morphometric Parameters		Control Group		Experimental Group		
		Mean	95% Confidence Interval	Mean	95% Confidence Interval	p Value
Parameters of morphometric characteristics of the midpiece and the tail	Midpiece area (μm^2)	19.42	19.29–19.55	19.48	19.35–19.60	0.52
	Midpiece outline (μm)	30.44	30.32–30.56	30.35	30.23–30.46	0.27
	Midpiece min. radius (μm)	0.39	0.37–0.39	0.39	0.38–0.40	0.66
	Midpiece max. radius (μm)	6.78	6.75–6.81	6.76	6.73–6.78	0.25
	Midpiece convex (μm)	24.58	24.32–24.84	24.92	24.67–25.18	0.06
	Midpiece length (μm)	13.25	13.19–13.30	13.2	13.15–13.26	0.27
	Midpiece width (μm)	2.05	2.03–2.08	2.08	2.05–2.11	0.21
	Tail length (μm)	34.05	33.84–34.26	33.92	33.71–34.12	0.37

Table 3. Boar spermatozoa morphometric head and tail parameters ratios in the control and experimental semen sample groups.

Spermatozoa Morphometric Parameters		Control Group		Experimental Group		
		Mean	95% Confidence Interval	Mean	95% Confidence Interval	p Value
Different ratios of morphometric parameters	Total length *	44.36	44.13–44.59	44.22	43.99–44.45	0.4
	Head length/Total length	0.23	0.232–0.234	0.23	0.23–0.24	0.52
	Head length/Tail length	0.31	0.30–0.31	0.31	0.30–0.31	0.53
	Tail length/Total length	0.77	0.766–0.768	0.77	0.765–0.767	0.52
	Head outline/Total length	0.62	0.615–0.621	0.62	0.616–0.621	0.78
	Head area/Total length	1.06	1.05–1.06	1.06	1.05–1.07	0.75
	Head length and width/Total length *	1.27	1.26–1.28	1.27	1.27–1.28	0.53

* Total length = head length + tail length.

3.3. Spermatozoa Subpopulations Based on Morphometric Parameters of Spermatozoa Head and Tail

Analysing the main components before grouping, four components (factor 1, 2, 3 and 4) with a characteristic value ($\lambda \geq 1$) were retained. All four components in total explained 84.6% of the variance of the morphometric parameters of the spermatozoa head, midpiece and tail (Table 4).

Table 4. Eigenvalues of boar spermatozoa morphometric head and tail parameters in the analysis of the main components. Four components (factor 1, 2, 3, 4) with a characteristic root $\lambda \geq 1$ were retained—Kaiser's criterion.

Spermatozoa Indicators	Factor 1	Factor 2	Factor 3	Factor 4
Head length	0.93 *			
Head width		0.97 *		
Head area	0.62			
Head outline	0.83			
Ellipticity	0.72			
Rugosity	−0.69			
Elongation	0.72			
Regularity				0.89 *
Midpiece length			0.70	
Midpiece width		0.39		
Midpiece area		0.65		
Midpiece outline			0.72 *	
Tail length	0.39			
Characteristic root (λ) and explained variance (%)	4.78 (36.8)	3.66 (28.2)	1.53 (11.8)	1.01 (7.8)

* The most important parameters for each factor.

Using Table 4, the most important parameters were selected from each component (head length, head width, midpiece outline and regularity of the spermatozoa head). The final number of subpopulations was obtained using the value of the equalised box criterion This analysis determined that the two subpopulations are the most optimal because the value of the equalised box criterion was the highest (Figure 2).

The grouping analysis revealed well-defined differences between head length, width and regularity and the midpiece outline in the two spermatozoa subpopulations (S1 subpopulation—smaller length and larger head width, more regular head shape and smaller midpiece outline, and S2 subpopulation—longer length and smaller head width, more elongated head shape and larger midpiece outline) (Table 5).

Table 5. Subpopulations of boar spermatozoa (S1 and S2) obtained using the analysis of grouping of spermatozoa morphometric head and midpiece parameters.

Spermatozoa Morphometric Head and Midpiece Parameters Mean ± SD	Spermatozoa Subpopulation	
	S1	**S2**
n (%)	942 (55.71)	749 (44.29)
Head length (μm)	10.16 ± 0.65	10.77 ± 0.68
Head width (μm)	5.44 ± 0.38	5.45 ± 0.35
Midpiece outline (μm)	29.67 ± 1.17	32.66 ± 1.09
Regularity	0.943 ± 0.02	0.941 ± 0.02

S1—Spermatozoa of smaller length and larger head width, more regular head shape and smaller midpiece outline; S2—Spermatozoa of longer length and smaller head width, more elongated head shape and larger midpiece outline.

Statistical analysis of the obtained subpopulations between the control and experimental groups of semen samples revealed that the experimental semen samples group had a higher percentage of spermatozoa of the S1 subpopulation (68.2% vs. 64.4%) and a lower percentage of the S2 subpopulation (31.8% vs. 35.6%) which is close to statistical significance ($p = 0.09$) (Figure 3).

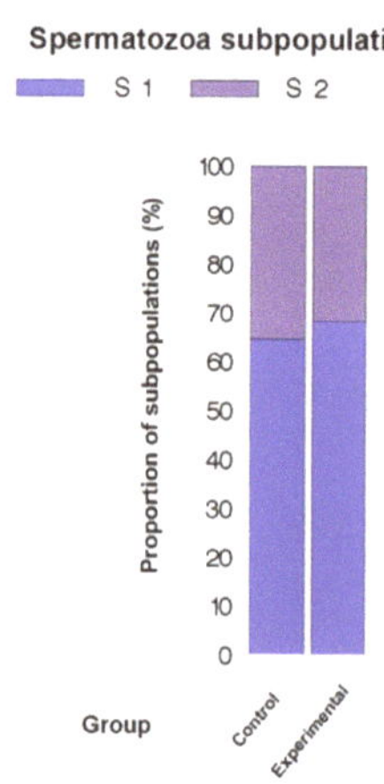

			Subpopulations S1	Subpopulations S2	Total (N)
Group	Control	N	540	299	839
		Row %	64.36	35.64	
		Column %	48.17	52.46	
	Experimental	N	581	271	852
		Row %	68.19	31.81	
		Column %	51.83	47.54	
	Total	N	1121	570	
	p value		0.09		

Figure 3. Proportion of spermatozoa subpopulations in control and experimental semen samples groups.

4. Discussion

This study showed that the exposure of semen of breeding boars in vitro to RF-EMR at a frequency of 2500 MHz and an electric field strength of 10 V/m for a duration of 2 h did not cause changes of spermatozoa individual morphometric parameters and sperm motility, but it decreased progressive sperm motility.

Although it is known that RF-EMR has a harmful effect on the male reproductive system, by reducing the number of Leydig cells, motility, and number of spermatozoa, and altering spermatozoa morphology in humans and animals [9,15,28–32], the results presented here do not support this. This could be due to the application of different study designs (protocols) during the experiment, different species and ages of animals, and different analysis methods. Furthermore, to the extent of our knowledge, the effect of RF-EMR on spermatozoa morphometric parameters has not previously been investigated in humans or other species. In addition, spermatozoa morphometry in boars is performed using different software and on semen smears stained with different methods. Wysokińska et al. [33] performed spermatozoa morphometric analysis on spermatozoa samples collected from 35 boars of the Polish Landrace breed at the age of 7 to 8 months stained with the Bydgoszka method using a computer image analysis package (Screen Measurement v. 4.1, Laboratory Imaging S.r.o. LIM, Prague, Czech Republic). Their research showed a lower mean length, width, area and outline of the head and a higher mean value for the total length and tail length compared to the results for the control group in the present study. This study used the computer program SFORM (VAMSTEC, Zagreb, Croatia) for the morphometric analysis, and there is no information about its use in the morphometric analysis of boar spermatozoa in the literature, and therefore, it is possible that the difference in the stated values was result of the differences in the programs for morphometric analysis or differences in staining methods, age and breed of boars. Górski et al. [34] showed a lower mean value for the length, width, area and outline of the spermatozoa head and a higher mean value for the spermatozoa total length and the tail length as compared to the control group of this study. Those authors performed a spermatozoa morphometric analysis on semen samples collected from 12 boars of the Duroc breed, stained using the method according to Kondracki et al. [24], and using the computerised image analysis system Screen Measurement v. 4.1. for morphometric measurements. One reason for the differences between the results of our control group and that study could be due to the different staining methods [26], animal breed [23] or the analysis method used.

Other aspects of spermatozoa physiology and morphology may also need to be considered, as they may affect their ability to actively move through the female reproductive system. In many species, the first barrier is the cervical mucus, which allows only progressively motile spermatozoa with normal morphology to pass into the uterus and through which they progressively move (with the help of myometrial contractions) to the fallopian tube, where fertilization occurs [35]. Therefore, spermatozoa motility is crucial. This study showed a significant reduction in progressive spermatozoa motility after exposure of the semen of the boars in vitro to RF-EMR at a frequency of 2500 MHz. Mailankot et al. [10] reported similar results, showing that exposing rats to RF-EMR 1 h a day for 28 days at a frequency of 900 and 1800 MHz originating from mobile devices caused a drop in the number of motile spermatozoa, and reduced progressive sperm motility. Oni et al. [36] and Gorpinchenko et al. [37] investigated the effect of electromagnetic radiation at the frequency of mobile telecommunications (900/1800 MHz) and laptops (2.45 GHz) on in vitro samples of human semen, and found that exposure to these frequencies also reduced the number of motile and progressively motile spermatozoa. On the other hand, some studies have reported no changes in spermatozoa motility after rats had been long-term exposed to RF-EMR frequency of 2.4 GHz [17]. A possible mechanism that leads to a decrease in spermatozoa motility after exposure of semen to RF-EMR is a lowered mitochondrial potential or oxidative stress, and consequently impaired spermatozoa vitality [28,38]. Namely, some studies on the impact of RF-EMR on spermatozoa, though not on boars, have indicated that RF-EMR at the frequency of mobile telephony can cause the formation of reactive oxygen species (ROS) and thus oxidative stress. It is also known that in aerobic organisms, a balance between antioxidant processes and reactive compounds formed requires an oxidative-reduction balance, because otherwise, during oxidative stress, an excess of ROS leads to the damage of numerous molecules. Kumar et al. [39] and Meena et al. [40] showed that exposing rats to RF-EMR frequencies of 2.45 GHz and 10 GHz for 2 h a day for 45 days led to cell damage mediated by oxidative stress, i.e., caused an increase in the concentration of ROS, an increase in the percentage of spermatozoa apoptosis in testicles, and DNA damage.

This study has shown that RF-EMR at a frequency of 2500 MHz could have a negative effect on the success of egg fertilisation. We base the above assumption not only on the obtained results that RF-EMR reduces the number of progressively motile spermatozoa, but also on the obtained percentages of the spermatozoa subpopulation after exposure to RF-EMR. The obtained proportions of spermatozoa subpopulations using PC and cluster analysis according to morphometric head and tail parameters showed that the experimental group of semen had a higher percentage of the less desirable subpopulation, characterised by spermatozoa of smaller length and larger head width, more regular head shape and smaller midpiece outline as compared to the more desirable spermatozoa characterised by longer length and larger head width, more elongated shape of the head and larger midpiece outline. It has been empirically proven in different species that spermatozoa function is related to its morphometry, which includes the area of the head, midpiece and tail [41,42]. Ramon et al. [43] stated that deer ejaculate containing a high percentage of spermatozoa with fast and linear motility have small and elongated heads and achieve higher fertility. An elongated spermatozoa head can have an important function in that such sperm will be hydrodynamically more efficient due to less resistance in forward movement, which can affect the fertilising ability of sperm [20,44]. However, Barquero et al. [23] reported that boars with a larger litter size had significantly less elongated spermatozoa, and the mortality of piglets was greater in these males. Evolutionary biology is still debating which of the two spermatozoa components, head characteristics or midpiece traits of the spermatozoa, is more important in spermatozoa competition during the egg fertilisation process. Namely, the increase in the midpiece of the spermatozoa increases its energy due to the increased area housing the mitochondria [45], as more energy is needed for faster sperm. FIRMAN and SIMMONS [46] reported that midpiece size is a predictor of swimming speed of Mus musculus domesticus spermatozoa. It is known that RF-EMR can cause cell/sperm apoptosis, and mitochondria are the main initiators of apoptosis [47]. In addition, RF-EMR promotes increased mitochondrial ROS production and expression of mitochondrial apoptotic markers [48] with decreased mitochondrial membrane potential [47,49]. Although the exact mechanism of apoptotic changes in spermatozoa, and in somatic cells, is still unknown, one of the signs of apoptosis was described as typical cell shrinkage [50]. The decrease in the mitochondria outline with reduced progressive motility of spermatozoa of the experimental group in this study could indicate the initiation of spermatozoa apoptosis. Furthermore, exposure of cells to RF-EMR if the intensity of the fields increases beyond the threshold, causes electroporation, during which water pores are created in the membrane, disrupting the ion balance and leading to water ingress in the cell [18], which is likely to cause a change in spermatozoa shape in a higher proportion in the experimental group in this study. If this was the case, then the increased ROS production generated in these highly vulnerable cells could reasonably be expected to impose an oxidative stress environment upon the aforementioned the sperm population. Given that there is no literature on the effects of RF-EMR on the proportion of boars or other species spermatozoa subpopulations obtained based on morphometric parameters, the results of the present study cannot be compared. There are scarce studies on the proportions of boar spermatozoa subpopulations using PC and cluster analysis according to morphometric parameters, therefore it is not possible to compare the obtained results. However, Barquero et al. [23] investigated the spermatozoa morphometry of boars without exposure to RF-EMR, and found four subpopulations of spermatozoa using PC and cluster analysis according only to morphometric head parameters: subpopulation 1 with lower values for ellipticity and the widest heads, sub-population 2 with the highest values for head area and perimeter, subpopulation 3 with shorter length, with a smaller area and head perimeter values, and subpopulation 4 with the highest values for head length, ellipticity, and elongation. We obtained two subpopulations related to not only morphometric head but also tail parameters, giving a more complete picture of spermatozoa morphology, which is related to their function, i.e., fertilisation ability.

5. Conclusions

The effect of RF-EMR at 2500 MHz on in vitro exposed breeding boar semen for two hours was seen in the decreased progressive spermatozoa motility and proportion of the spermatozoa subpopulation with a more elongated head shape and larger midpiece outline. Further research on the effects of RF-EMR on different animal species and breeds, especially domestic animals, is important both for the quality of semen and fertilization and for production and breeding goals. The observed results could also be crucial for comparison with human reproductive medicine and potential adverse effects during the specific technological process of semen processing on breeding pigs farms.

Author Contributions: Conceptualization, I.Ž.Ž., S.V., I.B., K.M. and M.V.; methodology, M.L., I.B.; software, S.V., N.P.M., M.K. and I.Ž.Ž.; validation, M.L.; investigation, I.Ž.Ž., S.V., I.B., K.S., M.L., I.F., S.Č., N.Ž. and M.V.; writing—original draft preparation, I.Ž.Ž., I.B., M.K. and K.S.; writing—review and editing, I.Ž.Ž., M.V., M.S., S.M.T., M.O. and N.P.M. All authors have read and agreed to the published version of the manuscript.

Funding: This research received no external funding.

Institutional Review Board Statement: This research was approved by the Ethics Committee in Veterinary Medicine, Faculty of Veterinary Medicine, University of Zagreb, Zagreb, Croatia (record No.: 640-01/21-02/03; file No.: 251-61-01/139-21-39), approval date: 19 February 2020.

Informed Consent Statement: Not applicable.

Data Availability Statement: The data presented in this study are available on reasonable request from the corresponding author. The data are not publicly available due to planned research in the future.

Acknowledgments: The authors would like to thank Sead Džubur, Vams Tec d.o.o., Zagreb, Croatia for his help and assistance with SFORM Software.

Conflicts of Interest: The authors declare no conflicts of interest.

References

1. Bandara, P.; Carpenter, D.O. Planetary Electromagnetic Pollution: It Is Time to Assess Its Impact. *Lancet Planet. Health* **2018**, *2*, e512–e514. [CrossRef]
2. Kesari, K.K.; Agarwal, A.; Henkel, R. Radiations and Male Fertility. *Reprod. Biol. Endocrinol.* **2018**, *16*, 118. [CrossRef]
3. Yadav, H.; Rai, U.; Singh, R. Radiofrequency Radiation: A Possible Threat to Male Fertility. *Reprod. Toxicol.* **2021**, *100*, 90–100. [CrossRef]
4. Prlić, I.; Šiško, J.; Varnai, V.M.; Pavelić, L.; Macan, J.; Kobešćak, S.; Hajdinjak, M.; Jurdana, M.; Cerovac, Z.; Zauner, B.; et al. Wi-Fi Technology and Human Health Impact: A Brief Review of Current Knowledge. *Arch. Ind. Hyg. Toxicol.* **2022**, *73*, 94–106. [CrossRef]
5. La Vignera, S.; Condorelli, R.A.; Vicari, E.; D'Agata, R.; Calogero, A.E. Effects of the Exposure to Mobile Phones on Male Reproduction: A Review of the Literature. *J. Androl.* **2012**, *33*, 350–356. [CrossRef]
6. Liu, K.; Li, Y.; Zhang, G.; Liu, J.; Cao, J.; Ao, L.; Zhang, S. Association between Mobile Phone Use and Semen Quality: A Systemic Review and Meta-Analysis. *Andrology* **2014**, *2*, 491–501. [CrossRef] [PubMed]
7. Kaur, P.; Rai, U.; Singh, R. Genotoxic Risks to Male Reproductive Health from Radiofrequency Radiation. *Cells* **2023**, *12*, 594. [CrossRef] [PubMed]
8. Erogul, O.; Oztas, E.; Yildirim, I.; Kir, T.; Aydur, E.; Komesli, G.; Irkilata, H.C.; Irmak, M.K.; Peker, A.F. Effects of Electromagnetic Radiation from a Cellular Phone on Human Sperm Motility: An In Vitro Study. *Arch. Med. Res.* **2006**, *37*, 840–843. [CrossRef]
9. Agarwal, A.; Singh, A.; Hamada, A.; Kesari, K. Cell Phones and Male Infertility: A Review of Recent Innovations in Technology and Consequences. *Int. Braz. J. Urol.* **2011**, *37*, 432–454. [CrossRef] [PubMed]
10. Mailankot, M.; Kunnath, A.P.; Jayalekshmi, H.; Koduru, B.; Valsalan, R. Radio Frequency Electromagnetic Radiation (RF-EMR) from GSM (0.9/1.8 GHZ) Mobile Phones Induces Oxidative Stress and Reduces Sperm Motility in Rats. *Clinics* **2009**, *64*, 561–565. [CrossRef] [PubMed]
11. Kesari, K.K.; Behari, J. Effects of Microwave at 2.45 GHz Radiations on Reproductive System of Male Rats. *Toxicol. Environ. Chem.* **2010**, *92*, 1135–1147. [CrossRef]

12. Gutschi, T.; Mohamad Al-Ali, B.; Shamloul, R.; Pummer, K.; Trummer, H. Impact of Cell Phone Use on Men's Semen Parameters. *Andrologia* **2011**, *43*, 312–316. [CrossRef]
13. Rago, R.; Salacone, P.; Caponecchia, L.; Sebastianelli, A.; Marcucci, I.; Calogero, A.E.; Condorelli, R.; Vicari, E.; Morgia, G.; Favilla V.; et al. The Semen Quality of the Mobile Phone Users. *J. Endocrinol. Investig.* **2013**, *36*, 970–974. [CrossRef]
14. Yildirim, M.E.; Kaynar, M.; Badem, H.; Cavis, M.; Karatas, O.F.; Cimentepe, E. What Is Harmful for Male Fertility: Cell Phone or the Wireless Internet? *Kaohsiung J. Med. Sci.* **2015**, *31*, 480–484. [CrossRef]
15. Agarwal, A.; Deepinder, F.; Sharma, R.K.; Ranga, G.; Li, J. Effect of Cell Phone Usage on Semen Analysis in Men Attending Infertility Clinic: An Observational Study. *Fertil. Steril.* **2008**, *89*, 124–128. [CrossRef]
16. Avendaño, C.; Mata, A.; Sanchez Sarmiento, C.A.; Doncel, G.F. Use of Laptop Computers Connected to Internet through Wi-Fi Decreases Human Sperm Motility and Increases Sperm DNA Fragmentation. *Fertil. Steril.* **2012**, *97*, 39–45.e2. [CrossRef]
17. Dasdag, S.; Taş, M.; Akdag, M.Z.; Yegin, K. Effect of Long-Term Exposure of 2.4 GHz Radiofrequency Radiation Emitted from Wi-Fi Equipment on Testes Functions. *Electromagn. Biol. Med.* **2015**, *34*, 37–42. [CrossRef] [PubMed]
18. Koohestanidehaghi, Y.; Khalili, M.A.; Fesahat, F.; Seify, M.; Mangoli, E.; Kalantar, S.M.; Annarita Nottola, S.; Macchiarelli, G.; Grazia Palmerini, M. Detrimental Effects of Radiofrequency Electromagnetic Waves Emitted by Mobile Phones on Morphokinetics, Oxidative Stress, and Apoptosis in Mouse Preimplantation Embryos. *Environ. Pollut.* **2023**, *336*, 122411. [CrossRef] [PubMed]
19. Verstegen, J.; Iguer-Ouada, M.; Onclin, K. Computer Assisted Semen Analyzers in Andrology Research and Veterinary Practice. *Theriogenology* **2002**, *57*, 149–179. [CrossRef] [PubMed]
20. Yániz, J.L.; Soler, C.; Santolaria, P. Computer Assisted Sperm Morphometry in Mammals: A Review. *Anim. Reprod. Sci.* **2015**, *156*, 1–12. [CrossRef]
21. Wysokińska, A.; Wójcik, E.; Chłopik, A. Evaluation of the Morphometry of Sperm from the Epididymides of Dogs Using Different Staining Methods. *Animals* **2021**, *11*, 227. [CrossRef] [PubMed]
22. Peña, F.J.; Saravia, F.; García-Herreros, M.; Núñezmartínez, I.; Tapia, J.A.; Johannisson, A.; Wallgren, M.; Rodríguez-Martínez, H. Identification of Sperm Morphometric Subpopulations in Two Different Portions of the Boar Ejaculate and Its Relation to Postthaw Quality. *J. Androl.* **2005**, *26*, 716–723. [CrossRef] [PubMed]
23. Barquero, V.; Roldan, E.R.S.; Soler, C.; Yániz, J.L.; Camacho, M.; Valverde, A. Predictive Capacity of Boar Sperm Morphometry and Morphometric Sub-Populations on Reproductive Success after Artificial Insemination. *Animals* **2021**, *11*, 920. [CrossRef] [PubMed]
24. Kondracki, S.; Bonaszewska, D.; Mielnicka, C. The Effect of Age on the Morphometric Sperm Traits of Domestic Pigs (Sus Scrofa Domestica). *Cell. Mol. Biol. Lett.* **2005**, *10*, 3–13. [PubMed]
25. Banaszewska, D.; Andraszek, K. Assessment of the Morphometry of Heads of Normal Sperm and Sperm with the Dag Defect in the Semen of Duroc Boars. *J. Vet. Res.* **2021**, *65*, 239–244. [CrossRef] [PubMed]
26. Szablicka, D.; Wysokińska, A.; Pawlak, A.; Roman, K. Morphometry of Boar Spermatozoa in Semen Stored at 17 °C—The Influence of the Staining Technique. *Animals* **2022**, *12*, 1888. [CrossRef] [PubMed]
27. Žaja, I.Ž.; Vince, S.; Milinković-Tur, S.; Milas, N.P.; Samardžija, M.; Valpotić, H.; Berta, V.; Vilić, M.; Rakić, K. Exogenous Melatonin Influences Distribution of French Alpine Buck Spermatozoa in Morphometrically Distinct Subpopulations during the Non-Breeding Season. *Anim. Reprod. Sci.* **2018**, *192*, 154–163. [CrossRef]
28. De Iuliis, G.N.; Newey, R.J.; King, B.V.; Aitken, R.J. Mobile Phone Radiation Induces Reactive Oxygen Species Production and DNA Damage in Human Spermatozoa In Vitro. *PLoS ONE* **2009**, *4*, e6446. [CrossRef]
29. Saygin, M.; Caliskan, S.; Karahan, N.; Koyu, A.; Gumral, N.; Uguz, A. Testicular Apoptosis and Histopathological Changes Induced by a 2.45 GHz Electromagnetic Field. *Toxicol. Ind. Health* **2011**, *27*, 455–463. [CrossRef]
30. Gye, M.C.; Park, C.J. Effect of Electromagnetic Field Exposure on the Reproductive System. *Clin. Exp. Reprod. Med.* **2012**, *39*, 1–9. [CrossRef]
31. Al-Bayyari, N. The Effect of Cell Phone Usage on Semen Quality and Fertility among Jordanian Males. *Middle East Fertil. Soc. J.* **2017**, *22*, 178–182. [CrossRef]
32. Kıvrak, E.G.; Yurt, K.K.; Kaplan, A.A.; Alkan, I.; Altun, G. Effects of Electromagnetic Fields Exposure on the Antioxidant Defense System. *J. Microsc. Ultrastruct.* **2017**, *5*, 167–176. [CrossRef]
33. Wysokińska, A.; Kondracki, S.; Banaszewska, D. Morphometrical Characteristics of Spermatozoa in Polish Landrace Boars with Regard to the Number of Spermatozoa in an Ejaculate. *Reprod. Biol.* **2009**, *9*, 271–282. [CrossRef] [PubMed]
34. Górski, K.; Kondracki, S.; Wysokińska, A. Ejaculate Traits and Sperm Morphology Depending on Ejaculate Volume in Duroc Boars. *J. Vet. Res.* **2017**, *61*, 121–125. [CrossRef] [PubMed]
35. Druart, X. Sperm Interaction with the Female Reproductive Tract. *Reprod. Domest. Anim.* **2012**, *47*, 348–352. [CrossRef] [PubMed]
36. Oni, O.; Amuda, D. Effects of radiofrequency radiation from wifi devices on human ejaculated semen. *Int. J. Res. Rev. Appl. Sci.* **2011**, *9*, 292–294.
37. Gorpinchenko, I.; Nikitin, O.; Banyra, O.; Shulyak, A. The Influence of Direct Mobile Phone Radiation on Sperm Quality. *Cent. Eur. J. Urol.* **2014**, *67*, 65–71. [CrossRef]
38. Ghanbari, M.; Mortazavi, S.B.; Khavanin, A.; Khazaei, M. The Effects of Cell Phone Waves (900 MHz-GSM Band) on Sperm Parameters and Total Antioxidant Capacity in Rats. *Int. J. Fertil. Steril.* **2013**, *7*, 21–28.
39. Kumar, S.; Kesari, K.K.; Behari, J. Influence of Microwave Exposure on Fertility of Male Rats. *Fertil. Steril.* **2011**, *95*, 1500–1502. [CrossRef]

40. Meena, R.; Kumari, K.; Kumar, J.; Rajamani, P.; Verma, H.N.; Kesari, K.K. Therapeutic Approaches of Melatonin in Microwave Radiations-Induced Oxidative Stress-Mediated Toxicity on Male Fertility Pattern of Wistar Rats. *Electromagn. Biol. Med.* **2014**, *33*, 81–91. [CrossRef]
41. Immler, S.; Pryke, S.R.; Birkhead, T.R.; Griffith, S.C. Pronounced Within-Individual Plasticity in Sperm Morphometry Across Social Environments. *Evolution* **2010**, *64*, 1634–1643. [CrossRef] [PubMed]
42. Shahani, S.K.; Revell, S.G.; Argo, C.G.; Murray, R.D. Mid-Piece Length of Spermatozoa in Different Cattle Breeds and Its Relationship to Fertility. *Pak. J. Biol. Sci. PJBS* **2010**, *13*, 802–808. [CrossRef] [PubMed]
43. Ramón, M.; Soler, A.J.; Ortiz, J.A.; García-Alvarez, O.; Maroto-Morales, A.; Roldan, E.R.S.; Garde, J.J. Sperm Population Structure and Male Fertility: An Intraspecific Study of Sperm Design and Velocity in Red Deer. *Biol. Reprod.* **2013**, *89*, 110. [CrossRef] [PubMed]
44. Malo, A.F.; Garde, J.J.; Soler, A.J.; García, A.J.; Gomendio, M.; Roldan, E.R.S. Male Fertility in Natural Populations of Red Deer Is Determined by Sperm Velocity and the Proportion of Normal Spermatozoa1. *Biol. Reprod.* **2005**, *72*, 822–829. [CrossRef] [PubMed]
45. Gomendio, M.; Roldan, E.R.S. Implications of Diversity in Sperm Size and Function for Sperm Competition and Fertility. *Int. J. Dev. Biol.* **2008**, *52*, 439–447. [CrossRef] [PubMed]
46. Firman, R.C.; Simmons, L.W. Sperm Midpiece Length Predicts Sperm Swimming Velocity in House Mice. *Biol. Lett.* **2010**, *6*, 513–516. [CrossRef] [PubMed]
47. Costa, J.; Braga, P.C.; Rebelo, I.; Oliveira, P.F.; Alves, M.G. Mitochondria Quality Control and Male Fertility. *Biology* **2023**, *12*, 827. [CrossRef] [PubMed]
48. Liu, Q.; Si, T.; Xu, X.; Liang, F.; Wang, L.; Pan, S. Electromagnetic Radiation at 900 MHz Induces Sperm Apoptosis through Bcl-2, Bax and Caspase-3 Signaling Pathways in Rats. *Reprod. Health* **2015**, *12*, 65. [CrossRef]
49. Houston, B.J.; Nixon, B.; King, B.V.; Iuliis, G.N.D.; Aitken, R.J. The Effects of Radiofrequency Electromagnetic Radiation on Sperm Function. *Reproduction* **2016**, *152*, R263–R276. [CrossRef]
50. Talarczyk-Desole, J.; Kotwicka, M.; Jendraszak, M.; Pawelczyk, L.; Murawski, M.; Jędrzejczak, P. Sperm Midpiece Apoptotic Markers: Impact on Fertilizing Potential in in Vitro Fertilization and Intracytoplasmic Sperm Injection. *Hum. Cell* **2016**, *29*, 67–75. [CrossRef]

Disclaimer/Publisher's Note: The statements, opinions and data contained in all publications are solely those of the individual author(s) and contributor(s) and not of MDPI and/or the editor(s). MDPI and/or the editor(s) disclaim responsibility for any injury to people or property resulting from any ideas, methods, instructions or products referred to in the content.

animals

MDPI

Article

Resveratrol Improves the Frozen-Thawed Ram Sperm Quality

Zhendong Zhu [1,†], Haolong Zhao [1,†], Haixiang Cui [1], Adedeji O. Adetunji [2] and Lingjiang Min [1,*]

1 College of Animal Science and Technology, Qingdao Agricultural University, Qingdao 266109, China; zzd2020@qau.edu.cn (Z.Z.); zhl2368447140@163.com (H.Z.); chx1044944467@163.com (H.C.)
2 Department of Agriculture, University of Arkansas at Pine Bluff, Pine Bluff, AR 71601, USA; adetunjiadedeji.aa@gmail.com
* Correspondence: mljlab405@163.com
† These authors contributed equally to this work.

Simple Summary: Cryopreservation generates a substantial quantity of reactive oxygen species (ROS) in semen, leading to a decline in sperm quality and fertilization capacity. In this study, the influence of resveratrol on thawed ram sperm quality and antioxidant capacity was investigated. The study demonstrated that the supplementation of 50 μM resveratrol significantly enhances the sperm quality and antioxidant capacity of the ram sperm post-thawing. Furthermore, the impact of resveratrol on sirtuin 1 (SIRT1) in thawed sperm was investigated. We found that the activation of adenosine 5′-monophosphate (AMP)-activated protein kinase (AMPK) by 50 μM resveratrol protected ram sperm from ROS-induced stress. However, at 100 μM, resveratrol appears to reduce sperm motility, mitochondrial membrane potential, acrosomes, and plasma membrane integrity. Consequently, the protective effect of resveratrol on sperm quality or potential cytotoxicity towards ram sperm is contingent on the concentration of resveratrol and filler components, as well as the ability to regulate antioxidant uptake in sperm.

Abstract: Cryopreservation generates a substantial quantity of ROS in semen, leading to a decline in sperm quality and fertilization capacity. The objective of this study was to investigate the effects of resveratrol and its optimal concentration on ram sperm quality after cryopreservation. Ram semen was diluted with a freezing medium containing different concentrations of resveratrol (0, 25, 50, 75, and 100 μM). After thawing, various sperm parameters such as total motility, progressive motility, acrosome integrity, plasma membrane integrity, mitochondrial membrane potential, glutathione (GSH) content, glutathione synthase (GPx) activity, superoxide dismutase (SOD) activity, catalase (CAT) activity, lipid peroxidation (LPO) content, malondialdehyde (MDA) content, ROS level, SIRT1 level, DNA oxidative damage, and AMPK phosphorylation level were assessed. In addition, post-thaw sperm apoptosis was evaluated. Comparatively, the addition of resveratrol up to 75 μM significantly improved the sperm motility and sperm parameters of cryopreserved ram sperm. Specifically, 50 μM resveratrol demonstrated a notable enhancement in acrosome and plasma membrane integrity, antioxidant capacity, mitochondrial membrane potential, adenosine triphosphate (ATP) content, SIRT1 level, and AMPK phosphorylation levels compared to the control group ($p < 0.05$). It also significantly ($p < 0.05$) reduced the oxidative damage to sperm DNA. However, detrimental effects of resveratrol were observed at a concentration of 100 μM resveratrol. In conclusion, the addition of 50 μM resveratrol to the cryopreservation solution is optimal for enhancing the quality of cryopreserved ram sperm.

Keywords: resveratrol; antioxidant; cryopreservation; ram; sperm

Citation: Zhu, Z.; Zhao, H.; Cui, H.; Adetunji, A.O.; Min, L. Resveratrol Improves the Frozen-Thawed Ram Sperm Quality. *Animals* **2023**, *13*, 3887. https://doi.org/10.3390/ani13243887

Academic Editor: Anna Wysokińska

Received: 19 November 2023
Revised: 14 December 2023
Accepted: 15 December 2023
Published: 18 December 2023

Copyright: © 2023 by the authors. Licensee MDPI, Basel, Switzerland. This article is an open access article distributed under the terms and conditions of the Creative Commons Attribution (CC BY) license (https://creativecommons.org/licenses/by/4.0/).

1. Introduction

The success of artificial insemination (AI) crucially depends on the quality of semen preserved in vitro. Cryopreservation technology is pivotal in overcoming temporal and spatial constraints, allowing for the extended preservation of sperm and thereby enhancing the utilization rate of high-quality sheep breeds [1]. This technology facilitates the full exploitation of the reproductive potential of sheep [2]. However, the morphology of sheep's cervix is a constraint to the use of AI in sheep [3]. Consequently, quality sperm is essential to ensuring successful colonization and migration of sperm through the cervix [4]. One significant challenge during cryopreservation is oxidative stress, primarily induced by elevated levels of ROS, which is the leading cause of sperm damage [5]. It was well known that oxidative stress resulted in a decline in sperm quality and fertilization capacity [6]. The sperm plasma membrane is rich in polyunsaturated fatty acids (PUFA), but low PUFA content in the cytoplasm compromises the antioxidant defense system, making sperm susceptible to lipid peroxidation and subsequent damage [7]. To counteract this, antioxidant compounds are added to cryoprotective solutions [8]. Previous studies have demonstrated that certain oxidants, such as trehalose [9], melatonin [10], and spermine [11], can protect sperm from oxidative stress by enhancing antioxidant capacity. In recent years, polyphenols like tannin extract [12] and tea polyphenol-t [13] have also been shown to exhibit antioxidant properties in sperm.

Among the polyphenols, resveratrol, derived from grape skins and seeds [14], stands out for its antioxidant activity. Resveratrol scavenges free radicals and inhibits lipid peroxidation, thereby safeguarding cellular functionality [15]. Moreover, resveratrol has been demonstrated to enhance the activities of SOD, CAT, GPx, and other antioxidant enzymes within cells [16]. Significant increases in the antioxidant capacity of sperm from humans [17], boars [18], buffaloes [19], and roosters [20] have been observed in cryoprotective fluids containing resveratrol. Research also indicates that resveratrol can protect sperm from oxidative stress by activating AMPK [21]. Furthermore, resveratrol has been shown to improve sperm motility, acrosome integrity, and mitochondrial activity in frozen-preserved giant pandas [22]. In sheep, Al-Mutary et al. [16] demonstrated that resveratrol enhances the quality and fertilization capacity of sheep semen stored at 5 °C. Despite these promising findings, the impact of resveratrol on the cryopreservation of ram semen remains understudied. Therefore, this study is aimed at investigating the effects of resveratrol supplementation in ram semen cryopreservation. It will contribute valuable insights into optimizing the cryopreservation process for ram semen, ultimately enhancing the efficiency of artificial insemination in sheep breeding programs.

2. Materials and Methods

2.1. Chemicals

All chemicals and reagents were purchased from Sigma-Aldrich (Shanghai, China) unless specified otherwise.

2.2. Ethical Approval

All animals and experimental procedures were approved by the Qingdao Agriculture University Institutional Animal Care and Use Committee (QAU1121010, 1 October 2019–30 December 2023).

2.3. Animals

Eight healthy and fertile rams (small-tailed Han sheep) aged approximately 2 years, utilized for routine artificial insemination in a commercial herd, were used in this study. The rams were kept in an enclosed facility with an evaporative cooling system and separate enclosures. Each ram was fed a commercial diet, and water was freely supplied through tanks.

2.4. Collection of Semen

Semen was collected from eight healthy and fertile rams (small-tailed Han sheep) (aged approximately 2 years) two times in one week using an artificial vagina in December of 2022 at the Hongde livestock farm (Shouguang, China). Semen were totally collected 8 times from each ram in this study. A total of 64 ejaculates were obtained and transported to the laboratory in insulated buckets at 37 °C. Sperm motility was analyzed with a computer-assisted sperm analysis (CASA), and only samples with over 80% motility were used in this study. Similarly, a hemocytometer was used to estimate sperm concentration, and only semen with a concentration of more than 2×10^9 sperm/mL was used. The ejaculated semen from the rams was pooled to minimize individual differences, split into 5 parts, and cryopreserved in freezing medium supplemented with different concentrations of resveratrol (0, 25, 50, 75, and 100 μM). The resveratrol was dissolved in dimethyl sulfoxide to create a 200 mM resveratrol solution.

2.5. Semen Freezing and Thawing

The semen samples were diluted in freezing extenders containing 250 mM Tris, 83 mM citric acid, 69 mM fructose, along with 5% (*v*/*v*) glycerol, and 20% (*v*/*v*) egg with varied concentrations of resveratrol (0, 25, 50, 75, and 100 μM) to achieve a sperm concentration of 1×10^8 sperm/mL. Subsequently, the samples were cooled to 4 °C for 3 h and loaded into 0.25-mL straws. Then, the straws were placed horizontally at a height of 5 cm above the surface of liquid nitrogen for 10 min, and then plunged in liquid nitrogen. Thereafter, the straws were stored in a cryogenic storage tank. A week later, the frozen straws were thawed in 37 °C water for 12 s, and after that, the sperm quality was evaluated.

2.6. Evaluation of Sperm Motility

Computer-assisted sperm analysis (CASA) (SCA 20-06-01; Goldcyto, Barcelona, Spain) was performed. For detection, images were acquired using a digital camera (acA780-75gc, Basler, Germany) and a negative phase contrast microscope at 100× magnification, set to a standard parameter of 25 frames/s. The post-thaw sperm was diluted with Tris-citrate-fructose extender before motility evaluation. According to our previous study [23], after preheating the analyzer, semen sample aliquots of 5 μL were added to the Makler chamber. Sperm motility was then assessed in three randomly selected areas using CASA, and more than 500 sperm were evaluated. The percentage of sperm moving at a path speed of 12 μm/s was defined as total sperm motility. Forward movement denotes the percentage of sperm moving in a straight line at a path velocity of 45 μm/s for more than 80% of the time.

2.7. Evaluation of Sperm Acrosome Integrity and Plasma Membrane Integrity

According to our previous study [24], sperm acrosome integrity and membrane integrity were detected by fluorescein isothiocyanate-peanut lectin (L-7381, Sigma-Aldrich, Shanghai, China) and the live/dead sperm motility assay kit (L-7011, Thermo Fisher, Shanghai, China), respectively. Briefly, to evaluate sperm acrosome integrity, sperm samples were fixed with methanol solution and incubated with 100 μg/mL fluorescein isothiocyanate-peanut lectin solution and 2.4 mM PI solution for 30 min in the dark before being observed under the microscope. Stained sperm samples were observed and photographed with an epifluorescence microscope (ZEISS DM200LED, Oberkochen, Germany) with 488 nm excitation for FITC-PNA green fluorescence and 535 nm excitation for PI red fluorescence.

For membrane integrity detection, sperm samples were incubated with 100 nM SYBR-14 working solution and 2.4 mM PI solution for 10 min in the dark. The stained sperm were observed and photographed using an epifluorescence microscope (ZEISS DM200LED, Oberkochen, Germany) with 488 nm excitation for SYBR-14 green fluorescence and 535 nm excitation for PI red fluorescence.

2.8. Evaluation of Mitochondrial Activity

The JC-1 mitochondrial membrane potential assay kit (C2003S; Beyotime Institute of Biotechnology; Shanghai, China) was used to analyze sperm mitochondrial activity (ΔΨm) [25]. Additionally, 50 μL of semen was centrifuged at 800× *g* for 5 min, and the supernatant was discarded. The collected sample was washed twice with TCG. Thereafter, 200 μL of JC-1 working solution was added and incubated at 37 °C for 20 min. Sperm was collected by centrifugation at 800× *g* at 4 °C for 5 min and washed twice with JC-1 staining buffer (1×). An appropriate amount of JC-1 staining buffer was used to resuspend the sperm sample. Thereafter, it was placed on ice prior to detection using a flow cytometer. The excitation wavelength was set at 485 nm and the emission wavelength at 590 nm, and a total of 2×10^4 sperm were detected. The experimental data were analyzed using FlowJo-V10, and the whole experimental process was carried out in the dark. All experiments were performed in triplicate (n = 3).

2.9. Evaluation of ATP Content

The ATP content in sperm was measured using an ATP content assay kit (A095-1-1; Nanjing Jiancheng Bioengineering; Wuhan, China) [23]. Additionally, 30 μL of semen was centrifuged at 800× g for 5 min, and the supernatant was discarded while the bottom layer of precipitated cells was collected. Thereafter, 300 μL of cold double-steaming water was added to the cells and placed in ice water to break the homogenate. Then, the cell suspension was heated in boiling water for 10 min, followed by extraction and mixing for 1 min according to the manufacturer's instructions. Finally, the absorbance was measured at 636 nm with a microplate reader (TECAN, Infinite M Nano, Männedorf, Switzerland). All experiments were performed in triplicate (n = 3).

2.10. Evaluation of NADH/NAD$^+$

According to the NAD(H) assay kit (A114-1-1, Nanjing Jiancheng Bioengineering Institute, Wuhan, China) to analyze sperm NADH/NAD$^+$. After thawing 10μL of semen, alkaline extract and homogenate were added and boiled for 5 min. Then, the mixture was cooled in ice water and centrifuged at 10,000× g for 10 min at 4 °C, and the supernatant was collected. Equal volumes of acidic extraction solution were added to the supernatant for neutralization and centrifuged at 10,000× g for 10 min at 4 °C. Thereafter, the supernatant was collected and placed on ice prior to measurement. Finally, the absorbance was measured at 570 nm with a microplate reader (TECAN, Infinite M Nano, Männedorf, Switzerland). All experiments were performed in triplicate (n = 3).

2.11. Evaluation of Sperm ROS Content

The ROS content in sperm was measured using an ROS level assay kit (M36008, Thermo Fisher Scientific, Shanghai, China) [26]. Sperm samples were centrifuged and resuspended with 200 μL of working solution. The cells were incubated for 15 min in the dark at 37 °C. Thereafter, the cells were centrifuged and washed three times with 1× PBS. Stained sperm were resuspended in 1× PBS and evaluated by flow cytometry (FACS Aria III, BD Biosciences, San Jose, CA, USA) using a filter with a bandwidth of 574/26 nm, and the measurements denote the mean fluorescence intensity (MFI). All experiments were performed in triplicate (n = 3).

2.12. Evaluation of Sperm MDA Content

According to our previous study [23], MDA content was measured with a commercial MDA assay kit (S0131S, Beyotime Institute of Biotechnology, Shanghai, China). Briefly, sperm stored at 4 °C were lysed by sonication (20 kHz, 300 W, operating at 50%, 3 min for 10 s on and 5 s off) on ice. The sample was mixed with the preprepared reaction buffer reagent and boiled for 40 min, then centrifuged to collect the supernatant after cooling. The absorbance was measured at 532 nm with a microplate reader (TECAN, Infifinite M Nano, Männedorf, Switzerland). All experiments were performed in triplicate (n = 3).

2.13. Evaluation of Sperm LPO Content

According to our previous study [27], LPO content was measured with a commercial LPO assay kit (A160-1, Nanjing Jiancheng Bioengineering Institute, Wuhan, China). In brief, normal saline was added to the sperm samples, followed by homogenization in ice water, centrifugation, and mixing with prepared buffer according to the manufacturer's instructions. The absorbance was measured at 586 nm with a microplate reader (TECAN,

Infinite M Nano, Männedorf, Switzerland). All experiments were performed in triplicate ($n = 3$).

2.14. Evaluation of GSH Content, GPx, SOD, and CAT Activity

Sperm GSH content, GPx, SOD, and CAT activity were analyzed by the GSH (A006-2-1, Nanjing Jiancheng Bioengineering, China), GPx (A005-1-2, Nanjing Jiancheng Bioengineering, China), SOD (A001-3-2, Nanjing Jiancheng Bioengineering, China), and CAT (A007-1-1, Nanjing Jiancheng Bioengineering, China) assay kit [23]. The sperm samples were centrifuged at 800× g for 10 min, and the supernatant was discarded. Thereafter, the collected sperm was washed twice in PBS. Furthermore, precooled RIPA lysate was added to the sperm collected and ground using an electric grinder in ice water, followed by centrifugation at 12,000× g. Then, the supernatant was taken for analysis of GSH content, GPx, SOD, and CAT activity according to the manufacturer's instructions. All experiments were performed in triplicate ($n = 3$).

2.15. Evaluation of Sperm Oxidative DNA Damage

Precise quantification of 8-hydroxyguanosine (8-OHdG), a biomarker for oxidative DNA damage, was conducted in sperm samples. The sperm was washed with 1×PBS twice, suspended with 500 μL PBS, and analyzed by flow cytometry after 8-OHdG staining. All experiments were performed in triplicate ($n = 3$).

2.16. Western Blotting

Total sperm protein was extracted using sodium dodecyl sulfate (SDS) sample buffer. Moreover, total proteins (20 μg) from each sample were separated on a 10% SDS-PAGE gel (E303-01, Vazyme, Nanjing, China), and the separated proteins were transferred to a polyvinylidene difluoride (PVDF) membrane. Nonspecific binding of PVDF membrane was blocked by TBST containing 5% BSA. Then, 1% BSA (dissolved in TBST) was used to dilute primary antibodies such as anti-SIRT1(13161-1-AP, Proteintech, Wuhan, China) (1:800), anti-AMPK (bs-1115R, Bioss, Beijing, China) (1:800), anti-p-AMPK (AP0432, ABclonal, Wuhan, China) (1:800), anti-p53 (A5761, ABclonal, Wuhan, China) (1:800), caspase 3 (A2156, ABclonal, Wuhan, China) (1:1000), caspase 9 (A0281, ABclonal, Wuhan, China) (1:1000), and incubated for a total of 12 h at 4 °C. Then, the PVDF membranes were placed in a TBST solution for washing. Thereafter, the membranes were incubated with a secondary antibody (AS014, 1:1000, ABclonal, Wuhan, China) for 1 h. ECL plus (ED0016-B, Sparkjade, Jinan, China) was added to the membrane for detection prior to developing the image with a gel imaging analyzer (Alpha, Fluor Chem Q, Shanghai, China).

2.17. Statistical Analysis

Data from all replicates were compared using one-way analysis of variance followed by Tukey's post hoc test (Stat view; Abacus Concepts, Inc., Berkeley, CA, USA). All the values in this study are presented as the mean ± standard error of the mean (SEM). Moreover, treatments were considered to be statistically different from one another at $p < 0.05$.

3. Result

3.1. Addition of Resveratrol Improved Sperm Motility Parameters

As shown in Table 1, the motility patterns of sperm analyzed through the movement trajectories generated by CASA showed that the addition of 50 and 75 μM resveratrol to the extender significantly increased ($p < 0.05$) sperm total motility. In addition, the values of progressive motility, curvilinear velocity (VCL), straight-line velocity (VSL), average path velocity (VAP), and wobble (WOB) in the 50 μM resveratrol treatment were higher than those in other treatments. However, values for the progressive motility, VCL, and VSL parameters for other treatments were similar to the control. Moreover, there were no differences ($p > 0.05$) among treatments for the lateral head (ALH), straightness (STR), and linearity (LIN) parameters.

Table 1. Effects of different concentrations of resveratrol on post-thaw sperm motility parameters.

Sperm Parameters \ Concentration	0 μM	25 μM	50 μM	75 μM	100 μM
TM (%)	37.6 ± 1.1 [b]	40.8 ± 1.8 [b]	69.2 ± 4.2 [a]	59.6 ± 5.0 [a]	33.2 ± 2.4 [b]
PM (%)	23.6 ± 0.6 [b]	24.0 ± 3.5 [b]	48.4 ± 7.0 [a]	30.4 ± 3.4 [b]	19.1 ± 0.7 [c]
VCL (μm/s)	36.7 ± 6.8 [b]	44.1 ± 13.0 [b]	61.1 ± 13.1 [a]	47.1 ± 9.1 [b]	43.0 ± 5.4 [c]
VSL (μm/s)	19.3 ± 2.9 [b]	24.4 ± 7.3 [b]	34.1 ± 3.6 [a]	22.3 ± 2.6 [b]	22.3 ± 5.2 [b]
VAP (μm/s)	24.2 ± 3.5 [d]	30.8 ± 9.1 [b]	35.8 ± 8.8 [a]	25.2 ± 5.2 [c]	28.4 ± 5.6 [c]
BCF (HZ)	7.8 ± 1.7 [a]	6.4 ± 1.4 [b]	6.2 ± 1.2 [b]	6.1 ± 1.1 [b]	5.4 ± 0.8 [c]
ALH (μm)	3.0 ± 0.2	3.0 ± 0.4	3.4 ± 0.7	3.8 ± 0.6	3.7 ± 0.2
STR (%)	79.6 ± 0.7	79.1 ± 1.0	77.5 ± 1.3	75.4 ± 1.2	76.6 ± 3.9
LIN (%)	53.7 ± 2.2	56.4 ± 3.9	54.3 ± 3.9	51.1 ± 4.5	51.9 ± 5.3
WOB (%)	67.1 ± 3.1 [b]	71.1 ± 2.0 [a]	69.8 ± 3.8 [a]	67.6 ± 5.5 [b]	67.2 ± 4.1 [b]

Values are expressed as mean ± standard error. Different letters within the same row indicate significant differences ($p < 0.05$). VCL, curvilinear velocity; VSL, straight-line velocity; VAP, average path velocity; BCF, beat-cross frequency; ALH, lateral head; STR, straightness (VSL/VAP); LIN, linearity (VSL/VCL); WOB, wobble (VAP/VCL).

3.2. Addition of Resveratrol Improved the Sperm Acrosome Integrity and Plasma Membrane Integrity

The addition of resveratrol to the extender significantly improved the integrity of the sperm acrosome ($p < 0.05$) after cryopreservation (Figure 1A, Supplementary Figure S1A). Among them, the 50- and 75-μM resveratrol significantly improved sperm acrosome integrity ($p < 0.05$). The addition of resveratrol to the extender up to 75 μM significantly improved ($p < 0.05$) sperm plasma membrane integrity after cryopreservation, with the 75 μM resveratrol treatment having the highest value (Figure 1B, Supplementary Figure S1B).

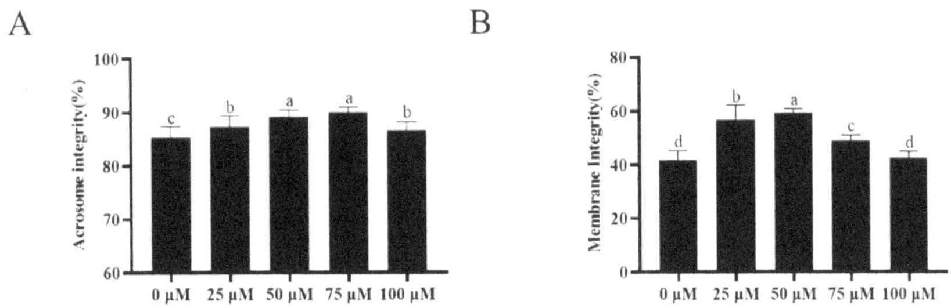

Figure 1. Effect of the addition of different concentrations of resveratrol to extender on acrosome integrity (A) and plasma membrane integrity (B) of ram sperm after cryopreservation. Values are presented as mean ± standard error of the mean (SEM). Columns with different lowercase letters were significantly different ($p < 0.05$), $n = 5$.

3.3. Addition of Resveratrol Improved Sperm Mitochondrial Activity

As shown in Figure 2 and Supplementary Figure S2, the addition of 25, 50, and 75 μM resveratrol to the extender significantly increased ($p < 0.05$) the sperm mitochondrial activity after cryopreservation, and the 75 μM resveratrol treatment presented the highest increase (Figure 2). However, 100 μM resveratrol treatment significantly reduced mitochondrial activity compared to the control group (Figure 2).

3.4. Addition of Resveratrol Improved Sperm ATP Content

As shown in Figure 3, it was observed that the addition of resveratrol to the extender significantly improved ($p < 0.05$) the sperm ATP content after cryopreservation. Moreover, the 50 μM resveratrol treatment showed the highest value for ATP content compared to the control (Figure 3).

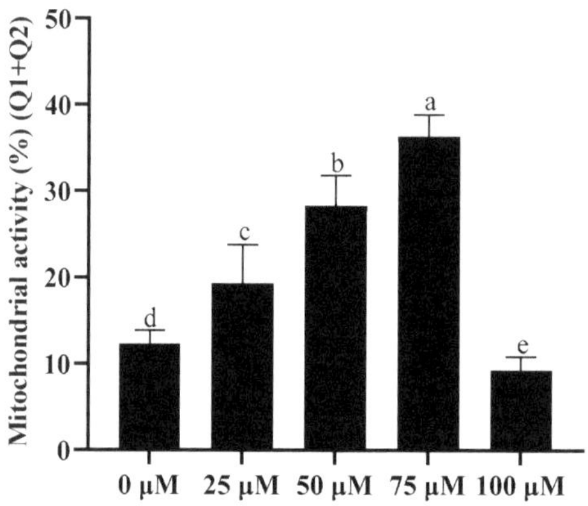

Figure 2. Effect of different concentrations of resveratrol on mitochondrial activity after cryopreservation. Values are presented as mean ± standard error of the mean (SEM). Columns with different lowercase letters are significantly different ($p < 0.05$), $n = 3$.

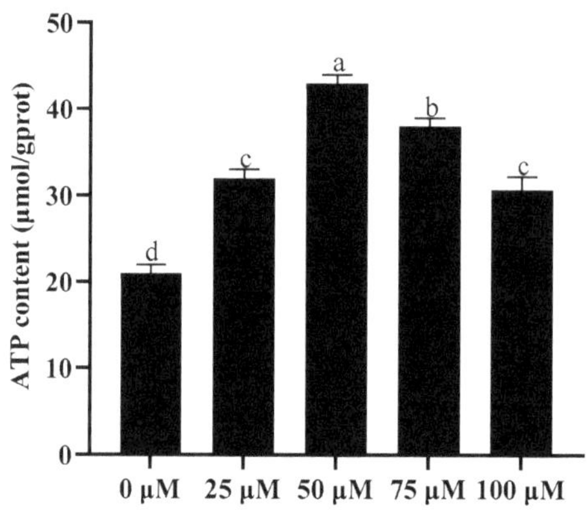

Figure 3. Effect of different concentrations of resveratrol on ATP content after cryopreservation. Values are presented as mean ± standard error of the mean (SEM). Columns with different lowercase letters are significantly different ($p < 0.05$), $n = 3$.

3.5. Addition of Resveratrol Improved Sperm NAD$^+$ Content and Reduced Sperm NADH/NAD$^+$

As shown in Figure 4A, it was observed that the addition of resveratrol to the extender significantly improved ($p < 0.05$) the sperm NAD+ content after cryopreservation. Moreover, the 75 μM resveratrol treatment showed the highest value of NAD+ content compared to the control (Figure 4A). Additionally, the addition of resveratrol to the extender significantly reduced ($p < 0.05$) the sperm NADH/NAD+ after cryopreservation compared to the control (Figure 4B).

3.6. Addition of Resveratrol Reduced Sperm LPO, ROS Level, and MDA Content

As shown in Figure 5A and Supplementary Figure S3, ROS levels were significantly decreased after the addition of resveratrol up to 75 μM compared with the control group ($p < 0.05$). Notably, the 50 μM resveratrol treatment showed the most significant decrease in ROS levels compared to other treatments. The addition of resveratrol up to 75 μM significantly decreased ($p < 0.05$) the contents of LPO (Figure 5B) and MDA (Figure 5C), with the most significant decrease observed in the 50 μM resveratrol treatment. However, there was no significant difference in LPO content between the 100 μM resveratrol treatment group and the control group ($p > 0.05$). Furthermore, the content of MDA for the 100 μM resveratrol treatment was significantly higher than the control group.

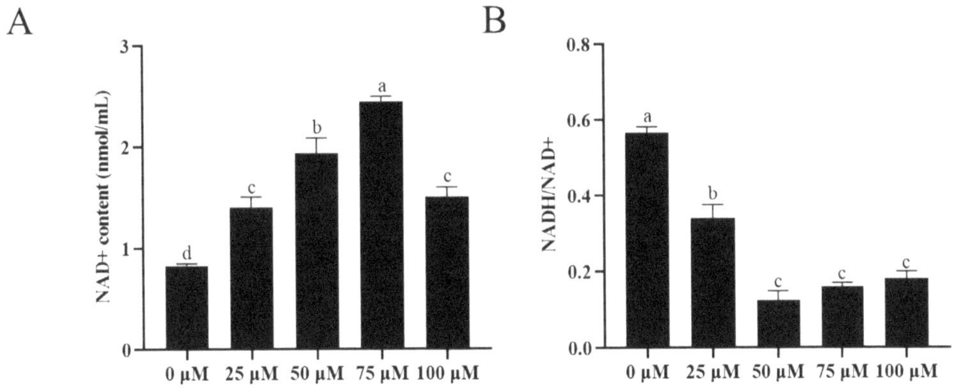

Figure 4. Effect of addition of different concentrations of resveratrol to extender on NAD+ content (**A**) and NADH/NAD+ (**B**) of ram sperm after cryopreservation. Values are presented as mean ± standard error of the mean (SEM). Columns with different lowercase letters were significantly different ($p < 0.05$), $n = 5$.

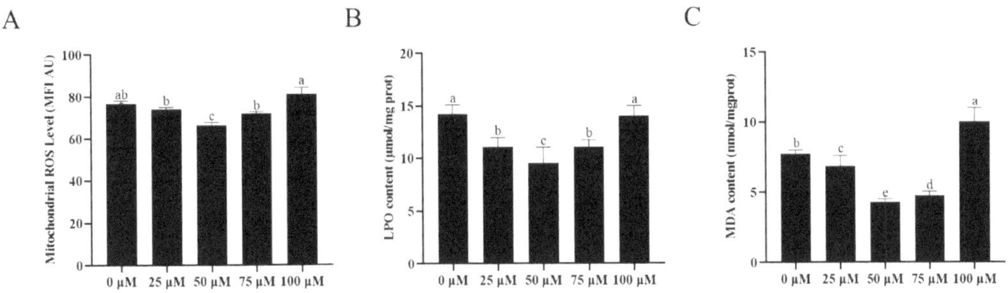

Figure 5. Effect of different concentrations of resveratrol on sperm ROS Level (**A**), LPO content (**B**), and MDA content (**C**) after cryopreservation. Values are presented as mean ± standard error of the mean (SEM). Columns with different lowercase letters were significantly different ($p < 0.05$), $n = 3$.

3.7. Addition of Resveratrol Improved the Sperm Antioxidative Ability

To investigate the effect of resveratrol on the antioxidant capacity of sperm cryopreservation, the SOD activity, CAT activity, GPx activity, and GSH content were measured. As shown in Figure 6A, the highest SOD activity was observed in the 50 and 75 μM resveratrol treatments ($p < 0.05$). However, the SOD activity for the 25 μM and 100 μM resveratrol treatments and the control were similar. In addition, the addition of resveratrol significantly increased ($p < 0.05$) GPx activity (Figure 6B). The CAT activity in the 50 and 75 μM resveratrol treatment groups significantly increased compared to the other treatments (Figure 6C). Similarly, an increase in the GSH content was also noted with the addition of 50 and 75 μM resveratrol ($p < 0.05$) (Figure 6D).

3.8. Addition of Resveratrol Reduced the Oxidative DNA Damage

Analysis of oxidative DNA damage in post-thawed sperm by 8-OHdG staining (Figure 7A–F) showed that the levels of 8-OHdG in post-thawed sperm were significantly different from those in the control ($p < 0.05$) after the addition of 25, 50, and 75 μM resveratrol. The addition of 50 μM resveratrol had a significant difference with other treatment groups ($p < 0.05$).

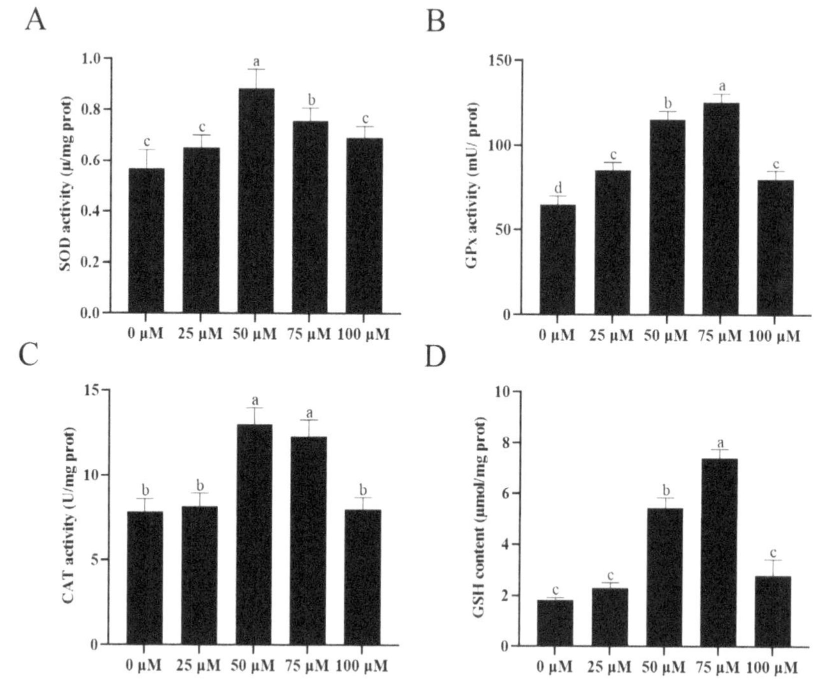

Figure 6. Effect of different concentrations of resveratrol on sperm SOD activity (**A**), GPx activity (**B**), CAT activity (**C**), and GSH content (**D**) after cryopreservation. Values are presented as mean ± standard error of the mean (SEM). Columns with different lowercase letters were significantly different ($p < 0.05$), $n = 3$.

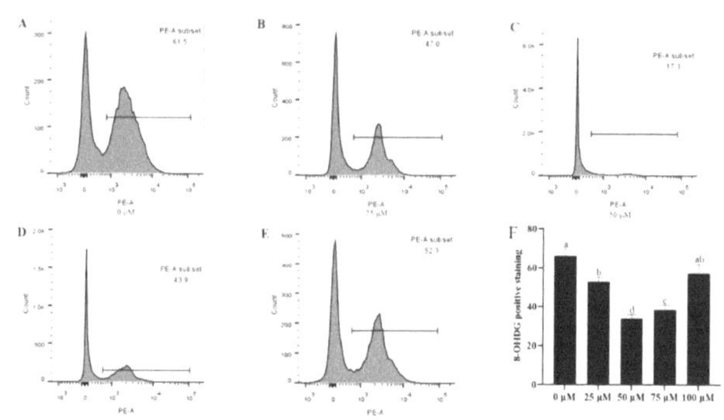

Figure 7. Effect of different concentrations of resveratrol on sperm DNA integrity (**A–F**) after cryopreservation. Values are presented as mean ± standard error of the mean (SEM). Columns with different lowercase letters were significantly different ($p < 0.05$), $n = 3$.

3.9. Addition of Resveratrol Promotes AMPK Phosphorylation against ROS Damage by Activating SIRT1

To investigate the mechanism of how resveratrol improves sperm quality, SIRT1 protein expression and AMPK phosphorylation were detected in thawed sperm (Figure 8A–D, Supplementary Figure S4). Compared with the control group, the SIRT1 expression increased after the addition of 25, 50, and 75 μM resveratrol ($p < 0.05$); however, SIRT1 expression for the 100 μM resveratrol treatment was not different from the control ($p > 0.05$) (Figure 8B). Furthermore, while there was no significant change in total AMPK level with the addition of resveratrol (Figure 8C), the addition of 25, 50, and 75 μM resveratrol significantly increased AMPK phosphorylation after thawing sperm ($p < 0.05$). The highest degree of AMPK phosphorylation was observed in the 50 μM resveratrol treatment group ($p < 0.05$) (Figure 8D).

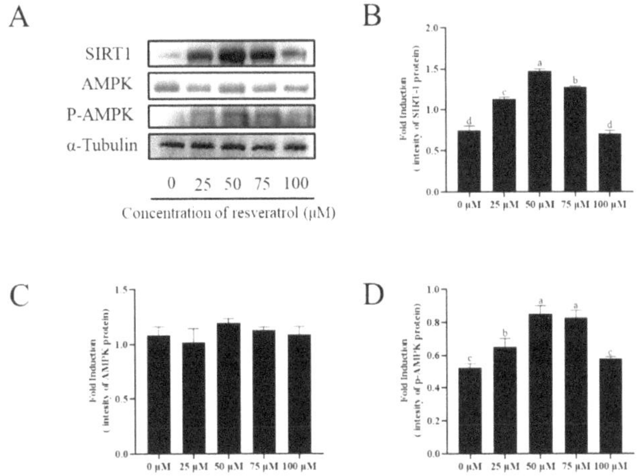

Figure 8. Effect of different concentrations of resveratrol on SIRT1, AMPK, and p-AMPK after cryopreservation. Detection of the expression of SIRT1, AMPK, and p-AMPK by Western blot (**A**–**D**). Columns with different lowercase letters were significantly different ($p < 0.05$), $n = 3$.

3.10. Addition of Resveratrol Attenuated the Sperm Apoptosis

Western blotting analysis shows that the expression of apoptosis proteins (caspase-3, caspase-9, and p53) significantly decreased in the resveratrol treatment groups (Figure 9A–D, Supplementary Figure S5). Among the treatments, the expression of apoptotic proteins was the lowest in the 50 μM resveratrol group for caspase-3 and p53, and the expression of apoptotic proteins was the lowest in the 75 μM resveratrol group for caspase-9.

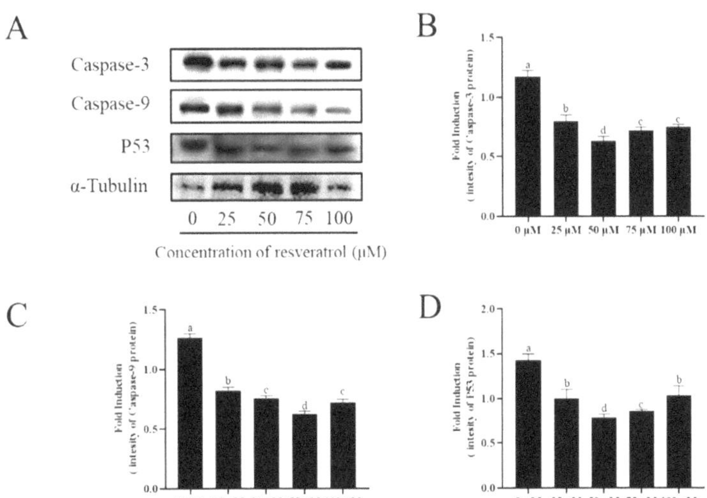

Figure 9. Effect of different concentrations of resveratrol on sperm apoptosis after cryopreservation. Detection of the expression of caspase3, caspase9, and P53 by Western blot (**A**–**D**). Values are presented as mean ± standard error of the mean (SEM). Columns with different lowercase letters were significantly different ($p < 0.05$), $n = 3$.

4. Discussion

Artificial insemination in sheep is limited by the morphology of the sheep cervix [28] The quality of thawed semen is instrumental in determining whether sperm can successfully colonize and migrate within the uterus [29]. The objective of this study was to examine the influence of resveratrol and its optimal concentration on the cryopreservation of ram sperm, focusing on its protective effects against freeze-thaw damage. Results show that the addition of 50 μM resveratrol in a frozen medium significantly improved sperm motility parameters, membrane integrity, acrosome integrity, mitochondrial activity, GPx, CAT, SOD activity, GSH content, SIRT1 expression level, and AMPK phosphorylation level after thawing. Similarly, the levels of LPO, MDA, DNA damage, and apoptosis in sperm decreased.

Sperm motility is important for sperm to swim from the site of ejaculation toward the oviduct for fertilization to take place [30]. In this study, post-thaw sperm motility parameters such as the total motility, progressive motility, VCL, VSL, VAP, and WOB of the 50 μM resveratrol treatment were significantly higher than those of the control. These findings are consistent with Zhu et al. (2019), who reported that the addition of resveratrol improved post-thaw boar sperm motility parameters [31]. In addition, supplementation with 50 μM resveratrol also increased post-thaw ram sperm mitochondrial activity. This result aligns with the findings of Kaeoket et al. (2023), who demonstrated that resveratrol improved sperm mitochondrial bioenergetics [18]. It is well known that the mitochondria are regarded as powerhouses due to their central role in ATP production [32]. It is also notable that sperm motility is dependent on ATP generation [33]. In this study, we found that the post-thaw ram sperm ATP level in the 50-μM resveratrol treatment was higher than that of the control. Thus, we can suggest that the addition of 50 μM resveratrol improved the post-thaw ram sperm motility parameters by maintaining mitochondrial membrane potentials, which are required for ATP generation.

The overproduction of ROS as a result of sperm cryopreservation procedures is a well-documented phenomenon, which significantly increases the risk of oxidative damage and potentially impairs sperm functionality [34]. ROS attack can lead to lipid peroxidation and DNA oxidative damage in sperm, which leads to changes in the structure and function of sperm and ultimately leads to the decline of sperm motility and fertilization ability [35]. Therefore, some studies have explored the addition of antioxidants to diluted semen to reduce ROS accumulation in sperm during cryopreservation to improve sperm motility and fertilization ability [8]. In the present study, the addition of 50 μM resveratrol to the extender protected sperm from oxidative stress by reducing ROS levels and lipid peroxidation. Similar to the present study, Ahmed et al. (2020) showed that the addition of 50–100 μM resveratrol to the extender reduced LPO in sperm [36]. Oxidation [37], heat stress [38], and osmotic stress [39] have been reported to damage the sperm plasma membrane and acrosome during cryopreservation. Nevertheless, the addition of resveratrol to sperm freezing fillers significantly improved the integrity of sperm acrosomes and plasma membranes, which is consistent with the findings of previous studies in humans [17], boar [18], buffalo [19], and rooster [20]. Furthermore, contrary to our findings, Garcez et al. (2010) [40] and Silva et al. (2012) [41] found that resveratrol supplementation did not enhance sperm motility and mitochondrial activity of human and ram sperm after thawing. It has also been documented that 50 μM resveratrol mitigated movement loss but failed to prevent the oxidative damage induced by the cryopreservation of buck sperm [42]. The difference in results may suggest that the protective effect of resveratrol on thawed semen may depend on resveratrol concentration, filler composition, animal species, storage procedures, and the entity under stress conditions.

The present investigation further elucidated that the incorporation of resveratrol into the extender diminished the concentrations of LPO and MDA, concurrently augmenting the GPx, CAT, SOD activity, and GSH content. These results are consistent with studies from other species that suggest that adding resveratrol protects sperm from oxidative damage by boosting their antioxidant capacity [20,31,43]. It is well known that the cellular antioxidant system (scavenging enzymes) is located in the cytoplasm, which is found in sperm in a minimal amount [44]. In addition, cryopreservation decreases the activity of sperm antioxidant enzymes, indicating that sperm had decreased antioxidant capacity and were more susceptible to oxidative damage. Interestingly, our study showed that the addition of resveratrol increased the activity of GPx, CAT, and SOD. These findings are consistent with Zhang et al. (2022) [22], who found that 100 μM of resveratrol significantly improved the activity of GPx and SOD in giant panda sperm.

The metabolism and function of sperm require energy to be maintained during sperm cryopreservation [45]. Resveratrol is an indirect SIRT1 activator [46,47]. The findings of previous research have demonstrated that SIRT1 reduction leads to mitochondrial dysfunction by increasing ROS, LPO, and DNA damage in sperm, resulting in decreased sperm fertilization ability [48]. Similarly, studies by Cuerda et al. have shown that insufficient SIRT1 function in men leads to overactivation of sperm and decreased fertilization ability [49]. Thus, SIRT1 expression was significantly positively correlated with sperm concentration, total motility, and normal sperm morphology. The deficiency of SIRT1 increases the effect of reduced antioxidant defense [50]. In addition, SIRT1 protects sperm from hydrogen peroxide apoptosis through ubiquitination and subsequent degradation of the transcription factor Foxo3a [51]. SIRT1 also inhibits microglia-derived factors through a p53-caspase-3-dependent mechanism, thereby eliminating caspase-mediated apoptosis [52]. Additionally, SIRT1 increases Bcl-2 expression and decreases BAX expression, thereby regulating mitochondrial membrane permeability, mitochondrial function, and cytochrome c release [53]. SIRT1 inhibition caused oocytes to fail to upregulate SOD2 and offset the ROS increase in the presence of increased ROS [54]. Similarly, this study found that adding resveratrol to the freezing additive activated the expression of SIRT1, increased the activity of GPx, CAT, and SOD, reduced the damage of ROS, LPO, and DNA in sperm, and inhibited the expression of P53, caspase-3, and caspase-9. In goat sperm, Price et al. (2012) found that SIRT1 plays a crucial role in activating AMPK and improving mitochondrial function at moderate doses of resveratrol, which is consistent with this study [55]. AMPK is a key kinase that regulates cellular REDOX status by altering metabolic pathways under stress conditions [56]. Additionally, AMPK is documented as a key regulator of sperm physiological function, particularly sperm motility, plasma membrane integrity, and mitochondrial activity [57]. Furthermore, Cantó et al. (2009) indicated that AMPK regulates energy expenditure by modulating NAD^+ metabolism and SIRT1 activity [58]. NAD^+ is a metabolic product associated with improved mitochondrial function under somatic stress [59]. Moreover, NAD^+ levels decline with mitochondrial dysfunction or mitochondrial disease [60]. Activation of AMPK compensates for ATP loss by stimulating catabolic mechanisms and inhibiting synthesis mechanisms [61]. This study also found that adding resveratrol increased ATP content and NAD^+ levels by activating AMPK. Recently, numerous studies have demonstrated the presence of AMPK in boar [62], ram [45], rooster [63], and mouse [64] sperm. In addition, we showed in a previous study that increased phosphorylation of AMPK in boars improves semen quality [31]. Similarly, in the present study, it was also found that adding resveratrol to the freezing additive improved sperm motility by increasing the phosphorylation level of AMPK. A recent study showed that in boars, the phosphorylated AMPK form is mainly located in the acrosome and equatorial subsegments of the head, as well as in the middle of sperm [65]. However, in roosters, AMPK is present in the acrosome, the middle, and the entire flagella [63]. The distribution of AMPK in different parts of the sperm suggests that, depending on the species, it may promote motility and acrosomal responses. As a result of this variability, the distinct roles of SIRT1 and AMPK in the reproductive activity of different species should be investigated.

5. Conclusions

This study showed that 50 μM resveratrol can effectively alleviate the decrease in sperm motility, sperm acrosome integrity, and membrane integrity damage that occurs during cryopreservation of ram semen while maintaining high mitochondrial activity. Additionally, the addition of resveratrol activated AMPK phosphorylation and SIRT1 expression, which is known to decrease ROS production. It also enhanced sperm antioxidant defense systems (e.g., GSH content and activities of GPx, SOD, and CAT) and reduced cell apoptosis and DNA damage.

Supplementary Materials: The following supporting information can be downloaded at: https://www.mdpi.com/article/10.3390/ani13243887/s1, Figure S1: Photographs of sheep sperm stained with fluorescein SYBR-14/PI (A) and LIVE/DEAD sperm viability kit (B), respectively. In Figure S1A, red arrows indicate sperm with intact acrosome, blue arrows indicate sperm with damaged acrosome. In Figure S1B, the red arrow indicates the sperm with intact plasma membrane, blue arrows indicate sperm with damaged plasma membrane. Figure S2: Different concentrations of resveratrol on the sheep sperm mitochondrial activity after cryopreservation. Figure S3: Effect of different concentrations of resveratrol on sperm ROS after cryopreservation. Values are presented as mean ± standard error of the mean (SEM). Columns with different lowercase letters were significantly different ($p < 0.05$), $n = 3$. Figure S4: Different concentrations of resveratrol on the expression of proteins (SIRT1, P-AMPK, AMPK, α-Tubulin) in sheep sperm after cryopreservation. Figure S5: Different concentrations of resveratrol on the expression of proteins (Caspase3, Caspase9 and P53) in sheep sperm after cryopreservation.

Author Contributions: Conceptualization, Z.Z. and H.Z.; Methodology, Z.Z. and H.Z.; Software, H.C.; Validation, Z.Z., H.C. and H.Z; Formal Analysis, H.C.; Investigation, H.C.; Resources, Z.Z. and L.M.; Data Curation, H.Z.; Writing-Original Draft Preparation, Z.Z., H.C. and H.Z.; Writing-Review & Editing, A.O.A.; Visualization, Z.Z.; Supervision, L.M.; Project Administration, Z.Z.; Funding Acquisition, Z.Z. and L.M.; All authors have read and agreed to the published version of the manuscript.

Funding: This work was supported in part by the Technology System of Modern Agricultural Industry in Shandong Province (SDAIT-10-08) for L Min, the Start-up Fund for High-Level Talents of Qingdao Agricultural University for Z Zhu (1121010), and the Young Innovation Team Plan Program of Higher Education of Shandong Province Project for Z Zhu (2022KJ170).

Institutional Review Board Statement: All animals and experimental procedures were approved by the Qingdao Agriculture University Institutional Animal Care and Use Committee (QAU1121010, 1 October 2019~30 December 2023).

Informed Consent Statement: Not applicable.

Data Availability Statement: Data are contained within the article and supplementary materials.

Conflicts of Interest: The authors declare no conflict of interest.

References

1. Yanez-Ortiz, I.; Catalan, J.; Rodriguez-Gil, J.E.; Miro, J.; Yeste, M. Advances in sperm cryopreservation in farm animals: Cattle, horse, pig and sheep. *Anim. Reprod. Sci.* **2022**, *246*, 106904. [CrossRef] [PubMed]
2. Lv, C.; Wu, G.; Hong, Q.; Quan, G. Spermatozoa Cryopreservation: State of Art and Future in Small Ruminants. *Biopreserv. Biobank.* **2019**, *17*, 171–182. [CrossRef] [PubMed]
3. Alvarez, M.; Anel-Lopez, L.; Boixo, J.C.; Chamorro, C.; Neila-Montero, M.; Montes-Garrido, R.; de Paz, P.; Anel, L. Current challenges in sheep artificial insemination: A particular insight. *Reprod. Domest. Anim.* **2019**, *54* (Suppl. 4), 32–40. [CrossRef] [PubMed]
4. Evans, G.; Hollinshead, F.K.; Maxwell, W.M. Preservation and artificial insemination of sexed semen in sheep. *Reprod. Fertil. Dev.* **2004**, *16*, 455–464. [CrossRef] [PubMed]
5. Khosravizadeh, Z.; Khodamoradi, K.; Rashidi, Z.; Jahromi, M.; Shiri, E.; Salehi, E.; Talebi, A. Sperm cryopreservation and DNA methylation: Possible implications for ART success and the health of offspring. *J. Assist. Reprod. Genet.* **2022**, *39*, 1815–1824. [CrossRef] [PubMed]
6. Peris, S.I.; Bilodeau, J.F.; Dufour, M.; Bailey, J.L. Impact of cryopreservation and reactive oxygen species on DNA integrity, lipid peroxidation, and functional parameters in ram sperm. *Mol. Reprod. Dev.* **2007**, *74*, 878–892. [CrossRef] [PubMed]
7. Dutta, S.; Majzoub, A.; Agarwal, A. Oxidative stress and sperm function: A systematic review on evaluation and management. *Arab. J. Urol.* **2019**, *17*, 87–97. [CrossRef]
8. Beygi, Z.; Forouhari, S.; Mahmoudi, E.; Hayat, S.M.G.; Nourimand, F. Role of Oxidative Stress and Antioxidant Supplementation in Male Fertility. *Curr. Mol. Med.* **2021**, *21*, 265–282. [CrossRef]

9. Nazari, P.; Farshad, A.; Hosseini, Y. Protective Effects of Trehalose and Pentoxifylline on Goat Sperm Exposed to Chilling-Freezing Process. *Biopreserv. Biobank.* **2022**, *20*, 540–550. [CrossRef]
10. Makris, A.; Alevra, A.I.; Exadactylos, A.; Papadopoulos, S. The Role of Melatonin to Ameliorate Oxidative Stress in Sperm Cells. *Int. J. Mol. Sci.* **2023**, *24*, 15056. [CrossRef]
11. Rubinstein, S.; Breitbart, H. Role of spermine in mammalian sperm capacitation and acrosome reaction. *Biochem. J.* **1991**, *278 Pt 1*, 25–28. [CrossRef]
12. Liman, M.S.; Hassen, A.; McGaw, L.J.; Sutovsky, P.; Holm, D.E. Potential Use of Tannin Extracts as Additives in Semen Destined for Cryopreservation: A Review. *Animals* **2022**, *12*, 1130. [CrossRef] [PubMed]
13. Parameswari, R.; Rao, K.A.; Manigandan, P.; Vickram, A.S.; Archana, A.; Sridharan, T.B. Tea polyphenol-T. arjuna bark as sperm antioxidant extender in infertile smokers. *Cryoletters* **2017**, *38*, 95–99. [PubMed]
14. Galiniak, S.; Aebisher, D.; Bartusik-Aebisher, D. Health benefits of resveratrol administration. *Acta Biochim. Pol.* **2019**, *66*, 13–21. [CrossRef] [PubMed]
15. Pasquariello, R.; Verdile, N.; Brevini, T.A.L.; Gandolfi, F.; Boiti, C.; Zerani, M.; Maranesi, M. The Role of Resveratrol in Mammalian Reproduction. *Molecules* **2020**, *25*, 4554. [CrossRef] [PubMed]
16. Al-Mutary, M.G.; Al-Ghadi, M.Q.; Ammari, A.A.; Al-Himadi, A.R.; Al-Jolimeed, A.H.; Arafah, M.W.; Amran, R.A.; Aleissa, M.S.; Swelum, A.A. Effect of different concentrations of resveratrol on the quality and in vitro fertilizing ability of ram semen stored at 5 degrees C for up to 168 h. *Theriogenology* **2020**, *152*, 139–146. [CrossRef] [PubMed]
17. Li, S.J.; Su, W.D.; Qiu, L.J.; Wang, X.; Liu, J. Resveratrol protects human sperm against cryopreservation-induced injury. *Zhonghua Nan Ke Xue* **2018**, *24*, 499–503. [PubMed]
18. Kaeoket, K.; Chanapiwat, P. The Beneficial Effect of Resveratrol on the Quality of Frozen-Thawed Boar Sperm. *Animals* **2023**, *13*, 2829. [CrossRef]
19. Longobardi, V.; Zullo, G.; Salzano, A.; De Canditiis, C.; Cammarano, A.; De Luise, L.; Puzio, M.V.; Neglia, G.; Gasparrini, B. Resveratrol prevents capacitation-like changes and improves in vitro fertilizing capability of buffalo frozen-thawed sperm. *Theriogenology* **2017**, *88*, 1–8. [CrossRef]
20. Najafi, A.; Daghigh Kia, H.; Hamishehkar, H.; Moghaddam, G.; Alijani, S. Effect of resveratrol-loaded nanostructured lipid carriers supplementation in cryopreservation medium on post-thawed sperm quality and fertility of roosters. *Anim. Reprod. Sci.* **2019**, *201*, 32–40. [CrossRef]
21. Shabani Nashtaei, M.; Amidi, F.; Sedighi Gilani, M.A.; Aleyasin, A.; Bakhshalizadeh, S.; Naji, M.; Nekoonam, S. Protective features of resveratrol on human spermatozoa cryopreservation may be mediated through 5′ AMP-activated protein kinase activation. *Andrology* **2017**, *5*, 313–326. [CrossRef] [PubMed]
22. Zhang, R.; Dong, H.; Zhao, P.; Shang, C.; Qi, H.; Ma, Y.; Gao, C.; Zhang, D.; Shen, J.; Lei, Y.; et al. Resveratrol and lycium barbarum polysaccharide improve Qinling giant panda (*Ailuropoda melanoleuca Qinlingensis*) sperm quality during cryopreservation. *BMC Vet. Res.* **2022**, *18*, 23. [CrossRef] [PubMed]
23. Zhang, W.; Li, Y.; Zhu, Z. Carboxylated epsilon-Poly-L-Lysine Supplementation of the Freezing Extender Improves the Post-Thawing Boar Sperm Quality. *Animals* **2022**, *12*, 1726. [CrossRef] [PubMed]
24. Zhang, W.; Min, L.; Li, Y.; Lang, Y.; Hoque, S.A.M.; Adetunji, A.O.; Zhu, Z. Beneficial Effect of Proline Supplementation on Goat Spermatozoa Quality during Cryopreservation. *Animals* **2022**, *12*, 2626. [CrossRef] [PubMed]
25. Zhu, Z.; Li, R.; Ma, G.; Bai, W.; Fan, X.; Lv, Y.; Luo, J.; Zeng, W. 5′-AMP-Activated Protein Kinase Regulates Goat Sperm Functions via Energy Metabolism In Vitro. *Cell Physiol. Biochem.* **2018**, *47*, 2420–2431. [CrossRef] [PubMed]
26. Sun, L.; Fan, X.; Zeng, Y.; Wang, L.; Zhu, Z.; Li, R.; Tian, X.; Wang, Y.; Lin, Y.; Wu, D.; et al. Resveratrol protects boar sperm in vitro via its antioxidant capacity. *Zygote* **2020**, *28*, 417–424. [CrossRef] [PubMed]
27. Li, R.; Wu, X.; Zhu, Z.; Lv, Y.; Zheng, Y.; Lu, H.; Zhou, K.; Wu, D.; Zeng, W.; Dong, W.; et al. Polyamines protect boar sperm from oxidative stress in vitro. *J. Anim. Sci.* **2022**, *100*, skac069. [CrossRef] [PubMed]
28. Wulster-Radcliffe, M.C.; Wang, S.; Lewis, G.S. Transcervical artificial insemination in sheep: Effects of a new transcervical artificial insemination instrument and traversing the cervix on pregnancy and lambing rates. *Theriogenology* **2004**, *62*, 990–1002. [CrossRef]
29. Gogol, P. The quality and fertilizing ability of the ram semen cryopreserved around 40 years ago. *Pol. J. Vet. Sci.* **2022**, *25*, 631–633. [CrossRef]
30. Suarez, S.S.; Pacey, A.A. Sperm transport in the female reproductive tract. *Hum. Reprod. Update* **2006**, *12*, 23–37. [CrossRef]
31. Zhu, Z.; Li, R.; Fan, X.; Lv, Y.; Zheng, Y.; Hoque, S.A.M.; Wu, D.; Zeng, W. Resveratrol Improves Boar Sperm Quality via 5′AMP-Activated Protein Kinase Activation during Cryopreservation. *Oxid. Med. Cell Longev.* **2019**, *2019*, 5921503. [CrossRef] [PubMed]
32. Amaral, A.; Lourenco, B.; Marques, M.; Ramalho-Santos, J. Mitochondria functionality and sperm quality. *Reproduction* **2013**, *146*, R163–R174. [CrossRef] [PubMed]
33. Du Toit, D.; Bornman, M.S.; Van Der Merwe, M.P.; Du Plessis, D.J.; Oosthuizen, J.M. Differential sperm motility scoring and sperm ATP concentrations. *Arch. Androl.* **1993**, *30*, 69–71. [CrossRef] [PubMed]
34. Aitken, R.J. Reactive oxygen species as mediators of sperm capacitation and pathological damage. *Mol. Reprod. Dev.* **2017**, *84*, 1039–1052. [CrossRef] [PubMed]
35. Chianese, R.; Pierantoni, R. Mitochondrial Reactive Oxygen Species (ROS) Production Alters Sperm Quality. *Antioxidants* **2021**, *10*, 92. [CrossRef] [PubMed]

36. Ahmed, H.; Jahan, S.; Ullah, H.; Ullah, F.; Salman, M.M. The addition of resveratrol in tris citric acid extender ameliorates post-thaw quality parameters, antioxidant enzymes levels, and fertilizing capability of buffalo (*Bubalus bubalis*) bull spermatozoa *Theriogenology* **2020**, *152*, 106–113. [CrossRef] [PubMed]
37. Guo, H.T.; Wang, J.R.; Sun, L.Z.; Jin, X.H.; Shi, X.Y.; Lin, J.Y.; Yue, S.L.; Zhou, J.B. Effects of astaxanthin on plasma membrane function and fertility of boar sperm during cryopreservation. *Theriogenology* **2021**, *164*, 58–64. [CrossRef]
38. Shahat, A.M.; Thundathil, J.C.; Kastelic, J.P. Melatonin improves post-thaw sperm quality after mild testicular heat stress in rams *Reprod. Domest. Anim.* **2023**, *58*, 423–430. [CrossRef]
39. Oldenhof, H.; Gojowsky, M.; Wang, S.; Henke, S.; Yu, C.; Rohn, K.; Wolkers, W.F.; Sieme, H. Osmotic stress and membrane phase changes during freezing of stallion sperm: Mode of action of cryoprotective agents. *Biol. Reprod.* **2013**, *88*, 68. [CrossRef]
40. Garcez, M.E.; dos Santos Branco, C.; Lara, L.V.; Pasqualotto, F.F.; Salvador, M. Effects of resveratrol supplementation on cryopreservation medium of human semen. *Fertil. Steril.* **2010**, *94*, 2118–2121. [CrossRef]
41. Silva, E.C.; Cajueiro, J.F.; Silva, S.V.; Soares, P.C.; Guerra, M.M. Effect of antioxidants resveratrol and quercetin on in vitro evaluation of frozen ram sperm. *Theriogenology* **2012**, *77*, 1722–1726. [CrossRef] [PubMed]
42. Falchi, L.; Pau, S.; Pivato, I.; Bogliolo, L.; Zedda, M.T. Resveratrol supplementation and cryopreservation of buck semen. *Cryobiology* **2020**, *95*, 60–67. [CrossRef] [PubMed]
43. Tvrda, E.; Kovacik, A.; Tusimova, E.; Massanyi, P.; Lukac, N. Resveratrol offers protection to oxidative stress induced by ferrous ascorbate in bovine spermatozoa. *J. Environ. Sci. Health A Toxic/Hazard. Subst. Environ. Eng.* **2015**, *50*, 1440–1451. [CrossRef] [PubMed]
44. Kowalczyk, A. The Role of the Natural Antioxidant Mechanism in Sperm Cells. *Reprod. Sci.* **2022**, *29*, 1387–1394. [CrossRef]
45. Ofosu, J.; Nartey, M.A.; Mo, X.; Ye, J.; Zhang, Y.; Zeng, C.; Zhang, M.; Fang, Y.; Zhou, G. Ram sperm cryopreservation disrupts metabolism of unsaturated fatty acids. *Theriogenology* **2023**, *204*, 8–17. [CrossRef] [PubMed]
46. Tatone, C.; Di Emidio, G.; Vitti, M.; Di Carlo, M.; Santini, S., Jr.; D'Alessandro, A.M.; Falone, S.; Amicarelli, F. Sirtuin Functions in Female Fertility: Possible Role in Oxidative Stress and Aging. *Oxid. Med. Cell Longev.* **2015**, *2015*, 659687. [CrossRef]
47. Morita, Y.; Wada-Hiraike, O.; Yano, T.; Shirane, A.; Hirano, M.; Hiraike, H.; Koyama, S.; Oishi, H.; Yoshino, O.; Miyamoto, Y.; et al. Resveratrol promotes expression of SIRT1 and StAR in rat ovarian granulosa cells: An implicative role of SIRT1 in the ovary. *Reprod. Biol. Endocrinol.* **2012**, *10*, 14. [CrossRef] [PubMed]
48. Tang, B.L. Sirt1 and the Mitochondria. *Mol. Cells* **2016**, *39*, 87–95. [CrossRef]
49. Iniesta-Cuerda, M.; Havrankova, J.; Rimnacova, H.; Garcia-Alvarez, O.; Nevoral, J. Male SIRT1 insufficiency leads to sperm with decreased ability to hyperactivate and fertilize. *Reprod. Domest. Anim.* **2022**, *57* (Suppl. 5), 72–77. [CrossRef]
50. Han, J.; Wang, S.; Wang, H.; Zhang, T.; Yang, Y.; Zhao, T.; Chen, Z.; Xia, G.; Wang, C. SIRT1 reduction contributes to doxorubicin-induced oxidative stress and meiotic failure in mouse oocytes. *Toxicol. Appl. Pharmacol.* **2023**, *476*, 116671. [CrossRef]
51. Wang, Y.Q.; Cao, Q.; Wang, F.; Huang, L.Y.; Sang, T.T.; Liu, F.; Chen, S.Y. SIRT1 Protects Against Oxidative Stress-Induced Endothelial Progenitor Cells Apoptosis by Inhibiting FOXO3a via FOXO3a Ubiquitination and Degradation. *J. Cell Physiol.* **2015**, *230*, 2098–2107. [CrossRef] [PubMed]
52. Ye, J.; Liu, Z.; Wei, J.; Lu, L.; Huang, Y.; Luo, L.; Xie, H. Protective effect of SIRT1 on toxicity of microglial-derived factors induced by LPS to PC12 cells via the p53-caspase-3-dependent apoptotic pathway. *Neurosci. Lett.* **2013**, *553*, 72–77. [CrossRef] [PubMed]
53. Zhang, W.; Huang, Q.; Zeng, Z.; Wu, J.; Zhang, Y.; Chen, Z. Sirt1 Inhibits Oxidative Stress in Vascular Endothelial Cells. *Oxid. Med. Cell Longev.* **2017**, *2017*, 7543973. [CrossRef] [PubMed]
54. Piras, A.R.; Ariu, F.; Maltana, A.; Leoni, G.G.; Martino, N.A.; Mastrorocco, A.; Dell'Aquila, M.E.; Bogliolo, L. Protective effect of resveratrol against cadmium-induced toxicity on ovine oocyte in vitro maturation and fertilization. *J. Anim. Sci. Biotechnol.* **2022**, *13*, 83. [CrossRef] [PubMed]
55. Price, N.L.; Gomes, A.P.; Ling, A.J.; Duarte, F.V.; Martin-Montalvo, A.; North, B.J.; Agarwal, B.; Ye, L.; Ramadori, G.; Teodoro, J.S.; et al. SIRT1 is required for AMPK activation and the beneficial effects of resveratrol on mitochondrial function. *Cell Metab.* **2012**, *15*, 675–690. [CrossRef] [PubMed]
56. Zhao, Y.; Hu, X.; Liu, Y.; Dong, S.; Wen, Z.; He, W.; Zhang, S.; Huang, Q.; Shi, M. ROS signaling under metabolic stress: Cross-talk between AMPK and AKT pathway. *Mol. Cancer* **2017**, *16*, 79. [CrossRef] [PubMed]
57. Martin-Hidalgo, D.; Hurtado de Llera, A.; Calle-Guisado, V.; Gonzalez-Fernandez, L.; Garcia-Marin, L.; Bragado, M.J. AMPK Function in Mammalian Spermatozoa. *Int. J. Mol. Sci.* **2018**, *19*, 3293. [CrossRef]
58. Canto, C.; Gerhart-Hines, Z.; Feige, J.N.; Lagouge, M.; Noriega, L.; Milne, J.C.; Elliott, P.J.; Puigserver, P.; Auwerx, J. AMPK regulates energy expenditure by modulating NAD+ metabolism and SIRT1 activity. *Nature* **2009**, *458*, 1056–1060. [CrossRef]
59. Chini, C.C.S.; Zeidler, J.D.; Kashyap, S.; Warner, G.; Chini, E.N. Evolving concepts in NAD(+) metabolism. *Cell Metab.* **2021**, *33*, 1076–1087. [CrossRef]
60. Waddell, J.; Khatoon, R.; Kristian, T. Cellular and Mitochondrial NAD Homeostasis in Health and Disease. *Cells* **2023**, *12*, 1329. [CrossRef]
61. Ke, R.; Xu, Q.; Li, C.; Luo, L.; Huang, D. Mechanisms of AMPK in the maintenance of ATP balance during energy metabolism. *Cell Biol. Int.* **2018**, *42*, 384–392. [CrossRef]
62. Feng, T.Y.; Lv, D.L.; Zhang, X.; Du, Y.Q.; Yuan, Y.T.; Chen, M.J.; Xi, H.M.; Li, Y.; Han, N.; Hu, J.H. Rosmarinic acid improves boar sperm quality, antioxidant capacity and energy metabolism at 17 degrees C via AMPK activation. *Reprod. Domest. Anim.* **2020**, *55*, 1714–1724. [CrossRef]

63. Rezaie, F.S.; Hezavehei, M.; Sharafi, M.; Shahverdi, A. Improving the post-thaw quality of rooster semen using the extender supplemented with resveratrol. *Poult. Sci.* **2021**, *100*, 101290. [CrossRef]
64. Lu, Y.; Pan, Y.; Sheng, N.; Zhao, A.Z.; Dai, J. Perfluorooctanoic acid exposure alters polyunsaturated fatty acid composition, induces oxidative stress and activates the AKT/AMPK pathway in mouse epididymis. *Chemosphere* **2016**, *158*, 143–153. [CrossRef]
65. Hurtado de Llera, A.; Martin-Hidalgo, D.; Gil, M.C.; Garcia-Marin, L.J.; Bragado, M.J. AMPK up-activation reduces motility and regulates other functions of boar spermatozoa. *Mol. Hum. Reprod.* **2015**, *21*, 31–45. [CrossRef]

Disclaimer/Publisher's Note: The statements, opinions and data contained in all publications are solely those of the individual author(s) and contributor(s) and not of MDPI and/or the editor(s). MDPI and/or the editor(s) disclaim responsibility for any injury to people or property resulting from any ideas, methods, instructions or products referred to in the content.

animals

MDPI

Article

Effects of Different Diluents on Semen Quality of Hu Ram Stored at 4 °C

Liuming Zhang †, Yanhu Wang †, Xiaomei Sun, Yan Kang, Tariq Sohail, Jian Wang * and Yongjun Li *

Key Laboratory for Animal Genetics & Molecular Breeding of Jiangsu Province, College of Animal Science and Technology, Yangzhou University, Yangzhou 225009, China; 18352767281@163.com (L.Z.)

* Correspondence: jianwang1223@163.com (J.W.); yzliyongjun@163.com (Y.L.)

† These authors contribute equally to this work.

Simple Summary: Diluent plays an important role in sperm storage. The various components of the diluent provide nutrition for sperm survival, maintain the stability of the preserved environment, and prolong the survival time of sperm. However, for the preservation of semen at 4 °C, diluent formulations are currently lacking or existing diluent formulations are not sufficient. Therefore, the effect of four diluents on semen preservation was studied through an assessment of sperm motility and functional integrity. Diluent D (Tris–Fructose–Citric acid–Egg yolk) was found to greatly improve semen quality during storage at 4 °C.

Abstract: This study aimed to investigate the effects of various diluents on the quality of *Hu* ram sperm stored at 4 °C. Semen samples were collected from three *Hu* rams and diluted with diluents A (Sodium citrate–Glucose–Egg yolk), B (Sodium citrate–Glucose), C (Fructose–Skimmed milk powder–Soy lecithin), and D (Tris–Fructose–Citric acid–Egg yolk). Total motility (TM), straight-line velocity (VSL), average path velocity (VAP), curvilinear velocity (VCL), average motion degree (MAD), acrosome integrity, membrane integrity, and reactive oxygen species (ROS) were evaluated. The results showed that diluent D had better preservation in terms of the sperm TM, VSL, VCL, VAP, MAD, and membrane and acrosome integrity. On the third day of the storage, the sperm PM of diluent D was higher than that of other diluents ($p < 0.05$). The ROS level of diluent D was lower than that of other diluents on the fifth day ($p < 0.05$). On the seventh day of the storage, the sperm TM in diluent D reached 50%, which was the highest in all diluent groups. On the seventh day of the storage, the integrity of the sperm membrane and the integrity of the acrosome of the sperm in diluent D were the highest in all diluent groups ($p < 0.05$). In conclusion, these results indicated that diluent D improved the semen quality during storage at 4 °C. In this study, diluent D was the best diluent formula for *Hu* ram semen stored at 4 °C.

Keywords: diluent; motility parameters; membrane integrity; acrosome integrity; ROS

Citation: Zhang, L.; Wang, Y.; Sun, X.; Kang, Y.; Sohail, T.; Wang, J.; Li, Y. Effects of Different Diluents on Semen Quality of Hu Ram Stored at 4 °C. *Animals* **2023**, *13*, 2823. https://doi.org/10.3390/ani13182823

Academic Editors: Anna Wysokińska and Giorgio Antonio Presicce

Received: 24 July 2023
Revised: 23 August 2023
Accepted: 1 September 2023
Published: 6 September 2023

Copyright: © 2023 by the authors. Licensee MDPI, Basel, Switzerland. This article is an open access article distributed under the terms and conditions of the Creative Commons Attribution (CC BY) license (https://creativecommons.org/licenses/by/4.0/).

1. Introduction

Hu sheep are a local breed under first-class protection in China and are also the only breed of sheep with lambs with white skin [1,2]. *Hu* sheep have excellent reproductive characteristics such as early sexual maturity and being highly prolific [3,4]. Therefore, improving the large-scale and standardized breeding level of *Hu* sheep can effectively increase farmers' economic benefits and improve people's living standards. However, the result of the preservation of ram semen is not satisfactory. In Bhalothia's [5] study of the preservation of ram semen at 4 °C, the TM decreased to 30% on the third day of preservation. In a study of ram semen, O'Hara [6] found that after three days of storage, the fertilization rate plummeted to just reached 20%. According to Druart [7], the pregnancy rate was significantly lower after insemination with semen stored for one day. With the extension of storage time, the quality of semen will decrease, which will further reduce the pregnancy rate [8].

Semen preservation is very important to prolong survival time, expand semen volume, effectively contribute to promoting the reproductive potential of excellent males, and overcome geographical limitations [9]. Semen preservation mainly includes room-temperature preservation (15–25 °C), low-temperature preservation (0–5 °C), and cryopreservation (−196 °C) [10,11]. Compared with fresh semen, frozen–thawed semen will have reduced sperm motility and structural integrity [12]. Cervical insemination with frozen–thawed semen generally resulted in low pregnancy rates [13]. The total motility (TM) of sperm stored at room temperature will decline in a short time, and the storage time is too short to satisfy the demand for extensive usage of semen [14]. At present, sheep semen is mainly preserved at low temperatures. Selecting an efficient and stable diluent formula for sheep semen is a prerequisite to improving the effect of low-temperature semen preservation. Low-temperature preservation of sheep semen refers to the method in which semen is diluted according to a certain proportion and stored at 4 °C to extend survival time [15]. At low temperatures, the metabolic rate of sperm is slowed down, even in a dormant state, and the accumulation of metabolites decreases significantly [16,17]. When stored at 4 °C, the reproduction and growth of microorganisms in the preserved semen were also inhibited, improving the quality of low-temperature preservation of sheep semen [18]. It should be noted that one study found that slow cooling during low-temperature preservation of sheep semen is the key to ensuring semen quality, and a sharp decrease in temperature will cause a cold shock to the sperm [19,20].

Currently, there are still many shortcomings in the low-temperature preservation of sheep semen, such as the uncertainty of basic diluent, the short survival time of sperm, and poor quality of semen. With the diluent formula for low-temperature preservation of sheep semen, the effective survival time of the sperm is only about three days, far from meeting the practical production needs. As the diluent formula in studies of low-temperature preservation of sheep semen does not give good results at present, the diluent formula selected in this study is based on the diluent formula of other species that are better preserved at low temperatures. Therefore, it is very important to select an efficient and stable diluent formula for low-temperature preservation of sheep semen. In this experiment, four kinds of diluent formulas for the low-temperature preservation of sheep semen were selected, and the best diluent formula was designated through the detection of semen storage quality. The purpose of this research was to assess the effects of different diluent formulas on the quality parameters (TM, straight-line velocity (VSL), average path velocity (VAP), curvilinear velocity (VCL), average motion degree (MAD), membrane integrity, acrosome integrity, and reactive oxygen species (ROS)) of *Hu* sheep semen when stored at 4 °C.

2. Materials and Methods

2.1. Animals and Semen Collection

Semen samples from three sexually mature and healthy *Hu* sheep (about three years old) were used in this research. After reproductive examination, there was no problem with the testes of the three rams, and the testicles were normal in size according to visual inspection. The rams were housed at the Experimental sheep farm of Yangzhou University (Jiangsu, China). The rams were free to drink water and provided with high-quality hay and concentrate.

A total of 60 ejaculates (20 ejaculates per ram) were collected from the three rams with an artificial vagina two times a week from February to April 2022. The collected semen was brought to the laboratory within 30 min. First, the collected semen was preliminarily selected, and samples with a milky white color and about 1 mL volume of semen were selected. The quality of semen was evaluated using a computer-aided sperm analyzer (CASA). In this study, only semen samples with a concentration $\geq 2.0 \times 10^9$ sperm/mL, motility $\geq$ 80%, and normal sperm morphology $\geq$ 85% were used. To overcome individual differences and balance the sperm contribution, the checked semen was pooled together.

2.2. Diluent Preparation

Four different diluents, A, B, C, and D, were prepared on the basis of composition and dose as shown in Table 1. In the laboratory, analytical balances (XSE105DU, Mettler Toledo Zurich, Switzerland) were used to accurately weigh reagents from different formulations. The weighed substances were dissolved in 100 mL of sterilized ultrapure water and stirred with a glass rod to fully dissolve. The diluent was filtered through a 0.2 μm filter membrane for later use. After filtration, 10% egg yolk was added to diluents A and D, and they were stored overnight in a refrigerator at 4 °C. The final diluent was the supernatant obtained after centrifugation at 12,000× *g* for 10 to 15 min.

Table 1. Four different diluent formulations.

Constituent	A	B	C	D
Sodium citrate	2.00 g	1.40 g	-	-
Glucose	3.00 g	3.00 g	-	-
Citric acid	-	-	-	1.64 g
Fructose	-	-	2.50 g	2.00 g
Tris	-	-	-	3.07 g
Skimmed milk powder	-	-	10.00 g	-
Soy Lecithin	-	-	0.375 g	-
Egg yolk	10%	-	-	10%
Penicillin sodium	100,000 IU	100,000 IU	100,000 IU	100,000 IU
Streptomycin sulfate	100,000 IU	100,000 IU	100,000 IU	100,000 IU
Total volume	100 mL	100 mL	100 mL	100 mL
Osmotic pressure	377 m Osm/L	316 m Osm/L	410 m Osm/L	375 m Osm/L

2.3. Semen Processing

After mixing the semen, it was split into four equal fractions and diluted with A, B, C, and D to 2.0×10^8 sperm/mL. Finally, the semen samples were wrapped with eight layers of cotton and kept in a 4 °C refrigerator. During the storage period, the plastic tubes containing the semen were gently shaken every 12 h to prevent sperm from depositing. In addition, semen quality parameters (TM, VSL, VCL, VAP, MAD, and membrane and acrosome integrity) were measured and analyzed once a day. The ROS level was measured on the fifth day.

2.4. Evaluation of Semen Quality

2.4.1. Analysis of Sperm Motility Parameters

Sperm TM, VSL, VCL, VAP, and MAD were assessed using the CASA (ML-608JZ II, Mailang, Nanning, China) equipped with a warm stage. The software used by the CASA is an automatic analysis and detection system for animal sperm belonging to Mailang. The microscope used in the detection system is an ML-800 (Mailang, Nanning, China). The microscope used a Mailang conjugate high-contrast sperm optical imaging device (10× PH). The CASA software (ML-800II) captured 30 frames per second. The diluted semen was incubated at 37 °C for 4 min. Then, each semen sample was placed on the Mailang computer sperm counting board (ML-CASA20-4) and estimated at 37 °C. The counting board was 20 μm deep. At least 200 sperm cells were observed per slide. The TM represents the proportion of active sperm. The VSL represents the speed of the straight-line movement distance of the sperm head. The VAP indicates the movement speed of the sperm head along its spatial average trajectory. The VCL indicates the velocity of the sperm head along its actual walking curve.

2.4.2. Analysis of Sperm Functional Integrity

The functional integrity of the sperm membrane was evaluated using HOST, as described by Wang Y [21]. It was evaluated every day. Measurement was carried out by mixing 20 μL semen with 200 μL hypo-osmotic solution. After incubation at 37 °C for

40 min, sperm swelling was observed under a 400× phase-contrast microscope. At least 200 sperm cells were observed on each slide.

The Coomassie brilliant blue stain was used to assess the integrity of the acrosome as described by Zhang L, and the Coomassie brilliant blue staining solution was prepared according to their method [22]. These parameters were assessed every day. Briefly, 50 μL of the preserved semen sample was fixed with 1 mL of 4% paraformaldehyde at room temperature for 10 min. The sample was centrifuged at 422× *g* for 5 min, the supernatant was discarded, and 10 μL was taken to make a smear. After air-drying, the smear was stained with Coomassie brilliant blue solution for 40 min and washed with distilled water. After natural drying, a minimum of 200 sperm cells on each slide were examined under a 1000× phase-contrast microscope. Sperm heads have two distinct varieties. The acrosome was intact if the sperm head appeared blue. The acrosome was not intact if the sperm head appeared unstained. At the same time, the spontaneous acrosome reaction (SAR) was evaluated in order to reflect the real influence of diluent on the sperm acrosome.

2.4.3. Analysis of the ROS Level of the Sperm

The ROS level of the sperm was measured with an ROS Assay kit (Beyotime Institute of Biotechnology, Shanghai, China) according to the instructions. First, a volume of 100 μL diluted semen was mixed with 300 μL of 4 °C pre-cooled PBS and centrifuged at 422× *g* for 5 min. This procedure was repeated two more times to remove the supernatant. Second, 300 μL DCFH-DC working solution was added to the sperm. After staining at 37 °C for 30 min in dark conditions, the samples were centrifuged at 422× *g* for 5 min, the supernatant was discarded, and the samples were washed with PBS two times. Finally, a volume of 600 μL PBS was added to the sperm. In the negative control group, DCFH-DC working solution was not added. In the positive control group, when 300 μL of CFH-DC working solution was added, 0.3 μL of Rosup reagent was added at the same time. The level of ROS was expressed as absorbance at an excitation wavelength of 488 nm and an emission wavelength of 525 nm for a microporous multimodal detection system (PerkinElmer Inc., NYSE: PKI, Waltham, MA, USA). The experiment was repeated three times.

2.5. Statistical Analysis

Statistical analyses were performed using SPSS 24.0. The Shapiro–Wilk test was used to test whether the data obey a normal distribution. The data do show a normal distribution. Duncan's multiple range tests using one-way analysis of variance procedures were used to compare the mean values of TM, VSL, VCL, VAP, MAD, membrane integrity, acrosome integrity, and ROS level between diluents within the same time points. Changes in variables across different time points in a group were assessed using repeated measures ANOVA to reveal time effects. The significance level of $p < 0.05$ was chosen unless otherwise noted. The results are presented as mean ± SD. All groups were replicated three times.

3. Result

3.1. Effects of Different Diluents on the Sperm Motility Parameters

Table 2 shows the effects of several diluents on the TM and kinematic parameters of the *Hu* ram sperm during liquid storage at 4 °C. With the extension of storage time, sperm motility gradually decreased, and the sperm motility of diluent B and diluent C decreased more rapidly. On the third day of storage, the sperm TM in diluent C dropped to 0. On the seventh day of storage, the sperm TM in diluent B dropped to 0. After three days of preservation, the sperm TM in diluent D was the highest among all the diluents.

On the fifth and sixth days of preservation, the sperm VSL in diluent D was significantly higher than that in the other diluents ($p < 0.05$). On the fourth, fifth, and sixth day of storage, the VCL and VAP of sperm in the diluent D group was significantly higher than that in other diluents ($p < 0.05$). On the second day of preservation, the MAD of sperm in the diluent A and D groups was significantly higher than that in other diluents ($p < 0.05$).

From day 3 to day 7, the MAD of the sperm in the diluent D group was significantly higher than that in the other diluents ($p < 0.05$).

Table 2. Effects of different diluents on the motility parameters of the *Hu* ram sperm.

Motility Parameters	Preserved Period (Days)	Different Diluents			
		A	B	C	D
TM (%)	0	94.40 ± 0.02^{A}	94.95 ± 0.06^{A}	94.85 ± 0.40^{A}	94.11 ± 0.92^{A}
	1	93.56 ± 0.14^{Aa}	79.67 ± 0.24^{Bc}	88.64 ± 0.43^{Bb}	93.37 ± 0.41^{ABa}
	2	91.74 ± 0.10^{Ba}	74.28 ± 1.86^{Cb}	37.84 ± 0.72^{Cc}	91.31 ± 1.34^{BCa}
	3	85.92 ± 0.73^{Cb}	64.22 ± 1.31^{Dc}	0^{Dd}	90.11 ± 0.06^{CDa}
	4	83.06 ± 0.10^{Db}	49.90 ± 0.21^{Ec}	0^{Dd}	88.96 ± 0.38^{Da}
	5	80.86 ± 1.11^{Eb}	32.01 ± 0.30^{Fc}	0^{Dd}	86.44 ± 0.58^{Ea}
	6	69.53 ± 0.36^{Fb}	14.60 ± 0.31^{Gc}	0^{Dd}	80.53 ± 0.21^{Fa}
	7	34.45 ± 0.47^{Gb}	0^{Hc}	0^{Dc}	51.01 ± 0.69^{Ga}
VSL (μm/s)	0	41.34 ± 0.57^{AB}	41.36 ± 1.59^{B}	42.05 ± 1.64^{A}	41.15 ± 1.65^{BC}
	1	42.98 ± 1.64^{A}	38.66 ± 1.78^{B}	39.65 ± 1.23^{B}	43.02 ± 1.01^{AB}
	2	45.31 ± 1.27^{Aab}	47.77 ± 0.23^{Aa}	26.98 ± 0.45^{Cc}	44.70 ± 0.74^{Ab}
	3	44.69 ± 2.02^{Aa}	41.04 ± 2.61^{Ba}	0^{Db}	38.98 ± 0.66^{CDa}
	4	41.17 ± 0.07^{ABa}	43.23 ± 1.12^{Ba}	0^{Db}	43.00 ± 0.66^{ABa}
	5	34.45 ± 0.20^{Cb}	30.33 ± 0.35^{Cc}	0^{Dd}	37.26 ± 0.28^{Da}
	6	36.33 ± 0.25^{BCb}	19.70 ± 1.86^{Dc}	0^{Dd}	40.16 ± 1.03^{BCa}
	7	35.33 ± 3.40^{Ca}	0^{Eb}	0^{Db}	33.57 ± 0.13^{Ea}
VCL (μm/s)	0	80.82 ± 0.98^{A}	82.22 ± 0.08^{A}	82.36 ± 2.62^{A}	82.11 ± 0.86^{A}
	1	75.71 ± 2.19^{ABab}	69.84 ± 2.91^{Bb}	72.23 ± 2.12^{Bb}	81.09 ± 1.08^{Aa}
	2	78.38 ± 1.71^{Ab}	83.16 ± 1.23^{Aa}	43.30 ± 1.13^{Cc}	82.88 ± 1.22^{Aa}
	3	74.32 ± 3.56^{ABa}	69.48 ± 3.43^{Ba}	0^{Db}	75.85 ± 0.78^{BCa}
	4	69.94 ± 0.45^{BCb}	69.39 ± 1.15^{Bb}	0^{Dc}	78.62 ± 1.09^{ABa}
	5	61.67 ± 0.85^{Db}	52.06 ± 0.15^{Cc}	0^{Dd}	67.26 ± 0.47^{Da}
	6	64.39 ± 0.26^{CDb}	41.92 ± 2.68^{Dc}	0^{Dd}	73.76 ± 2.86^{Ca}
	7	65.86 ± 5.14^{CDa}	0^{Eb}	0^{Db}	62.93 ± 0.74^{Ea}
VAP (μm/s)	0	57.15 ± 0.69^{A}	58.14 ± 0.05^{A}	58.23 ± 1.85^{A}	57.97 ± 0.61^{A}
	1	53.54 ± 1.54^{ABab}	49.38 ± 2.06^{Bc}	51.08 ± 1.50^{Bb}	57.34 ± 0.77^{Aa}
	2	55.42 ± 1.21^{Ab}	58.80 ± 0.87^{Aa}	30.62 ± 0.80^{Cc}	58.61 ± 0.86^{Aa}
	3	52.56 ± 2.51^{ABa}	49.13 ± 2.43^{Ba}	0^{Db}	53.64 ± 0.55^{BCa}
	4	49.46 ± 0.31^{BCb}	49.07 ± 0.81^{Bb}	0^{Dc}	55.60 ± 0.77^{ABa}
	5	43.61 ± 0.60^{Db}	36.81 ± 0.10^{Cc}	0^{Dd}	47.56 ± 0.33^{Da}
	6	45.53 ± 0.18^{CDb}	29.64 ± 1.90^{Dc}	0^{Dd}	52.16 ± 2.02^{Ca}
	7	46.57 ± 3.63^{CDa}	0^{Eb}	0^{Db}	44.50 ± 0.53^{Ea}
MAD (°/s)	0	133.19 ± 7.61^{A}	133.19 ± 6.76^{A}	132.37 ± 8.70^{A}	132.50 ± 7.07^{A}
	1	101.36 ± 5.65^{Ba}	77.65 ± 2.43^{Bb}	106.97 ± 11.17^{Ba}	118.47 ± 5.20^{Ba}
	2	98.55 ± 2.06^{BCa}	74.15 ± 4.01^{Bb}	28.66 ± 0.75^{Cc}	107.33 ± 3.91^{BCa}
	3	88.46 ± 1.52^{BCDb}	55.34 ± 1.60^{Cc}	0^{Dd}	101.15 ± 3.07^{CDa}
	4	88.00 ± 1.68^{CDb}	40.07 ± 0.94^{Dc}	0^{Dd}	95.65 ± 0.87^{CDa}
	5	81.92 ± 4.64^{Db}	23.84 ± 0.58^{Ec}	0^{Dd}	93.60 ± 2.20^{Da}
	6	59.36 ± 3.56^{Eb}	11.65 ± 0.91^{Fc}	0^{Dd}	71.99 ± 0.50^{Ea}
	7	28.59 ± 1.52^{Fb}	0^{Gc}	0^{Dc}	34.18 ± 3.00^{Fa}

Note: Data are expressed as mean ± SD of *Hu* ram sperm in different diluents. Different superscripts (lowercase) in the same row show significant differences ($p < 0.05$). Different superscripts (uppercase) in the same column show significant differences ($p < 0.05$).

3.2. Effects of Different Diluents on the Sperm Functional Integrity

The HOST microscopy result is shown in Figure 1a. There are three types of sperm tails. Sperm with a functional membrane are type A and B sperm, which have a curled tail. Sperm with a non-functional membrane are type C sperm, whose tail is not curled. Table 3 shows the effects of several diluents on the membrane function of the *Hu* ram sperm during liquid storage at 4 °C. From day 1 to day 7, diluent D was superior in terms of preserving the function of the sperm membrane during storage at 4 °C. These results indicated that diluent D was better preserving the membrane functional integrity of the *Hu* ram sperm during storage at 4 °C.

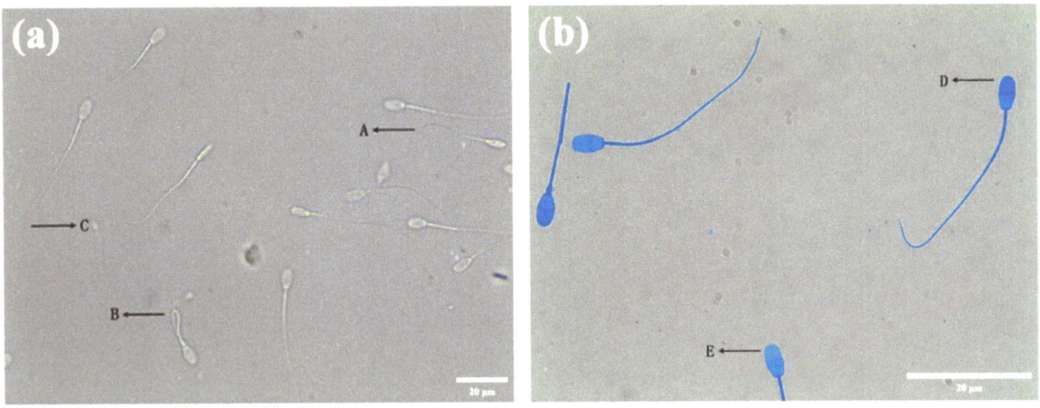

Figure 1. Detection of the sperm plasma membrane and acrosome. (**a**) The detection of sperm plasma membrane. The two types of tail curls A and B indicate intact-membrane sperm, and the tail non-curl type C is the sperm with a damaged membrane. (**b**) The detection of sperm acrosome. The sperm head is blue (D), which means that the acrosome is intact. The sperm head is unstained (E), which means that the acrosome is not intact.

Table 3. Effects of different diluents on the functional integrity of the *Hu* ram sperm.

Parameter	Preserved Period (Days)	Different Diluents			
		A	B	C	D
Sperm functional membrane integrity (%)	0	84.54 ± 0.59 ^{A}	84.51 ± 0.64 ^{A}	84.83 ± 0.63 ^{A}	85.10 ± 0.21 ^{A}
	1	73.65 ± 0.32 ^{Bb}	64.30 ± 1.29 ^{Bc}	65.77 ± 1.63 ^{Bc}	77.44 ± 0.33 ^{Ba}
	2	69.44 ± 0.76 ^{Cb}	60.79 ± 2.10 ^{Bc}	40.58 ± 1.66 ^{Cd}	75.00 ± 0.73 ^{Ca}
	3	63.47 ± 0.67 ^{Db}	54.85 ± 0.43 ^{Cc}	-	71.22 ± 0.35 ^{Da}
	4	57.67 ± 0.60 ^{Eb}	50.78 ± 0.29 ^{Dc}	-	67.62 ± 0.87 ^{Ea}
	5	52.32 ± 0.54 ^{Fb}	44.92 ± 0.79 ^{Ec}	-	61.26 ± 0.45 ^{Fa}
	6	38.22 ± 0.36 ^{Gb}	35.09 ± 0.38 ^{Fc}	-	54.12 ± 1.46 ^{Ga}
	7	34.23 ± 1.75 ^{Hb}	-	-	45.38 ± 0.75 ^{Ha}
Sperm acrosome integrity (%)	0	91.13 ± 0.23 ^{A}	90.84 ± 0.33 ^{A}	91.01 ± 0.56 ^{A}	90.87 ± 0.43 ^{A}
	1	89.83 ± 0.47 ^{Ba}	82.70 ± 0.31 ^{Bc}	84.53 ± 1.38 ^{Bbc}	85.92 ± 0.60 ^{Bb}
	2	80.75 ± 0.04 ^{Cb}	75.19 ± 0.17 ^{Cc}	83.08 ± 0.46 ^{Ba}	84.41 ± 0.67 ^{Ba}
	3	80.94 ± 0.64 ^{Ca}	74.17 ± 1.02 ^{Cb}	-	82.95 ± 0.74 ^{Ca}
	4	79.11 ± 0.92 ^{Ca}	73.61 ± 0.63 ^{Cb}	-	79.48 ± 1.37 ^{Da}
	5	70.26 ± 0.10 ^{Db}	58.60 ± 1.43 ^{Dc}	-	77.02 ± 0.47 ^{Ea}
	6	57.65 ± 3.06 ^{Ea}	42.72 ± 0.91 ^{Eb}	-	64.13 ± 0.27 ^{Fa}
	7	40.46 ± 0.88 ^{Fb}	-	-	45.95 ± 0.63 ^{Ga}

Note: Data are expressed as mean ± SD of *Hu* ram sperm in different diluents. Different superscripts (lowercase) in the same row show significant differences. ($p < 0.05$). Different superscripts (uppercase) in the same column show significant differences. ($p < 0.05$). "-" means that all the sperm had died, meaning loss of detection.

Figure 1b displays the result of the Coomassie brilliant blue staining. Table 3 shows the effects of several diluents on the acrosome integrity of the *Hu* ram sperm during liquid storage at 4 °C. The sAR of mixed semen was detected before semen diluents, and it was found that the value of SAR was only 4.55%, which was very low. The integrity of the acrosome of the preserved sperm in diluent C decreased the most rapidly. The integrity of the acrosome of the preserved sperm in diluent D was the highest among all the diluents ($p < 0.05$) on the seventh day of preservation.

3.3. *Effects of Different Diluents on the Sperm ROS Level*

The effects of different diluents on the ROS level of the *Hu* ram sperm during low temperature preservation are shown in Figure 2. The ROS level of sperm in diluent D was significantly lower than that of the groups of diluent A and diluent B on the fifth day ($p < 0.05$). The ROS level of sperm in diluent B was significantly higher than that of the groups of diluent A and diluent D on the fifth day ($p < 0.05$).

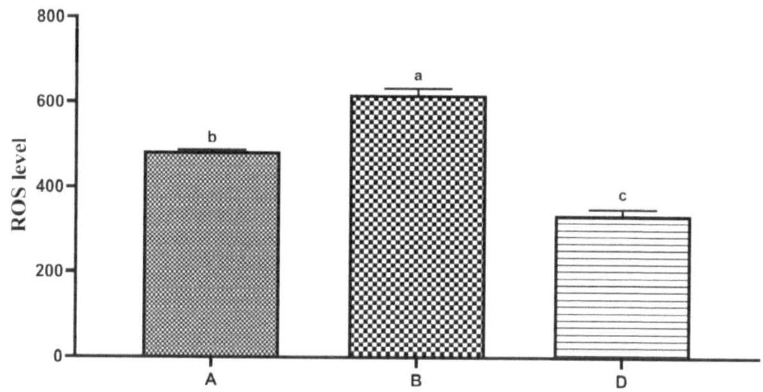

Figure 2. Effects of different diluents on the production of ROS in the sperm on the fifth day. Note: Different letters mean a significant difference ($p < 0.05$), but the same letter means no significant difference ($p > 0.05$).

4. Discussion

It is very important to detect the quality of semen before AI. Sperm motility parameters (such as sperm TM and kinematic parameters) and membrane and acrosome integrity are considered reliable indicators [23]. Therefore, in this study, the diluent formula suitable for the low-temperature preservation of *Hu* ram semen was identified by measuring the parameters of sperm TM, VSL, VCL, VAP, MAD, membrane integrity, acrosome integrity, and ROS level after different storage times. The experimental results show that diluent D has the best preservation effect. The sperm TM reached 51.01% on the seventh day after preservation and reached more than 86.44% on the fifth day. Followed by diluent A and diluent B, diluent C has the worst preservation impact on semen. On the fifth, sixth, and seventh s of preservation, the plasma membrane integrity and acrosome integrity of sperm in diluent D were significantly higher than those in other diluent groups. The basic diluent used by Merati [24] in the cryopreservation of sheep semen also contains the component of diluent D. After thawing with the optimized diluent formula, the sperm TM reached 45.93%. This study was consistent with Merati's findings, and diluent D had effective preservation properties. Sperm motility is an important component of male fertility because of its importance for migration in the genital tract [25]. Therefore, the composition of diluent D could be suitable for the preservation of sheep semen at low temperatures and cryopreservation. Under two conditions of preservation, it had a favorable effect on the preservation. The addition and concentration of other substances in the formula may be related to specific sheep breeds and storage conditions. Before diluting the mixed semen with each diluent, the sAR of sperm was detected, and it was found that the sAR rate was extremely low. If the sAR rate is high, it will affect the quality of semen preservation and the subsequent implementation of AI technology [26]. This shows that diluent D can prolong the low-temperature preservation time of semen and is the best diluent formula for the low-temperature preservation of *Hu* ram semen.

Sugars play three main roles in the diluent. Carbohydrates are the main nutrient substance in the diluent, providing energy for sperm. Sperm can use monosaccharides, such as glucose and fructose or their metabolites, for glycolysis and oxidative phosphorylation to provide energy for metabolism and activity [27]. In a study of sheep sperm, Molinia [28] found that sperm motility in a diluted solution containing glucose and fructose reached 35.3% and in solutions containing sucrose and lactose reached 24.7% and 25.7%, respectively. This may be because the efficiency of sperm utilization of monosaccharides is higher than that of disaccharides. Therefore, simple sugars such as fructose and glucose may be added to sheep sperm to provide energy. In this study, the nutrients added to the four diluents were monosaccharides. Yildiz [29] found that the effect of adding fructose to sperm was that motility reached 84.3% and acrosome integrity reached 98.4%, which was superior to other monosaccharides. This study was consistent with Yildiz's research results, and the sperm TM and acrosome integrity in diluent D were higher than that in diluent A. In addition to the effects of buffering substances, this may also be due to the higher utilization of fructose than glucose for sheep sperm.

At present, diluent containing milk, sodium citrate, and egg yolk is widely used in low-temperature preservation [30]. In this study, diluent A and D contained egg yolk, while diluent B did not contain any protective agent. The results indicated that the sperm TM in diluent A and D decreased slowly, while that in diluent B decreased rapidly. From the first day of semen preservation, the sperm TM in diluent A and D was significantly higher than that in diluent B. With the prolongation of semen preservation time, the sperm TM in diluent A and D was significantly higher than that in diluent B. On the seventh day after preservation, all the sperm stored in diluent B died, and the TM in diluent A and D, respectively, reached more than 30% and 50%. This suggests that egg yolk is essential to protect sperm from low-temperature effects. It may be that lecithin and unsaturated fatty acids in the yolk can alleviate the effect of free radicals, stabilize the cell membrane, promote lipoprotein synthesis, and improve sperm metabolism [31]. In addition, the addition of yolks can also improve the tolerance of sperm to osmosis. Egg yolk is a commonly used protective agent in low-temperature semen preservation diluents, although researchers suggest that adding egg yolk to the diluent may risk spreading animal diseases [32,33]. Some research is also devoted to the study of plant components that can replace egg yolks, such as soybean lecithin and others [34]. However, the egg yolk in the diluent can not only protect sperm from low-temperature effects but also provide nutrition for the sperm, which is not available in many protective agents [35]. Therefore, semen preservation still makes extensive use of egg yolks. In an experiment comparing the effects of untreated egg yolk and pasteurized egg yolk on semen, Selige [36] found that there was no significant difference in sperm PM, acrosome integrity, and membrane fluidity between the two groups, and the quality of sperm in both groups remained at a high level. This could be due to the fact that bacterial metabolism is effectively inhibited during cryopreservation, making any potential effect of bacterial metabolism unlikely. This study was consistent with Selige's research results. Sperm TM, acrosome integrity, and functional membrane were higher in diluents A and D, supplemented with egg yolk, than in other diluents. This shows that the addition of non-sterilized egg yolk to the diluent may not necessarily lead to a decline in semen quality. Under the condition of the low-temperature preservation, the influence of bacteria may also be limited. Soybean lecithin is a permeable protective agent that is equivalent to a low-density lipoprotein in the yolk and has a beneficial effect on sperm [37]. Skimmed milk powder is an impermeable protective agent that plays a role in the regulation of osmotic pressure outside of sperm [38]. Diluent C added two types of protective agents: skimmed milk powder and soybean lecithin. However, the quality of semen after storage decreased rapidly, and all sperm died on the third day after storage, likely because no buffer solution was added to dilute it.

The buffer substance is also an important component of the diluent. Its main function is to buffer the acidic products produced by sperm metabolism during preservation, mainly lactic acid. Due to the prolongation of semen preservation time, acid substances gradually accumulate, which can cause sperm acidosis [30]. It can be seen that adding buffer substances to the semen diluent is of great significance to maintain sperm TM [39].

Currently, the buffer substances commonly used in diluents are sodium citrate, sodium dihydrogen citrate, citric acid, Tris, and some others. It is found that the buffering effect of Tris is very good, mainly because the effective range of Tris buffer is neutral, the toxicity to sperm is low, and it can have a better buffering effect, which is more conducive to the preservation of sperm [40]. Sodium citrate is an inorganic salt buffer that can also maintain the pH value of the diluent. Citric acid can maintain the pH value of the diluent, chelate heavy metals, and adjust the osmotic pressure [41]. It has been found that the effect of adding various buffers to the diluent is better than that of a single buffer [42]. Diluent D has the longest sperm storage time and can maintain high sperm quality. This may be because diluent D contains Tris and citric acid, while diluent A and diluent B contain only sodium citrate as buffer. The detection of ROS content also proves this. ROS originate from endogenous and exogenous sources. The mitochondrial oxidative respiratory chain of normal sperm produces endogenous ROS, but damage to the mitochondria results in the release of a large amount of ROS, and the released ROS continue to damage the sperm, causally linked in a repeated cycle that leads to sperm apoptosis [43–45], while exogenous ROS are produced as a result of harmful substances and environmental changes [46]. The level of ROS in diluent D was lower than that in other diluent groups on the fifth day. The pH of the diluent can be affected by the by-products of sperm metabolism. If the pH level exceeds a certain threshold, it can damage the structure of the sperm and induce the programmed cell death [47]. Tris and citric acid, contained in diluent D, are commonly used buffer substances that can keep the diluent stable and reduce sperm apoptosis. On the one hand, this may be because the buffer substance of diluent D is the most suitable and can keep the pH and osmotic pressure of semen relatively stable. On the other hand, the sperm in diluent B had the lowest TM and the highest number of dead sperm, which leads to a higher production of ROS by the apoptotic sperm.

5. Conclusions

In this study, diluent D has a better low-temperature preserving effect on *Hu* ram semen, which can significantly improve the quality of semen preservation and prolong the semen preservation time. Comparing with other diluents, the best diluent formula suitable for the low-temperature conservation of *Hu* sheep semen was selected, which provided a theoretical reference for effectively prolonging the low-temperature preservation time of *Hu* ram semen and laid a foundation for AI technology.

Author Contributions: L.Z. and Y.W.: Conceptualization, Methodology, Investigation, Formal analysis, Writing—original draft. T.S.: Conceptualization, Writing—original draft, Writing—review and editing. Y.K.: Methodology, Investigation. X.S.: Formal analysis, Writing—review and editing. J.W. and Y.L.: Writing—review and editing, Visualization, Supervision, Project administration. All authors have read and agreed to the published version of the manuscript.

Funding: This work was supported by the Jiangsu Modern Agricultural Industrial Technology System Construction Project (JATS [2022]491) and the Postgraduate Research and Practice Innovation Program of Jiangsu Province (KYCX22_3531).

Institutional Review Board Statement: All animal procedures conform to the guidelines and regulatory standards of the Animal care committee of Yangzhou University (Approval ID: SYXK [Su]2017-0044).

Informed Consent Statement: Not applicable.

Data Availability Statement: All data sets collected and analyzed during the current study are available from the corresponding author on reasonable request.

Conflicts of Interest: The authors declare no conflict of interest.

References

1. Eer, H.; Ma, L.; Xie, X.; Ma, J.; Ma, X.; Yue, C.; Ma, Q.; Liang, X.; Ding, W.; Li, Y. Genetic polymorphism association analysis of SNPs on the species conservation genes of Tan sheep and Hu sheep. *Trop. Anim. Health Prod.* **2020**, *52*, 915–926. [CrossRef]

2. Chen, T.; Wang, L.; Li, Q.; Long, Y.; Lin, Y.; Yin, J.; Zeng, Y.; Huang, L.; Yao, T.; Abbasi, M.N.; et al. Functional probiotics of lactic acid bacteria from Hu sheep milk. *BMC Microbiol.* **2020**, *20*, 228. [CrossRef] [PubMed]
3. Wu, C.; Dai, J.; Zhang, S.; Sun, L.; Liu, Y.; Zhang, D. Effect of Thawing Rates and Antioxidants on Semen Cryopreservation in Hu Sheep. *Biopreserv. Biobank.* **2021**, *19*, 204–209. [CrossRef] [PubMed]
4. Yang, H.; Wan, Z.; Jin, Y.; Wang, F.; Zhang, Y. SMAD2 regulates testicular development and testosterone synthesis in Hu sheep. *Theriogenology* **2021**, *174*, 139–148. [CrossRef] [PubMed]
5. Bhalothia, S.K.; Mehta, J.S.; Kumar, T.; Prakash, C.; Talluri, T.R.; Pal, R.S.; Kumar, A. Melatonin and canthaxanthin enhances sperm viability and protect ram spermatozoa from oxidative stress during liquid storage at 4 °C. *Andrologia* **2022**, *54*, e14304. [CrossRef] [PubMed]
6. O'hara, L.; Hanrahan, J.; Richardson, L.; Donovan, A.; Fair, S.; Evans, A.; Lonergan, P. Effect of storage duration, storage temperature, and diluent on the viability and fertility of fresh ram sperm. *Theriogenology* **2010**, *73*, 541–549. [CrossRef]
7. Druart, X.; Cognié, J.; Baril, G.; Clément, F.; Dacheux, J.-L.; Gatti, J.-L. In vivo imaging of in situ motility of fresh and liquid stored ram spermatozoa in the ewe genital tract. *Reproduction* **2009**, *138*, 45–53. [CrossRef]
8. Arando, A.; Delgado, J.V.; León, J.M.; Nogales, S.; Navas-González, F.J.; Pizarro, M.G.; Pérez-Marín, C.C. Effect of three commercial extenders on sperm motility and fertility in liquid ram semen stored at 15 °C or 5 °C. *Acta Vet.-Hung.* **2019**, *67*, 430–444. [CrossRef]
9. Rizkallah, N.; Chambers, C.G.; de Graaf, S.P.; Rickard, J.P. Factors Affecting the Survival of Ram Spermatozoa during Liquid Storage and Options for Improvement. *Animals* **2022**, *12*, 244. [CrossRef]
10. Zhang, L.M.; Sohail, T.; Li, Y.J.; Wang, Y.H.; Chu, C.J.; Feng, Y.K.; Ma, J.L. Effect of different extenders on the sperm quality parameters of hu ram semen preserved at 16 °C. *Kafkas. Univ. Vet. Fak. Derg.* **2021**, *27*, 83–91.
11. Zhang, L.M.; Sohail, T.; Wang, Y.H.; Kang, Y.; Wang, X.Y.; Sun, X.M.; Li, Y.J. The effect of different storage temperature on Hu ram sperm parameters. *Kafkas. Univ. Vet. Fak. Derg.* **2022**, *28*, 201–209.
12. Paulenz, H.; Söderquist, L.; Pérez-Pé, R.; Berg, K.A. Effect of different extenders and storage temperatures on sperm viability of liquid ram semen. *Theriogenology* **2002**, *57*, 823–836. [CrossRef] [PubMed]
13. Câmara, D.; Mello-Pinto, M.; Pinto, L.; Brasil, O.; Nunes, J.; Guerra, M. Effects of reduced glutathione and catalase on the kinematics and membrane functionality of sperm during liquid storage of ram semen. *Small Rumin. Res.* **2011**, *100*, 44–49. [CrossRef]
14. Gillan, L.; Maxwell, W.M.; Evans, G. Preservation and evaluation of semen for artificial insemination. *Reprod. Fertil. Dev.* **2004**, *16*, 447–454. [CrossRef]
15. Eslami, M.; Jahan-Roshan, N.; Farrokhi-Ardabili, F. Influence of idebenone on ram semen quality stored at 4 °C. *Reprod. Domest. Anim.* **2018**, *54*, 486–497. [CrossRef]
16. Paulenz, H.; Kommisrud, E.; Hofmo, P. Effect of Long-Term Storage at Different Temperatures on the Quality of Liquid Boar Semen. *Reprod. Domest. Anim.* **2010**, *35*, 83–87. [CrossRef]
17. Karunakaran, M.; Dhali, A.; Mech, A.; Khate, K.; Rajkhowa, C.; Mishra, D. Preservation of mithun (*Bos frontalis*) semen at refrigeration temperature. *Anim. Reprod. Sci.* **2007**, *101*, 257–264. [CrossRef]
18. Meng, N.N. Effect of Skimmed-milk Powder and Soybean Lecithin Extender on Sheep Semen Preservation in Low Temperature. Master's Thesis, Hebei Agricultural University, Baoding, China, 2015.
19. Santymire, R.M.; Marinari, P.E.; Kreeger, J.S.; Wildt, D.E.; Howard, J.G. Slow cooling prevents cold-induced damage to sperm motility and acrosomal integrity in the black-footed ferret (*Mustela nigripes*). *Reprod. Fertil. Dev.* **2007**, *19*, 652–663. [CrossRef]
20. Ashrafi, I.; Kohram, H.; Naijan, H.; Bahreini, M.; Mirzakhani, H. Erratum to: Retraction: Effect of controlled and uncontrolled cooling rate on motility parameters of cryopreserved ram spermatozoa. *BMC Res. Notes* **2012**, *5*, 319. [CrossRef]
21. Wang, Y.; Zhang, L.; Sohail, T.; Kang, Y.; Sun, X.; Li, Y. Chlorogenic Acid Improves Quality of Chilled Ram Sperm by Mitigating Oxidative Stress. *Animals* **2022**, *12*, 163. [CrossRef]
22. Zhang, L.; Wang, Y.; Sohail, T.; Kang, Y.; Niu, H.; Sun, X.; Ji, D.; Li, Y. Effects of Taurine on Sperm Quality during Room Temperature Storage in Hu Sheep. *Animals* **2021**, *11*, 2725. [CrossRef]
23. Amann, R.P. Can the fertility potential of a seminal sample be predicted accurately? *J. Androl.* **1989**, *10*, 89–98. [CrossRef] [PubMed]
24. Merati, Z.; Farshad, A.; Farzinpour, A.; Rostamzadeh, J.; Sharafi, M. Anti-apoptotic effects of minocycline on ram epididymal spermatozoa exposed to oxidative stress. *Theriogenology* **2018**, *114*, 266–272. [CrossRef] [PubMed]
25. McDonough-Goldstein, C.E.; Whittington, E.; McCullough, E.L.; Buel, S.M.; Erdman, S.; Pitnick, S.; Dorus, S. Pronounced Postmating Response in the Drosophila Female Reproductive Tract Fluid Proteome. *Mol. Cell. Proteom.* **2021**, *20*, 100156. [CrossRef]
26. Albrizio, M.; Lacalandra, G.M.; Cinone, M. The role of bicarbonate in the modulation of capacitation, spontaneous acrosome reaction and motility of equine fresh and frozen spermatozoa. *Theriogenology* **2022**, *187*, 112–118. [CrossRef] [PubMed]
27. Hammerstedt, R.H.; A Lardy, H. The effect of substrate cycling on the ATP yield of sperm glycolysis. *J. Biol. Chem.* **1983**, *258*, 8759–8768. [CrossRef] [PubMed]
28. Molinia, F.; Evans, G.; Maxwell, W. In vitro evaluation of zwitterion buffers in diluents for freezing ram spermatozoa. *Reprod. Nutr. Dev.* **1994**, *34*, 491–500. [CrossRef]
29. Yildiz, C.; Kaya, A.; Aksoy, M.; Tekeli, T. Influence of sugar supplementation of the extender on motility, viability and acrosomal integrity of dog spermatozoa during freezing. *Theriogenology* **2000**, *54*, 579–585. [CrossRef]

30. Maxwell, W.; Salamon, S. Liquid storage of ram semen: A review. *Reprod. Fertil. Dev.* **1993**, *5*, 613–638. [CrossRef]
31. Subbaiah, P.V.; Horvath, P.; Achar, S.B. Regulation of the Activity and Fatty Acid Specificity of Lecithin-Cholesterol Acyltransferase by Sphingomyelin and Its Metabolites, Ceramide and Ceramide Phosphate. *Biochemistry* **2006**, *45*, 5029–5038. [CrossRef]
32. Bergeron, A.; Manjunath, P. New insights towards understanding the mechanisms of sperm protection by egg yolk and milk. *Mol Reprod. Dev.* **2006**, *73*, 1338–1344. [CrossRef] [PubMed]
33. Anzar, M.; Rajapaksha, K.; Boswall, L. Egg yolk-free cryopreservation of bull semen. *PLoS ONE* **2019**, *14*, e0223977. [CrossRef] [PubMed]
34. Gogol, P.; Bryła, M.; Trzcińska, M.; Bochenek, M. Quality parameters and fertility of ram semen cryopreserved in egg yolk and soybean lecithin supplemented extenders. *Pol. J. Vet. Sci.* **2019**, *22*, 177–179.
35. Drobnis, E.Z.; Crowe, L.M.; Berger, T.; Anchordoguy, T.J.; Overstreet, J.W.; Crowe, J.H. Cold shock damage is due to lipid phase transitions in cell membranes: A demonstration using sperm as a model. *J. Exp. Zool.* **1993**, *265*, 432–437. [CrossRef]
36. Selige, C.; Quirino, M.; Schröter, F.; Aepli, M.; Schulze, M. Effect of pasteurized egg yolk on the quality of cryopreserved boar semen. *Reprod. Domest. Anim.* **2021**, *56*, 1152–1155. [CrossRef]
37. Kmenta, I.; Strohmayer, C.; Müller-Schlösser, F.; Schäfer-Somi, S. Effects of a lecithin and catalase containing semen extender and a second dilution with different enhancing buffers on the quality of cold-stored canine spermatozoa. *Theriogenology* **2011**, *75*, 1095–1103. [CrossRef] [PubMed]
38. Diaz-Jimenez, M.; Dorado, J.; Pereira, B.; Consuegra, C.; Ortiz, I.; Hidalgo, M. Is sperm cryopreservation in absence of permeable cryoprotectants suitable for subfertile donkeys? *Reprod. Domest. Anim.* **2019**, *54*, 102–105. [CrossRef]
39. Miller, D.J.; Winer, M.A.; Ax, R.L. Heparin-Binding Proteins from Seminal Plasma Bind to Bovine Spermatozoa and Modulate Capacitation by Heparin1. *Biol. Reprod.* **1990**, *42*, 899–915. [CrossRef]
40. Osuagwuh, U.I.; Palomo, M.J. Effect of semen washing on thawed ram spermatozoa subjected to a four hour post-thawing thermal evaluation test. *Small Rumin. Res.* **2017**, *155*, 81–86. [CrossRef]
41. Lv, S.J.; Zhang, S.S.; Dai, J.J.; Sun, L.W.; Ge, L.; Liang, F.; Zhang, D.F. Effects of diluents and thawing conditions on the quality of Hu rams semen freezing. *Acta Agric. Shanghai* **2019**, *35*, 63–66.
42. Xing, Y. Prolonged the Liquid Preservation Time of Boar Semen by Keeping pH Stable. Master's Thesis, Northeast Agricultural University, Harbin, China, 2017.
43. Wang, E.; Huang, Y.; Du, Q.; Sun, Y. Silver nanoparticle induced toxicity to human sperm by increasing ROS (reactive oxygen species) production and DNA damage. *Environ. Toxicol. Pharmacol.* **2017**, *52*, 193–199. [CrossRef] [PubMed]
44. Gavella, M.; Lipovac, V. Protective effects of exogenous gangliosides on ROS-induced changes in human spermatozoa. *Asian J. Androl.* **2013**, *15*, 375–381. [CrossRef] [PubMed]
45. Sanocka, D.; Kurpisz, M. Reactive oxygen species and sperm cells. *Reprod. Biol. Endocrinol.* **2004**, *2*, 12. [CrossRef] [PubMed]
46. Bucak, M.N.; Tekin, N. Protective effect of taurine, glutathione and trehalose on the liquid storage of ram semen. *Small Rumin. Res.* **2007**, *73*, 103–108. [CrossRef]
47. Bonato, M.; Cornwallis, C.K.; Malecki, I.A.; Rybnik-Trzaskowska, P.K.; Cloete, S.W. The effect of temperature and pH on the motility and viability of ostrich sperm. *Anim. Reprod. Sci.* **2012**, *133*, 123–128. [CrossRef]

Disclaimer/Publisher's Note: The statements, opinions and data contained in all publications are solely those of the individual author(s) and contributor(s) and not of MDPI and/or the editor(s). MDPI and/or the editor(s) disclaim responsibility for any injury to people or property resulting from any ideas, methods, instructions or products referred to in the content.

animals

MDPI

Article

Effect of Ferulic Acid on Semen Quality of Goat Bucks during Liquid Storage at 17 °C

Feng Zhang [1], Shichang Han [1,2], Nian Zhang [1], Jin Chai [2,*] and Qi Xiong [1,*]

1 Hubei Key Laboratory of Animal Embryo Engineering and Molecular Breeding, Institute of Animal Husbandry and Veterinary, Hubei Academy of Agricultural Sciences, Wuhan 430064, China; zhangfeng0130@163.com (F.Z.); hanshic1235@163.com (S.H.); zhangnian@hbaas.com (N.Z.)

2 Key Laboratory of Swine Genetics and Breeding of the Agricultural Ministry, Key Laboratory of Agricultural Animal Genetics, Breeding and Reproduction of the Ministry of Education, College of Animal Science and Technology, Huazhong Agricultural University, Wuhan 430070, China

* Correspondence: chaijin@mail.hzau.edu.cn (J.C.); phenixxq@163.com (Q.X.)

Simple Summary: This study aims to explore whether ferulic acid (FA), as an exogenous antioxidant, could improve the quality of goat semen during liquid storage at 17 °C. Semen was collected from three black-headed goat bucks and stored at 17 °C with different concentrations of FA. The effect of FA on semen quality was evaluated via semen quality detection and antioxidant index analysis. The results showed that adding 50 µmol/L of FA significantly improved the semen quality from 1 to 5 days. Furthermore, supplementing semen with 50 µmol/L of FA preserved at 17 °C for 3 days had no significant effect on fertility. Overall, adding 50 µmol/L of FA in dilution improved the quality of goat semen stored at 17 °C.

Abstract: This study investigated the effect of different concentrations of ferulic acid (FA) on the quality of goat semen preserved at 17 °C. First, semen was collected from three black-headed goat bucks using an artificial vagina. Then, the mixed semen was diluted with basal dilutions containing different concentrations of FA (0, 25, 50, 100, and 200 µmol/L) and stored at 17 °C. Sperm total motility, plasma membrane integrity, acrosome integrity, reactive oxygen species (ROS) levels, malondialdehyde (MDA) content, and total antioxidant capacity (T-AOC) were measured during semen storage. The results showed that sperm total motility, plasma membrane integrity, and acrosome integrity were significantly improved in the 50 µmol/L FA group compared with the control group (0 µmol/L) on days 1–5, and the level of T-AOC significantly increased, while the contents of ROS and MDA significantly reduced. Meanwhile, the goats' conception rate showed that supplementing semen with 50 µmol/L FA preserved at 17 °C for 3 days had no significant effect on fertility. Taken together, our findings suggest that adding 50 µmol/L FA in dilution at 17 °C can improve goat bucks' semen quality.

Keywords: ferulic acid; goat; liquid storage; 17 °C; semen quality

Citation: Zhang, F.; Han, S.; Zhang, N.; Chai, J.; Xiong, Q. Effect of Ferulic Acid on Semen Quality of Goat Bucks during Liquid Storage at 17 °C. *Animals* **2023**, *13*, 2469. https://doi.org/10.3390/ani13152469

Academic Editor: Anna Wysokińska

Received: 29 June 2023
Revised: 28 July 2023
Accepted: 28 July 2023
Published: 31 July 2023

Copyright: © 2023 by the authors. Licensee MDPI, Basel, Switzerland. This article is an open access article distributed under the terms and conditions of the Creative Commons Attribution (CC BY) license (https://creativecommons.org/licenses/by/4.0/).

1. Introduction

With the development of science and technology, artificial insemination (AI) is seeing wider use in goat breeding. The success of AI is largely dependent on the quality of preserved semen, which is a key factor affecting its effectiveness [1,2]. However, compared with other livestock, semen preservation in goats is less studied, and the technology is still not mature. Thus, we are seeking methods that can be applied to the preservation of goat buck semen. Semen preservation technology allows semen to be preserved for longer by maintaining the functional, ultrastructural, and biochemical properties of the sperm after a specific treatment, followed by preservation in the corresponding environment [3]. It can be divided into liquid and frozen preservation, according to the preservation temperature [4]. Liquid preservation has the advantages of simple operation, low requirements for preservation conditions, and a good effect on fertility compared with frozen preservation. Therefore, the investigation of goat buck semen stored in a liquid state is important to goat AI.

Semen quality degradation is inevitable during storage, and one of the main reasons for this is oxidative damage. Reactive oxygen species (ROS) are products of the aerobic metabolism of sperm. Low concentrations of ROS are essential for sperm function, such as sperm capacitation, acrosome reaction, sperm–egg binding, and related signaling pathways [5]. However, high concentrations of ROS reduce sperm viability, plasma membrane integrity, and increase DNA damage through lipid peroxidation, leading to cell apoptosis, and reducing semen preservation quality [6,7]. In general, various enzyme systems involved in redox homeostasis regulate the ROS level of semen, whereas oxidative stress occurs as a result of oxidant excess, antioxidant deficiency, or both [8]. There are antioxidant defense balance systems in semen, including superoxide dismutase, catalase, glutathione peroxidase, glutathione reductase, and nonenzymatic antioxidants such as methionine, ascorbic acid, and α-tocopherol, etc. [9–12]. When semen is preserved in vitro, the balance is easily disrupted, so it is necessary to add antioxidants or other protective agents to semen to avoid oxidative stress damage [2].

Ferulic acid (FA), a natural phenolic phytochemical widely present in plants, is an important antioxidant. Previous studies have reported that FA plays a positive role in the preservation of rooster [13], boar [14], and Qezel ram [15] semen. However, to our knowledge, the effect of FA on the quality of goat buck semen during liquid storage has not been reported. Therefore, to test whether FA can improve the preservation quality of goat buck semen during liquid storage, FA of different concentrations was added to the semen diluent of goats and stored at 17 °C. Then, the effect of FA on semen quality was evaluated via semen quality detection and antioxidant index analysis. Finally, the optimal concentration of FA for semen preservation at 17 °C was determined.

2. Materials and Methods

2.1. Semen Collection and Liquid Preservation

The semen of three black-headed goat bucks (3–5 years old) was used in this study. The black-headed goat is a new breed based on Macheng Black Goat (Chinese local breed) and introduces Boer Goat lineage. It has a black head and a white body. Prior to this study, goat bucks were ejaculated two times per week during the breeding seasons. In this study, one ejaculate from each goat buck was collected using an artificial vagina in the presence of estrus goats. The semen volume of each goat buck collected was from 0.5~1.5 mL every time. The semen used was milky white and had no abnormal smell. The sperm concentration was >2 × 10^9 sperm/mL, and the total motility was >0.8. The semen was pooled to minimize individual differences between goat bucks.

Tris-based solution composed of Tris 250 mmol/L, fructose 100 mmol/L, sodium citrate 75 mmol/L, penicillin 50,000 IU, streptomycin 50,000 IU, and double-distilled water 100 mL (chemicals all from Sigma, St. Louis, MO, USA) was used as the base extender. The mixed semen was diluted 10-fold with diluent and divided equally into five aliquots. FA was added to the base extender at concentrations of 25, 50, 100, and 200 μmol/L, while the control was the base extender without FA. All samples were stored in a constant temperature refrigerator at 17 °C (shaken and turned over every 12 h) and used for the experiment.

2.2. Sperm Total Motility Analysis

Every 24 h, 10 μL of semen was taken, respectively, from four FA treatments and one control group, shaken slowly before use, diluted 10 times by adding isothermal base diluent, and incubated in a 37 °C water bath for 3 min. After incubation, 5 μL of the semen was added dropwise to a prewarmed slide, covered with a coverslip, and placed on a HT CASA II automatic sperm analyzer (Hamilton Thorne, Beverly, MA, USA) thermostatic loading table to assess sperm total motility. Five or more fields of view were randomly selected, and the total number of sperm observed was at least 800. There were three replicates per group.

2.3. Sperm Plasma Membrane Integrity Analysis

Plasma membrane integrity assay was performed with a commercial kit (Solarbio, Beijing, China). In brief, 20 μL of semen was taken, respectively, from four FA treatments and one control group and added to 200 μL of hypotonic solution. The solution was prewarmed, stirred gently, and incubated for 30 min at 37 °C. After incubation, 10 μL of the mixture was spread evenly on a slide for observation via light microscopy. Three or more fields of view were randomly selected, and the total number of sperm observed was at least 200. There were three replicates per group.

2.4. Sperm Acrosome Integrity Analysis

Semen from four FA treatments and one control group was used to analyze the sperm acrosome integrity. Fluorescein isothiocyanate-peanut agglutinin (FITC-PNA, Sigma, St. Louis, MO, USA) combined with 4,6-diamino-2-phenyl indole (DAPI, Beyotime (Shanghai, China)) was used to detect the sperm acrosome integrity. Additionally, 10 μL of semen was resuspended with phosphate-buffered saline (PBS) to a concentration of 1×10^6 sperm/mL, and then 10 μL of FITC-PNA staining solution was added and incubated in darkness for 10 min at 37 °C. After incubation, sperm cells were collected via centrifugation at 1500× *g* for 5 min; this process was repeated 1–2 times to wash off the floating color. The sperm cells were then fixed in 4% paraformaldehyde at room temperature for 30 min. After fixation, the cells were washed with PBS and incubated with DAPI. FITC-PNA staining was judged by the criteria that sperm with broken acrosomes were stained with green fluorescence, while sperm without green fluorescence had intact acrosomes. Three or more fields of view were randomly selected, and the total number of sperm observed was at least 200. There were three replicates per group.

2.5. ROS, Malondialdehyde (MDA) Content, and Total Antioxidant Capacity (T-AOC) Activity Assays

Semen from four FA treatments and one control group was used to analyze the ROS and MDA content and the T-AOC activity.

The ROS content was measured using the reactive oxygen species assay kit (Nanjing Jiancheng Bioengineering Institute, Jiangsu, China), following the manufacturer's protocol. In brief, 100 μL of semen was taken in a centrifuge tube and centrifuged at 1000× *g* for 10 min; the supernatant was discarded. Next, 500 μL of diluted 10 μmol/L DCFH-DA was added to resuspend the precipitate in a water bath at 37 °C for 30 min. At the end of the incubation period, the samples were centrifuged at 1000× *g* for 10 min, and the supernatant was discarded. The precipitate was washed twice with PBS, and the cell precipitate was collected via centrifugation; then, 600 μL PBS was added to resuspend the precipitate, and 20 μL suspension was added to the 96-well plate to measure the fluorescence value. A multifunctional enzyme standard was used; the excitation wavelength was set at 488 nm and the emission wavelength at 525 nm. There were three replicates per group.

The MDA content was measured using the MDA assay kit (Nanjing Jiancheng Bioengineering Institute, Jiangsu, China) according to the manufacturer's instructions. In brief, 100 μL of semen was mixed with 1 mL of MDA assay working solution and incubated in a water bath at 95 °C for 40 min. At the end of the incubation period, the samples were centrifuged at 4000× *g* for 10 min. Additionally, 200 μL of supernatant was added to the 96-well plate to measure the absorbance value at 530 nm and normalize the protein content (nmol/mg protein) using the BCA assay kit (Thermo Scientific, Waltham, MA, USA). There were three replicates per group.

The T-AOC activity was determined using the T-AOC assay kit (Nanjing Jiancheng Bioengineering Institute, Jiangsu, China) according to the manufacturer's instructions. In brief, 200 μL of semen was centrifuged at 800× *g* for 10 min, and the supernatant was discarded. The sperm pellets were collected and repeated twice using precooled PBS; 200 µL of precooled lysate was added, and the sperm pellets were shattered using ultrasonication operated on ice; the corresponding reagents were added sequentially according to the kit instructions. Finally, the absorbance value at 520 nm was measured and normalized to the protein content (U/mg protein) using the BCA assay kit (Thermo Scientific, Waltham, MA USA). There were three replicates per group.

2.6. Fertility Test

To evaluate the effect of FA on fertility, semen containing 0 µmol/L and 50 µmol/L FA was stored at 17 °C for 3 days and injected into synchronized estrus goats (aged 2–3 years) via artificial insemination; fresh semen was injected as the control. The dose of semen used was approximately 5×10^8 sperm/mL. A total of 54 estrus goats were inseminated at random and kept in the same feeding conditions and management environment. Early pregnancy was detected via ultrasound 35–40 days after artificial insemination.

2.7. Statistical Analysis

All the results are presented as the mean ± SE ($n = 3$). One-way ANOVA was used for statistical comparison using SPSS 23.0 software. $p \leq 0.05$ was considered statistically significant.

3. Results

To investigate whether FA affects semen quality and antioxidant capacity during storage at 17 °C, sperm total motility, sperm plasma membrane integrity, sperm acrosome integrity, and ROS, MDA, and T-AOC content were evaluated at different times and at different FA concentrations, respectively.

3.1. Effect of Different Concentrations of FA on Sperm Total Motility of Black-Headed Goats

The effects of different concentrations of FA on black-headed goat sperm total motility during liquid storage at 17 °C are shown in Table 1. On days 1 and 3, the sperm total motility was significantly improved in the 25 and 50 µmol/L FA concentration groups compared with the control group (0 µmol/L, $p < 0.05$); on day 2, the sperm total motility of the 25, 50, and 100 µmol/L groups was significantly higher than that of the control group ($p < 0.05$); on day 4, the sperm total motility in the 25, 50, and 100 µmol/L groups was significantly higher than that in the control group ($p < 0.05$); on day 5, the sperm total motility was highest in the 50 µmol/L group ($p < 0.05$), and the sperm total motility of the 200 µmol/L group was lower than that of the control group, but there was no significant difference. Generally, the sperm total motility was significantly improved in the 25 and 50 µmol/L FA concentration groups.

Table 1. Effect of FA on the sperm total motility of black-headed goats (%).

FA Concentrations	Time of Storage				
	1 d	2 d	3 d	4 d	5 d
0 µmol/L	88.26 ± 0.44^{b}	84.73 ± 2.18^{b}	83.27 ± 0.17^{c}	77.97 ± 0.67^{c}	77.13 ± 0.29^{d}
25 µmol/L	92.33 ± 0.24^{a}	87.40 ± 0.72^{a}	86.13 ± 0.29^{a}	83.60 ± 0.23^{a}	80.63 ± 0.66^{b}
50 µmol/L	92.90 ± 0.29^{a}	87.70 ± 0.20^{a}	85.10 ± 0.06^{ab}	84.20 ± 0.17^{a}	82.57 ± 0.29^{a}
100 µmol/L	88.17 ± 0.65^{b}	86.93 ± 0.44^{a}	83.70 ± 0.95^{bc}	81.40 ± 1.37^{ab}	79.07 ± 0.42^{c}
200 µmol/L	89.43 ± 1.57^{b}	85.73 ± 0.58^{ab}	83.30 ± 1.90^{c}	80.23 ± 0.52^{bc}	76.63 ± 0.35^{d}

Note: For the same column data, different lowercase superscript letters indicate significant differences ($p < 0.05$), while the same letters indicate no significant differences ($p > 0.05$).

3.2. Effect of Different Concentrations of FA on Sperm Plasma Membrane Integrity of Black-Headed Goats

The effect of different concentrations of FA on black-headed goat sperm plasma membrane integrity during liquid storage at 17 °C is shown in Table 2. On day 1, the 50 μmol/L group was significantly higher than that of the 0 and 100 μmol/L groups ($p < 0.05$); the plasma membrane integrity of the 50 μmol/L FA concentration group was significantly higher than that of the other groups on days 3 and 5 ($p < 0.05$), and there was no significant difference between the other groups; on day 2, the 50 μmol/L group was significantly higher than that of the 0, 100, and 200 μmol/L groups ($p < 0.05$). Meanwhile, there was no significant difference between the 50 μmol/L and 25 μmol/L groups. The plasma membrane integrity of the 50 μmol/L group was significantly higher than that of the 25 and 200 μmol/L groups on day 4 of semen preservation ($p < 0.05$). These results suggested that adding 50 μmol/L FA to the semen diluent could significantly improve the integrity of the sperm plasma membrane.

Table 2. Effects of FA on the sperm plasma membrane integrity of black-headed goats (%).

FA Concentrations	Time of Storage				
	1 d	2 d	3 d	4 d	5 d
0 μmol/L	78.34 ± 1.07^{b}	74.30 ± 0.86^{bc}	72.27 ± 0.73^{c}	70.50 ± 0.64^{abc}	64.18 ± 1.15^{bc}
25 μmol/L	80.37 ± 1.22^{ab}	76.71 ± 1.29^{ab}	73.52 ± 1.34^{b}	69.44 ± 1.03^{bc}	64.58 ± 0.52^{bc}
50 μmol/L	81.50 ± 1.32^{a}	78.93 ± 0.84^{a}	78.10 ± 0.54^{a}	72.79 ± 1.70^{a}	71.49 ± 0.87^{a}
100 μmol/L	78.33 ± 1.28^{b}	75.29 ± 0.91^{bc}	72.48 ± 1.14^{bc}	71.21 ± 0.49^{ab}	65.52 ± 0.59^{b}
200 μmol/L	79.74 ± 0.70^{ab}	73.61 ± 1.01^{c}	71.67 ± 0.70^{c}	68.48 ± 1.00^{c}	62.30 ± 1.13^{c}

Note: For the same column data, different lowercase superscript letters indicate significant differences ($p < 0.05$), while the same letters indicate no significant differences ($p > 0.05$).

3.3. Effects of Different Concentrations of FA on the Sperm Acrosome Integrity of Black-Headed Goats

The effect of different concentrations of FA on black-headed goat sperm acrosome integrity during liquid storage at 17 °C is shown in Table 3. Sperm acrosome integrity was significantly improved only in the 50 μmol/L FA concentrations group compared with the 0 μmol/L group on days 1, 2, and 5 ($p < 0.05$); on 3 day, the sperm acrosome integrity in the 25 and 50 μmol/L groups was significantly higher than in the control group ($p < 0.05$); on day 4, sperm acrosome integrity was significantly improved in the 25, 50, and 100 μmol/L groups compared with the 0 μmol/L group.

Table 3. Effects of FA on the sperm acrosome integrity of black-headed goats (%).

FA Concentrations	Time of Storage				
	1 d	2 d	3 d	4 d	5 d
0 μmol/L	92.11 ± 0.70^{b}	88.65 ± 0.83^{b}	86.59 ± 0.70^{c}	83.66 ± 0.66^{c}	81.39 ± 0.70^{bc}
25 μmol/L	93.05 ± 0.54^{ab}	90.51 ± 0.81^{ab}	89.21 ± 0.64^{ab}	86.37 ± 0.77^{b}	83.44 ± 0.73^{b}
50 μmol/L	94.22 ± 0.50^{a}	91.34 ± 0.66^{a}	89.62 ± 0.86^{a}	88.39 ± 0.79^{a}	86.34 ± 0.29^{a}
100 μmol/L	92.03 ± 0.68^{b}	89.32 ± 0.75^{ab}	87.42 ± 0.45^{bc}	85.63 ± 0.61^{b}	82.32 ± 0.81^{bc}
200 μmol/L	92.75 ± 0.52^{ab}	88.84 ± 0.95^{b}	87.27 ± 0.63^{c}	85.03 ± 0.45^{bc}	81.16 ± 0.54^{c}

Note: For the same column data, different lowercase superscript letters indicate significant differences ($p < 0.05$), while the same letters indicate no significant differences ($p > 0.05$).

3.4. Effects of Different Concentrations of FA on Sperm ROS Content in Black-Headed Goats

The effect of different concentrations of FA on black-headed goat sperm ROS content during liquid storage at 17 °C is shown in Table 4. The ROS content was significantly reduced in the 25, 50, and 100 μmol/L FA concentration groups compared with the control group ($p < 0.05$) and reached its lowest in the 50 μmol/L group on the third and fifth days of preservation. There was no significant difference between the 200 μmol/L group and the control group.

Table 4. Effects of FA on the sperm ROS content of black-headed goats.

FA Concentrations	Time of Storage	
	3 d	5 d
0 μmol/L	2631.99 ± 13.81 ^{a}	3092.3 ± 23.39 ^{a}
25 μmol/L	2258.46 ± 57.56 ^{b}	2763.69 ± 15.68 ^{b}
50 μmol/L	2026.3 ± 42.47 ^{c}	2044.18 ± 18.53 ^{c}
100 μmol/L	2148.54 ± 12.47 ^{bc}	2523.23 ± 22.8 ^{b}
200 μmol/L	2643.61 ± 47.78 ^{a}	3223.03 ± 36.11 ^{a}

Note: For the same column data, different lowercase superscript letters indicate significant differences ($p < 0.05$), while the same letters indicate no significant differences ($p > 0.05$).

3.5. *Effects of Different Concentrations of FA on Sperm MDA Content in Black-Headed Goats*

The effect of different concentrations of FA on black-headed goat sperm MDA content during liquid storage at 17 °C is shown in Table 5. The MDA content was significantly reduced in the 50 μmol/L group ($p < 0.05$), while there were no significant differences among the other groups on day 3; on day 5, the MDA content of the 25 and 50 μmol/L groups was significantly decreased, while that of the 200 μmol/L group was significantly increased ($p < 0.05$).

Table 5. Effects of FA on sperm MDA content of black-headed goats (nmol/mg protein).

FA Concentrations	Time of Storage	
	3 d	5 d
0 μmol/L	2.29 ± 0.08 ^{a}	2.68 ± 0.14 ^{b}
25 μmol/L	2.10 ± 0.08 ^{ab}	2.37 ± 0.13 ^{c}
50 μmol/L	1.92 ± 0.10 ^{b}	2.28 ± 0.08 ^{c}
100 μmol/L	2.21 ± 0.05 ^{a}	2.79 ± 0.09 ^{b}
200 μmol/L	2.35 ± 0.12 ^{a}	2.91 ± 0.09 ^{a}

Note: For the same column data, different lowercase superscript letters indicate significant differences ($p < 0.05$), while the same letters indicate no significant differences ($p > 0.05$).

3.6. *Effects of Different Concentrations of FA on Sperm T-AOC Levels in Black-Headed Goats*

The effects of different concentrations of FA on the T-AOC level of black-headed goat sperm during liquid storage at 17 °C are shown in Table 6. The T-AOC level was the highest in the 50 μmol/L FA concentration group either on the third or fifth day of preservation ($p < 0.05$).

Table 6. Effects of FA on the sperm T-AOC level of black-headed goats (U/mg protein).

FA Concentrations	Time of Storage	
	3 d	5 d
0 μmol/L	1.44 ± 0.01 ^{b}	1.22 ± 0.04 ^{ab}
25 μmol/L	1.63 ± 0.07 ^{ab}	0.94 ± 0.04 ^{c}
50 μmol/L	1.69 ± 0.08 ^{a}	1.42 ± 0.02 ^{a}
100 μmol/L	1.59 ± 0.04 ^{b}	1.19 ± 0.03 ^{abc}
200 μmol/L	1.45 ± 0.01 ^{b}	1.03 ± 0.02 ^{bc}

Note: For the same column data, different lowercase superscript letters indicate significant differences ($p < 0.05$), while the same letters indicate no significant differences ($p > 0.05$).

3.7. *Effects of Semen Stored at 17 °C for 3 Days on the Conception Rate of Black-Headed Goats*

The effects of different semen kept in 17 °C liquid storage for 3 days on the conception rate of black-headed goats are shown in Table 7. There was no significant difference between the fresh semen group and the 50 μmol/L FA group ($p < 0.05$), while the 0 μmol/L FA group showed a significant decrease.

Table 7. Effect of different semen on the conception rate of black-headed goats.

Group	Inseminated Goats (n)	Pregnant Goats (n)	Conception Rate (%)
Fresh semen	18	15	83.33 [a]
0 μmol/L FA	16	7	43.75 [b]
50 μmol/L FA	20	16	80.00 [a]

Note: Different lowercase letters among groups indicate significant differences ($p < 0.05$), while the same letters indicate no significant differences ($p > 0.05$).

4. Discussion

Semen preservation during liquid storage has the advantages of being a simple operation, reducing damage to sperm, and having low investment costs. However, the surface of the sperm cell membrane contains large amounts of polyunsaturated fatty acids (PUFA), which are susceptible to free radicals such as ROS attacks, shortening the sperm's survival time [16,17]. Generally, seminal plasma contains a large amount of enzymatic and nonenzymatic antioxidants [18], which act simultaneously to neutralize free radicals and prevent further oxidative reactions [12]. In the process of semen preservation in vitro, its own antioxidants are insufficient to maintain semen quality for long periods of time, so it is crucial to add effective exogenous antioxidants to the semen extender [19].

Ferulic acid (FA) is a natural phenolic phytochemical that exists in plants in free or conjugated forms and can also be covalently bound to plant cell wall polysaccharides. FA has low toxicity and extensive pharmacological activities, such as antioxidant, antiallergic, anticancer, antibacterial, and anti-inflammatory properties, and plays a regulatory role in cell signaling and gene expression [20]. Some studies have reported that FA could be used as an exogenous antioxidant for semen preservation. For instance, FA protected the mitochondrial, acrosome, and plasma membrane integrity of stallion sperm after 8 h of storage at 4 °C [21]. FA is beneficial to human sperm viability and motility in both fertile and infertile individuals [22]. Trans-FA could ameliorate the toxic effect of β-cyfluthrin via the reduction of peroxidative and nitrosative reactions during the cold preservation (4 °C) of rooster semen [13]. Pei et al. [14] found that FA significantly improved the quality of frozen and thawed boar sperm. In this study, to test whether FA added at different concentrations to the semen extender can improve the preservation quality of goat buck semen during liquid storage at 17 °C, motility parameters such as sperm total motility, functional integrity factors such as plasma membrane integrity and acrosome integrity rates, and oxidative status parameters such as ROS, MDA, and T-AOC were evaluated. Our results demonstrated that adding an appropriate concentration of FA significantly improved sperm total motility, plasma membrane integrity, and acrosome integrity while reducing the peroxidation rate (as evidenced by the levels of ROS, MDA, and T-AOC). These results indicated that FA could alleviate oxidative damage during the liquid storage of goat buck semen and improve the quality of semen preservation. Abnormal sperm morphology, DNA damage, and lipid peroxidation affect fertilization and early embryonic development [23,24]. To further evaluate the effect of FA on semen quality, the conception rate of estrus goats was tested. Our results showed that the conception rate of goats was not significantly affected by semen stored at 17 °C with 50 μmol/L FA for 3 days.

ROS, a class of single electrons of oxygen, comprises superoxide anions (O^{2-}), hydrogen peroxide (H_2O_2), and hydroxyl radicals (OH^-) [25]. The content of ROS increases significantly with preservation time, resulting in decreased sperm motility and both changes in membrane permeability and acrosome structure [26], reducing the quality of semen. MDA is the final product of free radicals and lipid peroxidation [27] and has been used in various biochemical assays to monitor the degree of peroxidative damage sustained by sperm [28]. Both ROS and MDA negatively affect semen quality. T-AOC activity could reveal the total antioxidant capacity of semen [29]. In this study, we chose ROS, MDA, and T-AOC as redox parameters to evaluate semen's oxidative status.

It was noteworthy that the change in sperm total motility, plasma membrane integrity, acrosome integrity, and T-AOC levels showed a trend of rising first and then falling with the increase in FA concentration, while the contents of ROS and MDA decreased first and then increased. On the one hand, the high concentration of FA may influence the osmotic pressure of the extender, affecting the permeability of the sperm membrane, destroying the sperm structure, and reducing sperm progressive motility [30,31]. On the other hand, the high concentration of FA may be toxic to and damage sperm, and it may cause excessive activation of antioxidant enzymes and mitochondria, which affects the physiological state of the sperm [32]. Additionally, high amounts of antioxidative substances disturb redox balances and act as a pro-oxidant, increasing proinflammatory mediators of free radicals, stimulating oxidative toxicity, and nitrosylation of proteins [33–36].

The protective ability of antioxidants depends on their concentration, and lipid peroxidation is induced when antioxidants and ROS production are imbalanced [37,38]. Some studies have shown that FA has a dose-dependent protective effect on semen: low doses increased cell survival, while high doses increased apoptosis. Shayan-Nasr et al. [13] reported that trans-FA at 10 and 25 mM doses significantly restored the motility and viability of rooster spermatozoa at 4 °C. When exceeding the optimal concentration, FA not only failed to provide adequate protection but also had adverse effects on sperm viability and metabolites of the diluent, inducing elevated MDA content and reduced T-AOC levels. Pei et al. [14] reported that supplementation of the freezing extender with 0.25 mM or 0.35 mM FA has a beneficial effect on frozen and thawed boar sperm. The sperm viability, plasma membrane integrity, and acrosome integrity of thawed spermatozoa reached a maximum at the optimal addition of FA, while these parameters remained stable or even showed a significant decrease as the concentration continued to increase [14]. Salimi et al. [15] showed that 5 mM and 10 mM trans-FA improved the forward progressive motility and curvilinear velocity of ram semen at 4 °C; samples treated with 25 mM trans-FA showed the lowest total motility, forward progressive motility, and viability. In this study, we found that the optimal amount of FA added during the storage of black-headed goat semen at 17 °C was 50 μmol/L, which differed from the results of previous studies. This effect may be the result of different extenders, different dilution ratios, different animal species, and different storage procedures.

5. Conclusions

The results of this study showed that when 50 μmol/L FA was added to the extender, the sperm total motility, plasma membrane integrity, acrosome integrity, and T-AOC were significantly higher than those of the control group throughout the preservation period. Comprehensive indicators of all aspects showed that the concentration of 50 μmol/L FA was the best for the preservation of black-headed goat bucks' semen at 17 °C; it could effectively alleviate the oxidative stress damage caused by ROS, prolong the survival time of sperm, improve the quality of semen preservation, and do so without affecting the goat conception rate within 3 days. This study provides an important reference value for the liquid preservation of goat semen and helps promote the application of AI technology in goat breeding.

Author Contributions: Conceptualization, F.Z., S.H., J.C. and Q.X.; methodology, F.Z. and S.H.; software, F.Z.; validation, F.Z.; formal analysis, F.Z.; investigation, S.H. and N.Z.; resources, S.H. and N.Z.; data curation, S.H.; writing—original draft preparation, F.Z. and S.H.; writing—review and editing, F.Z., S.H., J.C., N.Z. and Q.X.; visualization, F.Z.; supervision, J.C. and Q.X.; project administration, J.C. and Q.X.; funding acquisition, Q.X. All authors have read and agreed to the published version of the manuscript.

Funding: This research was funded by the Key Research and Development Program of Hubei Province, China (2021BBA086 and 2022BBA0049), the Innovation Team Project of Hubei Provincial Agricultural Science and Technology Innovation Center (2021-620-000-001-020).

Institutional Review Board Statement: The animal study protocol was approved by the Scientific Ethics Committee of the Hubei Academy of Agricultural Sciences (No. HBAASEC-2021-015 on 19 October 2021).

Informed Consent Statement: Not applicable.

Data Availability Statement: The data presented in this study are available upon request from the corresponding author.

Conflicts of Interest: The authors declare no conflict of interest.

References

1. Faigl, V.; Vass, N.; Jávor, A.; Kulcsár, M.; Solti, L.; Amiridis, G.; Cseh, S. Artificial insemination of small ruminants—A review. *Acta Vet. Hung.* **2012**, *60*, 115–129. [CrossRef] [PubMed]
2. Rizkallah, N.; Chambers, C.G.; de Graaf, S.P.; Rickard, J.P. Factors Affecting the Survival of Ram Spermatozoa during Liquid Storage and Options for Improvement. *Animals* **2022**, *12*, 244. [CrossRef] [PubMed]
3. Bucak, M.N.; Tekin, N. Protective effect of taurine, glutathione and trehalose on the liquid storage of ram semen. *Small Rumin. Res.* **2007**, *73*, 103–108. [CrossRef]
4. Salamon, S.; Maxwell, W.M. Storage of ram semen. *Anim. Reprod. Sci.* **2000**, *62*, 77–111. [CrossRef] [PubMed]
5. Agarwal, A.; Aponte-Mellado, A.; Premkumar, B.J.; Shaman, A.; Gupta, S. The effects of oxidative stress on female reproduction: A review. *Reprod. Biol. Endocrin.* **2012**, *10*, 49. [CrossRef]
6. Guthrie, H.D.; Welch, G.R. Effects of reactive oxygen species on sperm function. *Theriogenology* **2012**, *78*, 1700–1708. [CrossRef]
7. Moradi, M.; Karimi, I.; Ahmadi, S.; Mohammed, L.J. The necessity of antioxidant inclusion in caprine and ovine semen extenders: A systematic review complemented with computational insight. *Reprod. Domest. Anim.* **2020**, *55*, 1027–1043. [CrossRef]
8. Silvestre, M.A.; Yániz, J.L.; Peña, F.J.; Santolaria, P.; Castelló-Ruiz, M. Role of Antioxidants in Cooled Liquid Storage of Mammal Spermatozoa. *Antioxidants* **2021**, *10*, 1096. [CrossRef]
9. Gadea, J.; García-Vazquez, F.; Matás, C.; Gardón, J.C.; Cánovas, S.; Gumbao, D. Cooling and freezing of boar spermatozoa: Supplementation of the freezing media with reduced glutathione preserves sperm function. *J. Androl.* **2005**, *26*, 396–404. [CrossRef]
10. Weir, C.P.; Robaire, B. Spermatozoa have decreased antioxidant enzymatic capacity and increased reactive oxygen species production during aging in the Brown Norway rat. *J. Androl.* **2007**, *28*, 229–240. [CrossRef]
11. Câmara, D.R.; Mello-Pinto, M.M.C.; Pinto, L.C.; Brasil, O.O.; Nunes, J.F.; Guerra, M.M.P. Effects of reduced glutathione and catalase on the kinematics and membrane functionality of sperm during liquid storage of ram semen. *Small Rumin. Res.* **2011**, *100*, 44–49. [CrossRef]
12. Gibb, Z.; Blanco-Prieto, O.; Bucci, D. The role of endogenous antioxidants in male animal fertility. *Res. Vet. Sci.* **2021**, *136*, 495–502. [CrossRef]
13. Shayan-Nasr, M.; Ghaniei, A.; Eslami, M.; Zadeh-Hashem, E. Ameliorative role of trans-ferulic acid on induced oxidative toxicity of rooster semen by β-cyfluthrin during low temperature liquid storage. *Poult. Sci.* **2021**, *100*, 101308. [CrossRef]
14. Pei, Y.; Yang, L.; Wu, L.; He, H.; Geng, G.; Xu, D.; Chen, H.; Li, Q. Combined effect of apigenin and ferulic acid on frozen-thawed boar sperm quality. *Anim. Sci. J.* **2018**, *89*, 956–965. [CrossRef]
15. Salimi, A.; Eslami, M.; Farrokhi-Ardabili, F. Influence of trans-ferulic acid on the quality of ram semen upon cold preservation. *Vet. Med. Sci.* **2023**, *9*, 1369–1378. [CrossRef]
16. Sarlos, P.; Molnar, A.; Kokai, M.; Gábor, G.; Rátky, J. Comparative evaluation of the effect of antioxidants in the conservation of ram semen. *Acta Vet. Hung.* **2002**, *50*, 235–245. [CrossRef]
17. Maneesh, M.; Jayalekshmi, H. Role of reactive oxygen species and antioxidants on pathophysiology of male reproduction. *Indian J. Clin. Biochem.* **2006**, *21*, 80–89. [CrossRef]
18. Patil, S.; Kumar, P.; Singh, G.; Bala, R.; Jerome, A.; Patil, C.S.; Kumar, D.; Singh, S.; Sharma, R.K. 'Semen dilution effect' on sperm variables and conception rate in buffalo. *Anim. Reprod. Sci.* **2020**, *214*, 106304. [CrossRef]
19. Wang, Y.; Zhang, L.; Sohail, T.; Kang, Y.; Sun, X.; Li, Y. Chlorogenic Acid Improves Quality of Chilled Ram Sperm by Mitigating Oxidative Stress. *Animals* **2022**, *12*, 163. [CrossRef]
20. Kumar, N.; Pruthi, V. Potential applications of ferulic acid from natural sources. *Biotechnol. Rep.* **2014**, *4*, 86–93. [CrossRef]
21. Affonso, F.J.; Carvalho, H.F.; Lançoni, R.; Lemes, K.M.; Leite, T.G.; Oliveira, L.Z.; Celeghini, E.C.C.; Arruda, R.P. Addition of antioxidants myoinositol, ferulic Acid, and melatonin and their effects on sperm motility, membrane integrity, and reactive oxygen species production in cooled equine semen. *J. Equine Vet. Sci.* **2017**, *59*, 57–63. [CrossRef]
22. Zheng, R.L.; Zhang, H. Effects of ferulic acid on fertile and asthenozoospermic infertile human sperm motility, viability, lipid peroxidation, and cyclic nucleotides. *Free Radic. Biol. Med.* **1997**, *22*, 581–586. [CrossRef] [PubMed]
23. Aitken, R.J.; De Iuliis, G.N.; McLachlan, R.I. Biological and clinical significance of DNA damage in the male germ line. *Int. J. Androl.* **2009**, *32*, 46–56. [CrossRef] [PubMed]

24. Lone, F.A.; Naikoo, M.; Khatun, A.; Shah, R.A.; Pampori, Z.A.; Khan, H.M.; Ahanger, A.A. Idebenone improves quality of ram sperm by mitigating oxidative stress during cryopreservation. *Cryobiology* **2019**, *90*, 15–20. [CrossRef] [PubMed]
25. Carrière, A.; Fernandez, Y.; Rigoulet, M.; Pénicaud, L.; Casteilla, L. Inhibition of preadipocyte proliferation by mitochondrial reactive oxygen species. *FEBS Lett.* **2003**, *550*, 163–167. [CrossRef]
26. Taylor, K.; Roberts, P.; Sanders, K.; Burton, P. Effect of antioxidant supplementation of cryopreservation medium on post-thaw integrity of human spermatozoa. *Reprod. Biomed. Online* **2009**, *18*, 184–189. [CrossRef]
27. Tsikas, D. Assessment of lipid peroxidation by measuring malondialdehyde (MDA) and relatives in biological samples: Analytical and biological challenges. *Anal. Biochem.* **2017**, *524*, 13–30. [CrossRef]
28. Aitken, J.; Krausz, C.; Buckingham, D. Relationships between biochemical markers for residual sperm cytoplasm, reactive oxygen species generation, and the presence of leukocytes and precursor germ cells in human sperm suspensions. *Mol. Reprod. Dev.* **1994**, *39*, 268–279. [CrossRef]
29. Zhang, X.G.; Liu, Q.; Wang, L.Q.; Yang, G.S.; Hu, J.H. Effects of glutathione on sperm quality during liquid storage in boars. *Anim. Sci. J.* **2016**, *50*, 263–269. [CrossRef]
30. Hu, J.H.; Zan, L.S.; Zhao, X.L.; Li, Q.W.; Jiang, Z.L.; Li, Y.K.; Li, X. Effects of trehalose supplementation on semen quality and oxidative stress variables in frozen-thawed bovine semen. *J. Anim. Sci.* **2010**, *88*, 1657–1662. [CrossRef]
31. Li, H.; Zhang, X.G.; Fang, Q.; Liu, Q.; Du, R.R.; Yang, G.S.; Wang, L.Q.; Hu, J.H. Supplemental effect of different levels of taurine in Modena on boar semen quality during liquid preservation at 17 °C. *Anim. Sci. J.* **2017**, *88*, 1692–1699. [CrossRef]
32. Stephens, T.D.; Brooks, R.M.; Carrington, J.L.; Cheng, L.; Carrington, A.C.; Porr, C.A.; Splan, R.K. Effects of Pentoxifylline, Caffeine, and Taurine on Post-Thaw Motility and Longevity of Equine Frozen Semen. *J. Equine Vet. Sci.* **2013**, *33*, 615–621. [CrossRef]
33. Jiang, Y.; Kusama, K.; Satoh, K.; Takayama, E.; Watanabe, S.; Sakagami, H. Induction of cytotoxicity by chlorogenic acid in human oral tumor cell lines. *Phytomedicine* **2000**, *7*, 483–491. [CrossRef]
34. Du, W.Y.; Xiao, Y.; Yao, J.J.; Hao, Z.; Zhao, Y.B. Involvement of NADPH oxidase in high-dose phenolic acid-induced pro-oxidant activity on rat mesenteric venules. *Exp. Ther. Med.* **2017**, *13*, 17–22. [CrossRef]
35. Dutta, S.; Sengupta, P.; Roychoudhury, S.; Chakravarthi, S.; Wang, C.W.; Slama, P. Antioxidant paradox in male infertility: 'A blind eye' on inflammation. *Antioxidants* **2022**, *11*, 167. [CrossRef]
36. Eslami, M.; Ghaniei, A.; Rad, H.M. Effect of the rooster semen enrichment with oleic acid on the quality of semen during chilled storage. *Poult. Sci.* **2016**, *95*, 1418–1424. [CrossRef]
37. Dowling, D.K.; Simmons, L.W. Reactive oxygen species as universal constraints in life-history evolution. *Proc. Biol. Sci.* **2009**, *276*, 1737–1745. [CrossRef]
38. Câmara, D.R.; Silva, S.V.; Almeida, F.C.; Nunes, J.F.; Guerra, M.M.P. Effects of antioxidants and duration of pre-freezing equilibration on frozen-thawed ram semen. *Theriogenology* **2011**, *76*, 342–350. [CrossRef]

Disclaimer/Publisher's Note: The statements, opinions and data contained in all publications are solely those of the individual author(s) and contributor(s) and not of MDPI and/or the editor(s). MDPI and/or the editor(s) disclaim responsibility for any injury to people or property resulting from any ideas, methods, instructions or products referred to in the content.

animals

MDPI

Article

Glucose Starvation Inhibits Ferroptosis by Activating the LKB1/AMPK Signaling Pathway and Promotes the High Speed Linear Motility of Dairy Goat Sperm

Yu Li, Guangzhi Zhang, Fei Wen, Ming Xian, Songmao Guo, Xing Zhang, Xianzhou Feng, Zhangtao Hu and Jianhong Hu *

Key Laboratory of Animal Genetics, Breeding and Reproduction of Shaanxi Province, College of Animal Science and Technology, Northwest A&F University, Yangling 712100, China; yuli0912@126.com (Y.L.); z13514592253@163.com (G.Z.); 13087561604@163.com (F.W.); xianmingtz@126.com (M.X.); 18854883308@163.com (S.G.); axingzzd66@126.com (X.Z.); fengxianzhou0930@163.com (X.F.); huzhangtao0819@163.com (Z.H.)

* Correspondence: 2008114994@nwafu.edu.cn

Simple Summary: In mammals, sperm acquire fertilization ability after capacitation in vitro or when in the female reproductive tract. Adenosine triphosphate (ATP) is required for sperm capacitation through two main metabolic processes, oxidative phosphorylation (OXPHOS) and glycolysis. This study incubated dairy goat sperm with different concentrations of ROT, FCCP, TIG, and AMPK inhibitors. Sperm motility attributes, ATP content, pyruvate and lactate levels, mitochondrial permeability transition pore fluorescence intensity, MMP, protein synthesis, and ferroptosis were analyzed. The results showed that glucose starvation inhibited ferroptosis by activating the LKB1/AMPK signaling pathway and promoted the motility and linear motility of dairy goat sperm, thereby promoting fertilization.

Abstract: In mammals, sperm acquire fertilization ability after capacitation in vitro or when in the female reproductive tract. The motility patterns of sperm undergo continuous changes from the moment of ejaculation until fertilization in the female reproductive tract. In vitro, hyperactivated motility can be induced through high glucose mediums, while in vivo, it is induced by oviduct fluids. Conversely, sperm maintain linear motility in seminal plasma or uterine fluids that contain low glucose levels. In dairy goat sperm, energy metabolism associated with capacitation depends on the energy sources in vitro, seminal plasma, or the female reproductive tract, especially the glucose levels. However, there is little experimental knowledge that glucose levels affect sperm energy metabolism in dairy goats. To clarify these hypotheses, we incubated dairy goat spermatozoa with different concentrations of rotenone-glucose (ROT), carbonyl cyanide 4-(trifluoromethoxy) phenylhydrazone (FCCP), and tigecycline (TIG) in vitro. Sperm motility attributes, ATP content, pyruvate and lactate levels, mitochondrial permeability transition pore fluorescence intensity, mitochondrial membrane potential (MMP), and protein synthesis were analyzed. Sperm motility patterns changed from circular to linear under low glucose conditions compared with those in high glucose conditions and showed a significant improvement in progressive motility and straight line speed, whereas lactate and pyruvate levels and MMP decreased remarkably. Incubation of spermatozoa with ROT, FCCP, and TIG inhibited sperm mitochondrial activity, protein synthesis, oxidative phosphorylation, and ATP levels, thereby reducing sperm motility, including the progressive motility, straight line speed, and total motility. Simultaneously, incubation of spermatozoa with Compound C under low glucose conditions significantly decreased the ATP levels and MMP, as well as liver kinase B1 and AMPK protein expression. Under low glucose conditions, sperm mainly produce ATP through mitochondrial OXPHOS to achieve high speed linear movement, inhibit ferroptosis through the LKB1/AMPK signaling pathway, and further maintain energy metabolism homeostasis.

Keywords: dairy goat; sperm; glucose; mitochondrial oxidative phosphorylation; AMPK; ferroptosis

Citation: Li, Y.; Zhang, G.; Wen, F.; Xian, M.; Guo, S.; Zhang, X.; Feng, X.; Hu, Z.; Hu, J. Glucose Starvation Inhibits Ferroptosis by Activating the LKB1/AMPK Signaling Pathway and Promotes the High Speed Linear Motility of Dairy Goat Sperm. *Animals* **2023**, *13*, 1442. https://doi.org/10.3390/ani13091442

Academic Editor: Anna Wysokińska

Received: 7 March 2023
Revised: 12 April 2023
Accepted: 20 April 2023
Published: 23 April 2023

Copyright: © 2023 by the authors. Licensee MDPI, Basel, Switzerland. This article is an open access article distributed under the terms and conditions of the Creative Commons Attribution (CC BY) license (https://creativecommons.org/licenses/by/4.0/).

1. Introduction

Sperm are highly specialized cells with unique compositional, morphological, and functional properties. After ejaculation, mammalian sperm acquire fertilization ability after capacitation in the female reproductive tract. Fertilization in mammals is a complex process influenced by multiple factors, especially because the site of ejaculation is at a distance from the site of fertilization [1]. Therefore, sperm migration, capacitation, hyperactivation, and acrosome reaction are all necessary for successful fertilization. In addition, the components in the female reproductive tract change during different reproductive stages. Under normal physiological conditions, the vagina mainly contains glucose, lactic acid, and glycerol, whereas the uterus and oviduct mainly contain glucose, pyruvic acid, and lactic acid [2]. When sperm enter the female reproductive tract, the components in the female reproductive tract are converted into fructose, citrate, glucose, lactic acid, and free amino acids [3]. These substrates of energy metabolism are necessary for spermatogenesis, maturation, and fertilization. As a substrate for sperm preservation in vitro, the glucose concentration should be in accordance with that in the female reproductive tract. Besides glucose, sperm also utilize other substrates such as fructose and sorbitol via glycolysis to maintain motility [4,5]. Sperm motility is driven by ATP from cytosolic glycolysis, mitochondrial OXPHOS, or both [6].

Sperm motility depends on the movement of sperm flagella and is driven by ATP production. Mitochondria are the most important organelles for ATP production in sperm. In human sperm, the ATP required for sperm motility is mainly generated from glycolysis. Although OXPHOS can produce ATP, it is not enough to sustain sperm movement during high motility [7]. Both glycolysis and OXPHOS are necessary to maintain vigorous motility. Incubation of mouse sperm in glucose-free media with a mitochondrial OXPHOS inhibitor (FCCP) and mitochondrial respiratory chain inhibitor (ROT) significantly reduced ATP levels and progressive motility [5,8,9]. Sperm motility parameters and the mitochondrial membrane potential (MMP) decrease significantly in sheep after treatment with FCCP and the glycolytic inhibitor 2-DOG, whereas sensitivity to lipid peroxidation (LPO) increases remarkably [10]. However, ATP production in horse sperm relies on mitochondrial OXPHOS in contrast to that in other mammals [11–15]. Therefore, sperm produce ATP for successful fertilization via different metabolic pathways in different energy conditions. However, the changes in motility patterns and selection of energy metabolic pathways for ATP production in dairy goat sperm in different energy conditions are unknown.

5′-AMP-activated protein kinase (AMPK) is a key regulator of energy homeostasis, which is achieved by stimulating ATP production or inhibiting ATP consumption to maintain energy homeostasis [16]. Glucose starvation is a classical activating condition of AMPK that maintains cell survival and redox homeostasis through multiple pathways [17]. In rat or mouse Sertoli cells, AMPK activation induces an increase in glucose uptake and lactate production, whereas the expressions of glucose transporter 1 (GLUT1) and monocarboxylic acid transporter 4 (MCT4) reduce significantly [18]. Platycodin D inhibits ferroptosis induced by high glucose in HK-2 cells, downregulates ACSL4 and TFR1 expressions, and upregulates FTH-1 and SLC7A11 expressions [19]. Furthermore, glycolysis and pentose phosphate pathways are also inhibited by glucose starvation, which induces oxidative stress that is characterized by an increased production of reactive oxygen species (ROS). This leads to interference with the antioxidant system, redox dysregulation, and cell death [20]. Hence, in this study, sperm motility patterns at different concentrations of glucose were measured, and sperm motility parameters were evaluated using a computer-assisted sperm analysis (CASA) system. We also determined the MMP, the opening fluorescence intensity of mitochondrial permeability transition pores (mPTP), and the protein expression after incubation with a mitochondrial respiratory chain inhibitor (rotenone), a mitochondrial OXPHOS inhibitor (FCCP), and a mitochondrial translation inhibitor (TIG) to confirm the metabolic pathways at different glucose concentrations. Lastly, we explored the relationship between ferroptosis-induced glucose starvation and oxidative damage.

2. Materials and Methods

2.1. Chemicals and Reagents

All chemicals and reagents were purchased from Sigma-Aldrich unless specified otherwise. All antibodies were used for incubation following the manufacturers' instructions and were obtained from the following vendors: COX-1 (13393-1-AP, Proteintech, Wuhan, China), COXVB (11418-2-AP, Proteintech, Wuhan, China), ND3 (ab192306, Abcam, Cambridge, UK), NRF1 (12482-1-AP, Proteintech, Wuhan, China), SLC7A11 (26864-1-AP, Proteintech, Wuhan, China), GPX4 (67763-1-AP, Proteintech, Wuhan, China), DHODH (14877-1-AP, Proteintech, Wuhan, China), β-actin (20536-1-AP, Proteintech, Wuhan, China), horseradish peroxidase (HRP)-labeled goat anti-mouse immunoglobulin (Ig)G (CW0102S, CWBIO, Beijing, China), HRP-labeled goat anti-rabbit IgG (CW0103S, CWBIO, Beijing, China), and enhanced chemiluminescence (ECL) fluid (SuperSignal^{TM} West Femto Maximum Sensitivity Substrate, 34095, Thermo Fisher Scientific, Waltham, MA, USA).

2.2. Animals and Semen Collection

Semen was collected using an artificial vagina in Guan Zhong dairy goats (n = 5) from Shaanxi Aonick Dairy Goat Breeding Co., Ltd. (Weinan, China), that were raised under the same feeding condition as in Weinan, Shaanxi Province (34°76′ N, 109°17′ E). The semen of dairy goats was collected using the artificial vagina method and the collection frequency was three times a week on Monday, Wednesday, and Friday mornings. Semen samples with a deformity rate of less than 5%, normal morphology of sperm, motility greater than 0.8, and semen density exceeding 1.5 billion/mL were suitable for this study. Subsequently, the semen samples were pooled to eliminate individual variation and used for subsequent experiments.

2.3. Ethics Statement

All animal experiments were approved by the Animal Ethics Committee of Northwest A&F University. All experiments with dairy goats were implemented strictly according to the Guide for the Care and Use of Laboratory Animals.

2.4. Semen Processing

Dilute fresh sperm with a modified solution consisting of 172 mM glucose, 90 mM lactose, 58 mM sodium citrate, 1000 IU/mL penicillin G potassium, and 1 mg/mL streptomycin was passed through a 0.22 μm filter. The glucose concentration (172 mM) in the modified solution was determined to be 40%. A portion of lactose was used with glucose to prepare different doses of glucose bulking agents (0, 86, 172, 258, 344, and 430 mM, i.e., 0%, 20%, 40%, 60%, 80%, and 100%), as sperm do not use lactose as an energy substrate. Sperm were incubated with different concentrations of glucose medium at 37 °C for 1 h to evaluate whether dairy goat sperm were involved in the mitochondrial oxidative phosphorylation mode of energy supply. Moreover, sperm were incubated in a low glucose group containing 86 mM glucose and different concentrations of ROT, FCCP, and TIG at 37 °C for 5 h to evaluate the role of mitochondrial oxidative phosphorylation in the regulation of mitochondrial energy metabolism function in dairy goat spermatozoa. Finally, Compound C (AMPK inhibitor) was added to low and high glucose diluents and incubated in a water bath at 37 °C for 5 h to evaluate whether the low glucose dilutions activated AMPK to maintain energy homeostasis and inhibit sperm oxidative damage.

2.5. Sperm Motility

Sperm motility parameters were assessed using a CASA system (HVIEW-SSAV8.0, FuZhouHongShiYeSoftware Technology Co., Ltd., Fuzhou, China). Briefly, 10 μL of the samples was dropped onto a preheated glass slide and covered with a coverslip. Then, sperm motion was captured in at least five different regions using dynamic video acquisition of CASA on a heated platform at 37 °C. Each region was required to contain at least

200 sperm counts. Lastly, sperm motility parameters were assessed using the analysis software of CASA.

2.6. Biochemical Assays

All tests were conducted per the respective manufacturers' instructions for each kit and were obtained from the following vendors: reactive oxygen species (ROS) assay kit (CA1410, Solarbio, Beijing, China), malondialdehyde (MDA) assay kit (BC0025, Solarbio, Beijing, China), pyruvate (PA) content assay kit (BC2205, Solarbio, Beijing, China), lactic acid (LA) assay kit (A019-2-1, Nanjing Jiancheng, Nanjing, China), ATP assay kit (S0026, Beyotime, Shanghai, China), MMP assay kit (M8650, Solarbio, Beijing, China), and bicinchoninic acid (BCA) protein assay kit (PA115, TAINGEN, Beijing, China). Finally, the related indexes were detected by a multifunctional enzyme label instrument (Synergy H1, American Berten, VT, USA).

2.7. Mitochondrial Permeability Transition Pore (mPTP) Fluorescence Intensity

Sperm samples were centrifuged at $1000\times g$ at room temperature for 5 min and the supernatant was discarded. The sperm density was diluted to achieve a final concentration of 1×10^6 sperm/mL by adding an appropriate volume of Calcein AM (1000×) staining solution and fluorescence quenching solution or ionomycin (200×) control, and incubated in the dark for 30 min. Then, the supernatant was removed by centrifugation at $1000\times g$ at room temperature for 5 min and slowly resuspended twice at 37 °C. Subsequently, the samples were smeared and analyzed using flow cytometry (FACS Melody, BD Biosciences, Franklin Lakes, NJ, USA) within 1 h. The excitation wavelength was 494 nm and the emission wavelength was 517 nm.

2.8. Immunofluorescence Assays

Sperm samples were washed with phosphate-buffered saline (PBS), spread on a glass slide with poly-D-lysine, dried naturally, and fixed in 4% paraformaldehyde for 10 min. After drying, the sperm were penetrated with PBS containing 0.5% Triton X-100 for 30 min, washed three times for 5 min each, and blocked with 5% bovine serum albumin (BSA) in PBS-T for 30 min. Next, the slides were incubated with the primary antibody (1:100) overnight at 4 °C. Subsequently, the slides were incubated with the second antibody (1:200) at 37 °C for 1 h in the dark. The sperm nuclei were stained with DAPI at room temperature for 1 min in the dark and washed three times with PBS for 5 min each. Lastly, the samples were observed using a fluorescence microscope (LECIA-DM6 B, LECIA Co., Ltd., WETZLAR, Hessian, Germany).

2.9. Western Blotting

Protein concentrations were determined using a BCA protein assay kit (PA115, TAINGEN, Beijing, China). Proteins were separated using FuturePAGE™ 4–12% (ET15412Gel, ACE Biotechnology, Nanjing, China) sodium dodecyl sulfate–polyacrylamide gel electrophoresis, transferred to polyvinylidene fluoride membranes, and blocked with 5% BSA for 1 h at room temperature. The primary antibody was diluted and incubated with the membrane overnight at 4 °C. After incubation, the membrane was washed three times for 10 min each using Tris-buffered saline tween (TBST) and incubated with the second antibody at 37 °C for 1 h. After washing with TBST, detection was performed using enhanced chemiluminescence (ECL) (34095, Thermo Fisher Scientific, Waltham, MA, USA) using a Gel Doc XR System (BioRad, Hercules, CA, USA) according to the manufacturer's specifications. Lastly, the intensity of the protein bands was analyzed using ImageJ software.

2.10. Statistical Analysis

Data are expressed as the mean ± standard error of the mean from at least 3 independent experiments. A statistical analysis among groups was performed using PRISM (version 6, GraphPad), followed by post hoc test using the Student's *t*-test. The mean difference at $p < 0.05$ was considered statistically significant (* $p < 0.05$, ** $p < 0.01$, *** $p < 0.001$, and **** $p < 0.0001$. ns, with no apparent difference).

3. Results

3.1. Sperm Energy Metabolic Pathway and Sperm Motility Patterns Are Changed in Different Concentrations of Glucose

The CASA system can detect the motility and movement trajectory of sperm. Sperm motility tracks changed into circle-like tracks in high glucose media (>60%), and the progressive motility and straight line velocity reduced significantly. However, the total motility and straight line velocities in the low glucose groups (20% and 40%) were significantly higher than those in the other groups (Figure 1A–D). To further determine the energy metabolic pathway, the levels of pyruvate, lactic acid, ATP, and MMP in the sperm of dairy goats were assessed at different glucose concentrations. The levels of pyruvate, lactic acid, and MMP were significantly higher in the 40% glucose group after incubation for 1 h compared with those in the control group, whereas the levels decreased remarkably in the high glucose group (Figure 1E–H, $p < 0.001$).

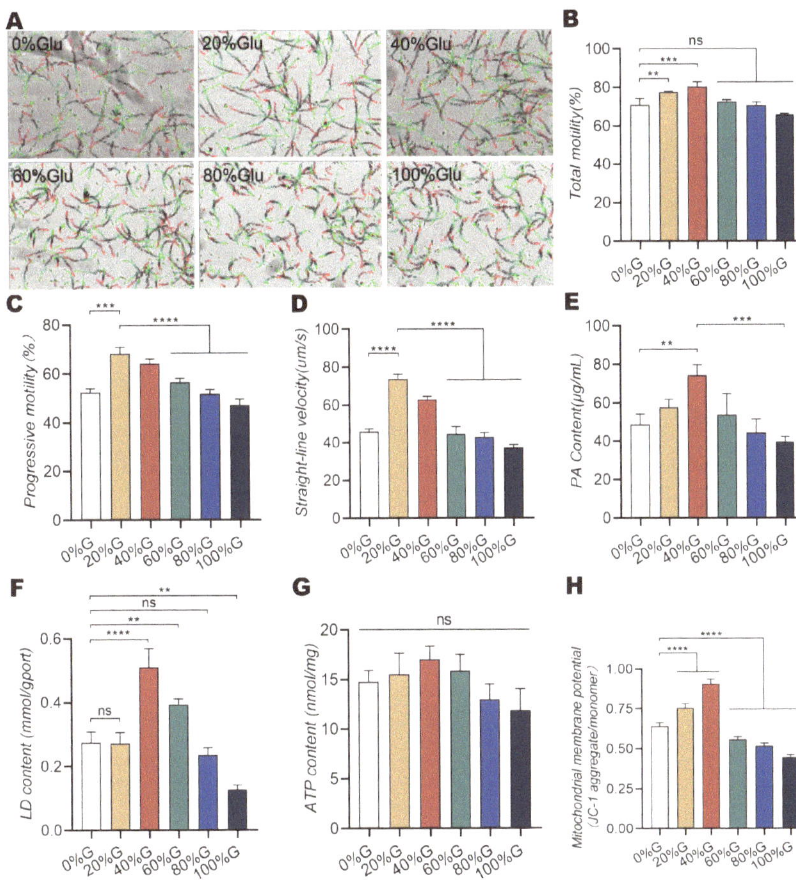

Figure 1. Effect of different concentrations of glucose on the motility and energy metabolism of dairy goat sperm. (**A**) Changes in sperm motility track using CASA. (**B**) Motility of sperm in dairy goats.

(**C**,**D**) Progressive motility and straight line velocity of sperm of dairy goats. (**E**) PA content. (**F**) LD content. (**G**) ATP content in sperm. (**H**) Mitochondrial membrane potential. Values are presented as the mean ± standard error of the mean (SEM) of three replicates. Asterisks indicate statistical significance compared with the respective control groups. All results are expressed as the mean ± SEM, with asterisks indicating statistical significance for the respective control group. ** $p < 0.01$, *** $p < 0.001$, and **** $p < 0.0001$. ns, no significant difference.

3.2. Addition of Rotenone to the Culture Reduces ATP Content and Straight Line Motility

To further validate the role of OXPHOS in energy metabolism in dairy goat sperm, rotenone (an inhibitor of mitochondrial respiratory chain complex I) was incubated with sperm samples for 5 h in this study. The results from CASA showed significant reductions in forward sperm motility and straight line velocity, but no significant difference was observed in the total motility of sperm (Figure 2A–D). Moreover, there was a considerable decrease in the fluorescence intensity in a dose-dependent manner with respect to mPTP and ATP levels with the addition of rotenone (Figure 2E–G, $p < 0.05$). Therefore, the expression of the marker proteins of ETC and OXPHOS was determined using immunofluorescence staining and Western blotting. Immunofluorescence staining results showed that COX-1 and COXVB were mainly distributed in the middle segment of the sperm tails (Figure 2H) and that the expression of COX-1, COXVB, ND3, and NRF1 decreased significantly after treatment with rotenone (Figure 2I–M, $p < 0.001$). Taken together, it could be inferred that rotenone inhibits the regulation of mitochondrial OXPHOS in dairy goat sperm.

3.3. The OXPHOS Uncoupler FCCP Reduces Sperm Motility and ATP Levels

To further explore the effect of OXPHOS on energy metabolism in dairy goat sperm, the samples were treated with a specific inhibitor of mitochondrial OXPHOS (FCCP). The sperm motility, forward motility, and linear velocity significantly decreased in a dose-dependent manner after treatment with FCCP for 5 h (Figure 3A–D, $p < 0.0001$). The fluorescence intensity of mPTP and the level of ATP in mitochondria also decreased significantly (Figure 3E–G, $p < 0.01$). Furthermore, we observed that COX-1, COXVB, ND3, and NRF1 expressions in mitochondria decreased obviously (Figure 3H–L, $p < 0.001$).

3.4. The Mitochondrial Translational Inhibitor TIG Reduces MMP and ATP Levels in a Dose-Dependent Manner

To further confirm that the straight line motility pattern in dairy goat sperm relies on mitochondrial transcription and translation, the mitochondrial translation inhibitor TIG was incubated with sperm. There were no significant differences in sperm motility at the 5 h point, but the progressive motility and straight line velocity of sperm decreased significantly (Figure 4A–C, $p < 0.0001$). Moreover, ATP levels and MMP decreased noticeably after incubation with TIG (Figure 4D,E, $p < 0.0001$), indicating that the inhibition of mitochondrial translation reduces the progressive motility and straight line velocity of sperm. Additionally, COX-1, COXVB, NRF1, and TFAM expressions in mitochondria decreased significantly (Figure 4F–J, $p < 0.0001$). Therefore, energy metabolism and the linear motility pattern are closely related to transcription and translation in mitochondria.

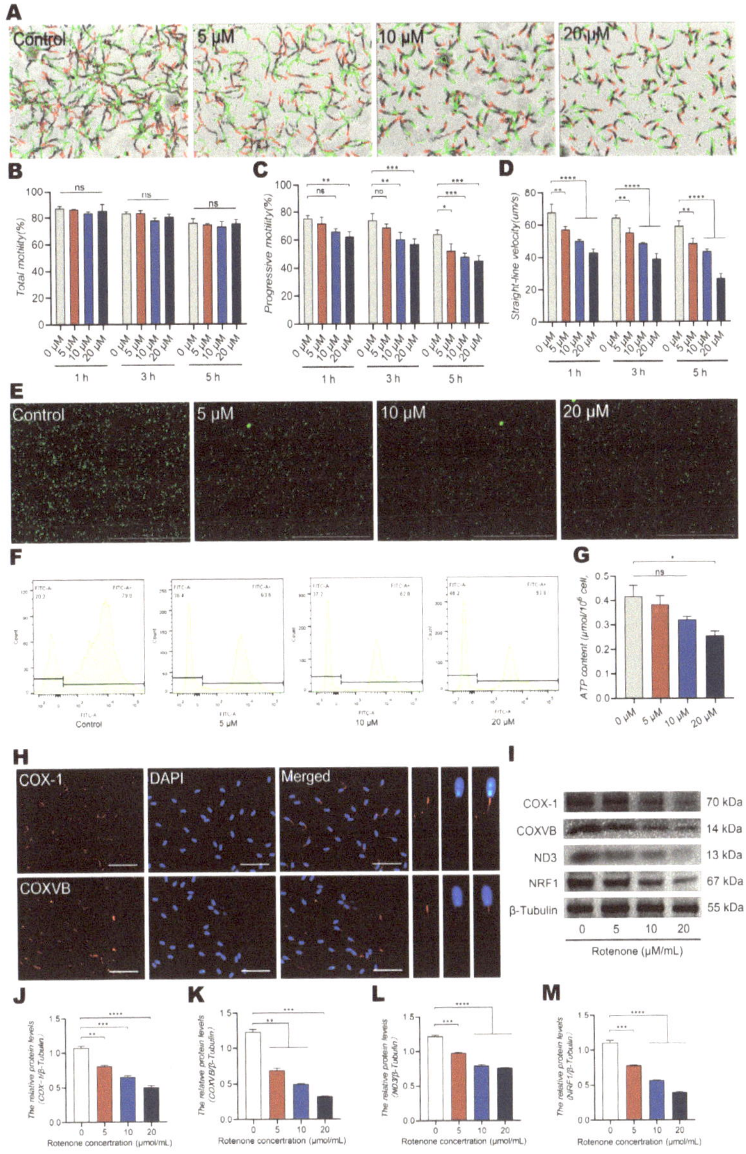

Figure 2. Incubation with rotenone reduces progressive motility and ATP levels. (**A–D**) Changes in sperm motility parameters evaluated using CASA: (**A**) sperm motility tracks, (**B**) motility of sperm of dairy goats. (**C,D**) Progressive motility and straight line velocity of sperm of dairy goats. (**E–H**) Effect of rotenone on sperm mitochondrial functions at the 5 h point: (**E,F**) opening fluorescence intensity of mitochondrial mPTP, (**G**) ATP content, and (**H**) immunolocalizations of COX-1 and COXVB in mitochondria. (**I–M**) Quantitative expression of COX-1, COXVB, ND3, and NRF1 using Western blotting after rotenone incubation for 5 h. All results are expressed as the mean ± standard error of the mean, with asterisks indicating statistical significance for the respective control group. * $p < 0.05$, ** $p < 0.01$, *** $p < 0.001$, and **** $p < 0.0001$. ns, no significant difference.

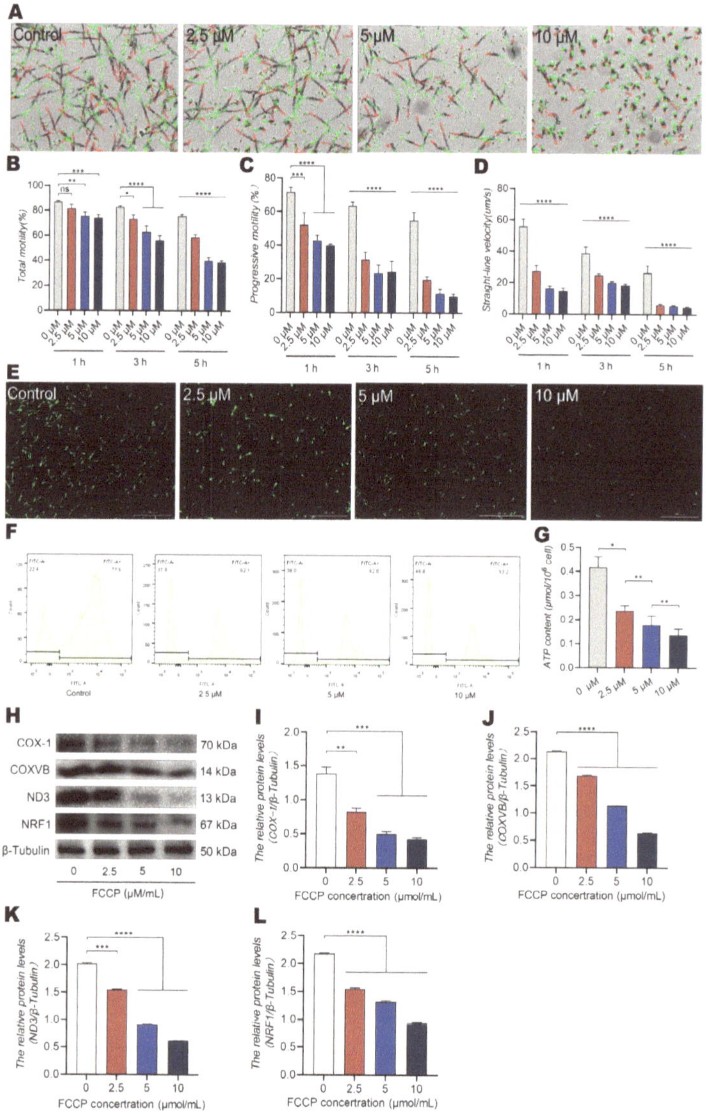

Figure 3. The oxidative phosphorylation uncoupler FCCP reduces sperm motility and ATP content. (**A–D**) Motility parameters of dairy goat sperm using CASA: (**A**) motility tracks were generated using CASA. (**B**) Motility of sperm of dairy goats. (**C**,**D**) Progressive motility and straight line velocity of sperm of dairy goats. (**E–G**) Effect of FCCP on sperm mitochondrial functions: (**E**,**F**) opening fluorescence intensity of mPTP in mitochondria and (**G**) ATP content. (**H–L**) Expressions of COX-1, COXVB, ND3, and NRF1 determined using Western blotting after treatment with FCCP for 5 h. All results are expressed as the mean ± standard error of the mean, with asterisks indicating statistical significance for the respective control group. * $p < 0.05$, ** $p < 0.01$, *** $p < 0.001$, and **** $p < 0.0001$. ns, no significant difference.

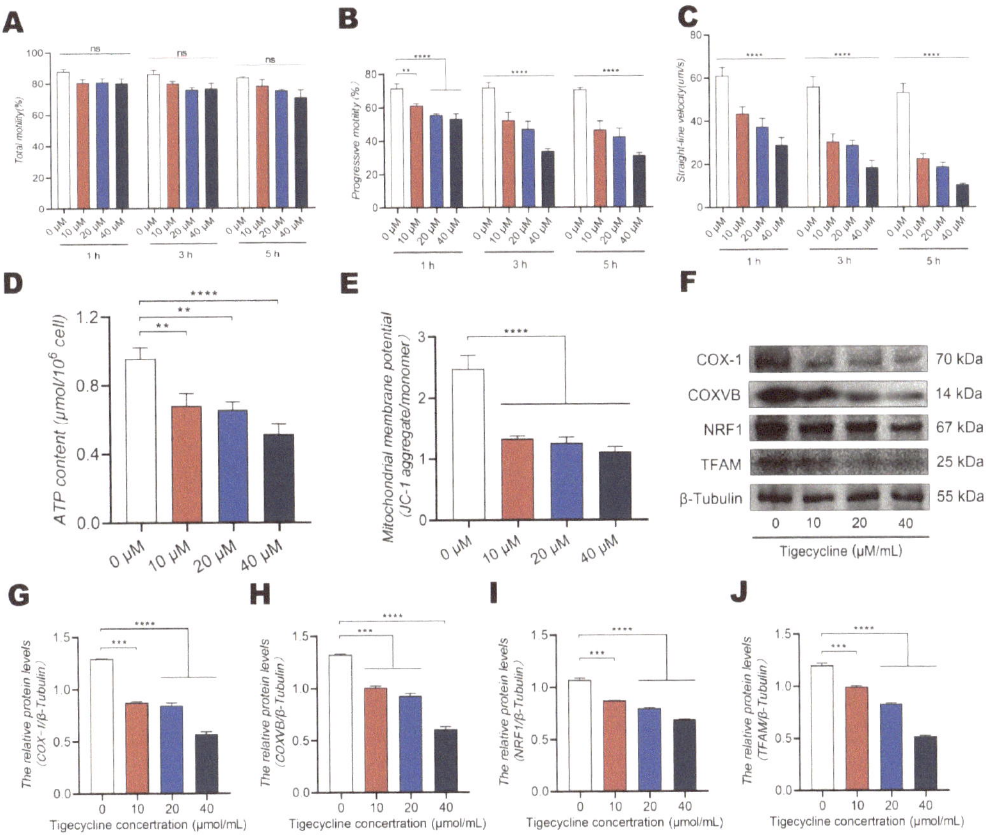

Figure 4. The mitochondrial translation inhibitor TIG decreases MMP and ATP levels in a dose-dependent manner. (**A–C**) Changes in sperm motility parameters using CASA: (**A**) motility of sperm of dairy goats. (**B**,**C**) Progressive motility and straight line velocity of sperm of dairy goats. (**D**) ATP levels in the mitochondria. (**E**) Mitochondrial membrane potential. (**F–J**) Protein levels of COX-1, COXVB, NRF1, and TFAM during 5 h of incubation with TIG determined using Western blotting. All results are expressed as the mean ± standard error of the mean, with asterisks indicating statistical significance for the respective control group. ** $p < 0.01$, *** $p < 0.001$, and **** $p < 0.0001$. ns, no significant difference.

3.5. Low Glucose Conditions Inhibit Ferroptosis-Induced Oxidative Damage by Regulating the Liver Kinase B1 (LKB1)/AMPK Signaling Pathway

High glucose media cause a reduction in cell viability as well as ROS accumulation, LPO, and outer mitochondrial membrane rupture, eventually inducing ferroptosis [21]. Therefore, low and high concentrations of glucose were used to determine whether low glucose conditions could inhibit ferroptosis and maintain the redox balance by activating AMPK in dairy goat sperm. Sperm motility in high glucose media did not change obviously compared with that in low glucose media at the 1 h point, as determined using CASA, but decreased significantly after high glucose or erastin incubation for 3 h or 5 h; moreover, ROS and MDA levels increased significantly (Figure 5A–C, $p < 0.01$). In addition, the intensities of LKB1, AMPK, glutathione peroxidase 4 (GPX4), and family 7 member 11 (SLC7A11) proteins decreased significantly in high glucose conditions or when treated with erastin (Figure 5D,E, $p < 0.001$). Therefore, high glucose levels could induce a reduction in sperm motility and lead to the accumulation of ROS and LPO in dairy goat sperm, which, in turn, caused ferroptosis. Glucose is the primary energy source for cell types and AMP-activated protein kinase (AMPK) is an energy sensor. In low energy conditions cells can be isomerically activated by adenosine monophosphate (AMP) and adenosine diphosphate [22]. Thus, an AMPK inhibitor (Compound C) was used to determine whether low glucose levels could inhibit ferroptosis in dairy goat sperm by regulating AMPK A significant decrease was observed in ATP levels and MMP in the low glucose group ($p < 0.01$) but not in the high glucose group (Figure 5F,G). Furthermore, Western blotting revealed that the expression of LKB1 and AMPK proteins decreased noticeably in high glucose conditions or after treatment with Compound C (Figure 5H–J, $p < 0.001$).

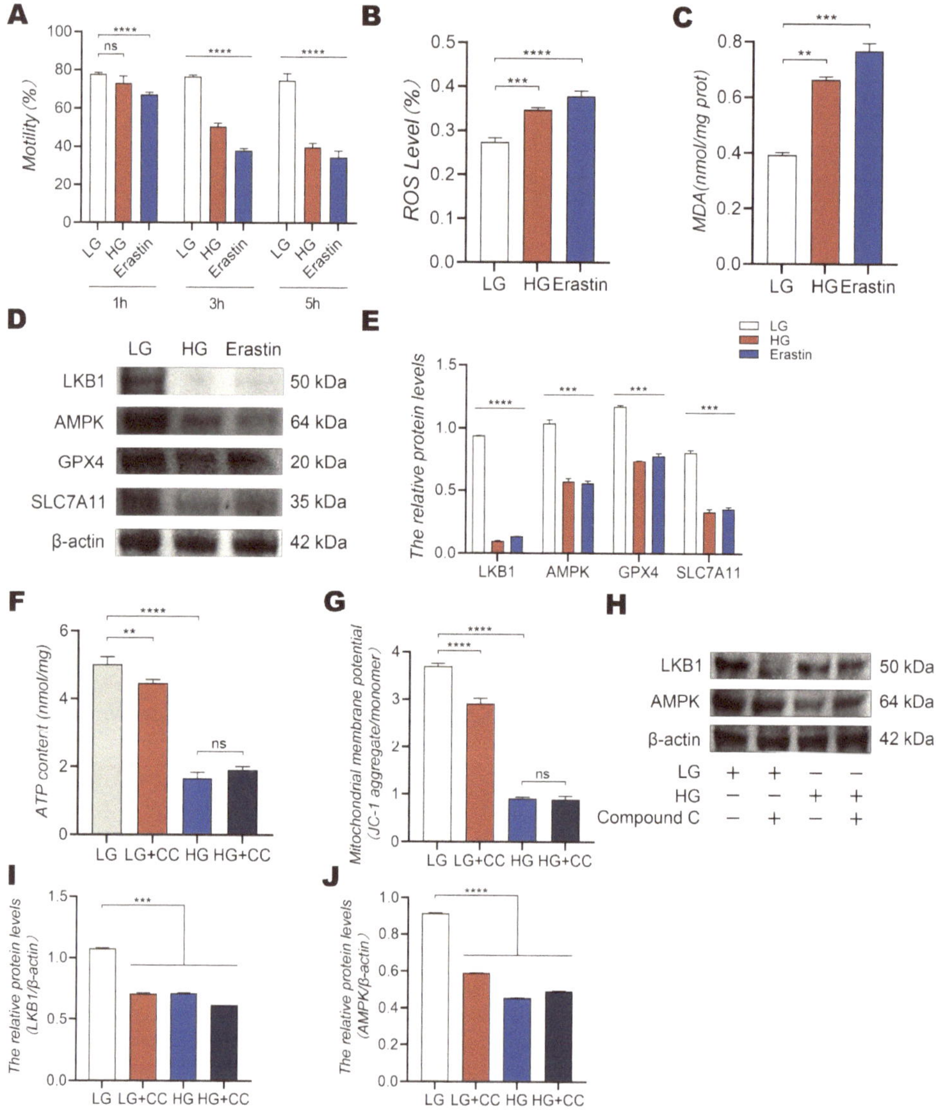

Figure 5. Low glucose conditions inhibit ferroptosis-induced oxidative damage by regulating the LKB1/AMPK signaling pathway. (**A**) Sperm motility. (**B**) ROS levels. (**C**) MDA levels. (**D**,**E**) Protein

levels of LKB1, AMPK, GPX4, and SLC7A11 during 5 h of incubation in high glucose medium or with erastin determined using Western blotting. (**F**) ATP levels. (**G**) Mitochondrial membrane potential. (**H–J**) Expressions of LKB1 and AMPK after incubation with low or high concentrations of glucose and Compound C for 5 h. All results are expressed as the mean ± standard error of the mean, with asterisks indicating statistical significance for the respective control group. ** $p < 0.01$, *** $p < 0.001$, and **** $p < 0.0001$. ns, no significant difference.

4. Discussion

Mammalian sperm are produced through spermatogenesis and serve the purpose of fertilizing an oocyte. After ejaculation, sperm require energy to meet their demands. This energy can be obtained through the importation of exogenous substrates or the use of endogenous sources. Adenosine triphosphate (ATP), an important energy source that supports sperm capacitation and migration in the female reproductive tract, is generated from two main metabolic pathways, namely glycolysis and oxidative phosphorylation (OXPHOS). They occur in different compartments in sperm flagella, both of which are located in the principal piece [23]. The metabolic pathways of energy in sperm change in the presence of different metabolic substrates. In addition to these changes, the sperm motility pattern associated with capacitation also changes and is called hyperactivation [24]. The hyperactivated status is always characterized by a high curvilinear velocity and a high lateral amplitude [25]. However, the relationship between the energy metabolic pathway and motility patterns in dairy goat sperm remains unclear. In this study, we established a model by incubation of sperm with different concentrations of glucose and found that not only the progressive motility and straight line velocity of sperm but also ATP levels and MMP increased significantly in low glucose media. This finding indicated that sperm regulate the energy metabolism pathway based on changes in the metabolic substrates and that low glucose levels activate OXPHOS in the mitochondria to maintain the normal functions of sperm. In terms of the changes in oxygen and substrate levels in the female reproductive tract, sperm adapt to the fluctuations in exogenous substrate levels via different energy metabolism pathways [26]. Qiu et al. found that among the substrates tested, glucose and pyruvate were better than lactate at maintaining goat sperm motility. Pyruvate enters goat spermatozoa through monocarboxylate transporters and is oxidized by the tricarboxylic acid cycle and electron transfer to sustain sperm motility [27]. Zhu et al. reported that GSK3α/β was expressed in the head acrosome and the mid, main end of the tail of goat sperm and that it regulated goat sperm motility and the acrosome response by mediating energy pathways in glycolysis and oxidative phosphorylation [28]. Thus, these different energy metabolic pathways are interconnected in sperm, and different concentrations of glucose can selectively activate glycolysis as well as OXPHOS in the mitochondria. Moreover, pyruvate and lactate are involved in the glycolysis pathway. In this study, the progressive motility of sperm was found to be regulated by OXPHOS in the mitochondria, which is similar to the physiological state of sperm in the female reproductive tract after ejaculation. The semen, uterus, and oviduct contain high levels of pyruvate, lactate, and amino acids, and sperm can utilize these metabolites through OXPHOS for energy metabolism, further enhancing their progressive motility and speed of access to reach the site of fertilization. Zhu et al. demonstrated that ATP generated from OXPHOS in the mitochondria in low glucose conditions induced progressive motility in boar sperm. On the other hand, incubation with CRP, a mitochondrial translation inhibitor, inhibited mitochondrial translation in sperm and decreased their progressive motility but not total motility, suggesting that motility patterns in sperm depend on the substrates of energy metabolism [29]. Moreover, high glucose concentrations led to a transient inefficiency in the late stages of both glycolysis and NADH accumulation and decreased the conversion flux from fructose-1,6-bisphosphate to pyruvate and the ATP content [30].

Mitochondria are called the motive source of cells and play a key role in energy metabolism through the tricarboxylic acid (TCA) cycle, β-oxidation, and OXPHOS [31]. During glycolysis, a glucose molecule produces two ATP molecules, whereas in the TCA cycle or OXPHOS pathway, a glucose molecule produces 32 ATP molecules. OXPHOS is composed of four respiratory complexes (complexes I to IV) and ATP synthase (complex V) and plays a key role in MMP across the mitochondrial inner membrane [32]. Rotenone is a strong inhibitor of complex I of the mitochondrial respiratory chain. Heo et al. demonstrated that rotenone induces mitochondrial dysfunction by inhibiting SIRT1 during oocyte maturation in pigs in vitro, thereby reducing mitochondrial activity and ATP generation and increasing ROS production [33]. There was no significant difference in sperm motility after incubation with rotenone, but there was a significant decrease in mitochondrial activity and ATP levels [29]. Besides, FCCP is a classic uncoupler of mitochondrial OXPHOS. Davila et al. showed that sperm motility, speed, and the number of live sperm decreased significantly in horses after incubation for 3 h with FCCP, but the MMP and ATP levels were unaffected [15]. Blanco-Prieto et al. have reported that FCCP significantly increases oxygen consumption and decreases the total motility of boar sperm [34]. In this study, dairy goat sperm were incubated with rotenone and FCCP separately. After treatment with rotenone, their total motility did not change much, but the progressive motility and straight linear speed reduced significantly. The addition of FCCP decreased the overall survival and mPTP permeability in a dose-dependent manner. Both these treatments significantly reduced the MMP and ATP content. Furthermore, the protein expression related to the mitochondrial respiratory chain determined using Western blotting showed that incubation with rotenone and FCCP resulted in the disruption of the ETC in the mitochondria and inactivation of ATP synthase [35]. On the other hand, the relationship between mitochondrial function and ATP generation from OXPHOS is especially close, and mitochondrial activity is influenced by proteins from the mitochondrial respiratory chain. Zhu et al. found that in a low glucose environment, 13 genes involved in mitochondrial transcription and translation were remarkably upregulated, which, in turn, improved mitochondrial OXPHOS and progressive motility in boar sperm [29]. Consistent with the above findings, tigecycline (a mitochondrial translation inhibitor) inhibited the expression of proteins in the mitochondrial respiratory chain in our study, decreased the MMP and ATP content, and reduced the progressive motility of dairy goat sperm. Glucose and fructose are essential components as a semen preservation diluent, but excessive glucose induces cellular damage [36]. High glucose concentrations induce cell apoptosis, ferroptosis, necrosis, or other types of cell death [37,38]. Ferroptosis is a new form of oxidative cell death that is induced by small molecules and especially by the imbalance of lipid ROS in cells [8]. Ferroptosis is also induced by other factors, including the inhibition of glutathione synthetase or GPX4, leading to the accumulation of LPO products and ROS and eventually resulting in cell death [39]. In our study, incubation with either high glucose levels or erastin for 3 h significantly reduced sperm motility, increased ROS and MDA levels, and decreased the expressions of LKB1/AMPK, SLC7A11, and GPX4. However, in low glucose media, there was no significant difference in sperm motility after incubation for 3 h, indicating that the normal functions of sperm could be maintained by accelerating ATP production through the AMPK pathway [40]. Glucose starvation is a typical activator of AMPK and an important regulator in energy homeostasis. The LKB1/AMPK signaling pathway is critical in maintaining cell survival under low glucose conditions [20]. Metformin (an activator of AMPK) was found to reduce the levels of LPO-stimulated steroid, restore spermatogenesis, and increase sperm motility in the testes of diabetic or obese rats [41]. During cryopreservation of chicken sperm, the addition of an AMPK activator, metformin or AICAR, improved AMPK phosphorylation after thawing, whereas LPO and ROS production was reduced [42]. Additionally, the activation of AMPK in goat sperm improved sperm motility, plasma membrane integrity, and the acrosome reaction; maintained normal levels of MMP, lactate acid, and ATP; and enhanced the activity of AMPK, PK, and lactate dehydrogenase. However, the addition of Compound C inhibited the AMPK pathway and induced the opposite effects [43]. In the present study, the addition of Compound C to the low glucose media led to reductions in ATP levels and MMP as well as a decrease in LKB1/AMPK expression compared with that in the high glucose group. These results suggested that the LKB1/AMPK signaling pathway can be activated to maintain energy homeostasis in sperm during energy stress.

5. Conclusions

Energy metabolism in dairy goat sperm is realized through the mitochondrial OXPHOS pathway under low glucose conditions, and the high speed linear motility and straight line velocity improved significantly. Moreover, in this study, the results of incubation with rotenone and FCCP confirmed that the progressive motility of sperm mainly relies on OXPHOS in the mitochondria and that ATP generation from OXPHOS is closely related to mitochondrial functions. Low glucose conditions promote transcription and translation in the mitochondria and activate mitochondrial OXPHOS to supply energy to sperm. Moreover, the LKB1/AMPK pathway is activated to maintain energy homeostasis and inhibit ferroptosis-induced oxidative damage. Therefore, the high speed linear motion induced by low sugar is a new factor to improve insemination in artificial insemination of livestock and human fertility treatments. As a convenient and inexpensive method, sperm dilution with a low-sugar insemination solution improves the rapid linear movement of sperm within the female reproductive tract, thereby facilitating fertilization.

Author Contributions: Y.L. and J.H. designed and conceived the study. S.G. organized the database. M.X. conducted the statistical analysis. G.Z. wrote the first draft of the manuscript. F.W., X.Z., X.F. and Z.H. wrote parts of the manuscript. All the authors were involved in revising the manuscript and reading and endorsing the submitted version. All authors have read and agreed to the published version of the manuscript.

Funding: This work was supported by the Shaanxi Province Technology Innovation Guide Project (2022QFY11-04) and the Fuping City Open Competition Project (202200423100512).

Institutional Review Board Statement: All animal experiments used in the experiment were raised in accordance with the No. 5 Announcement of the Ministry of Agriculture, P. R. China. All experiments with dairy goats were implemented strictly according to the Guide for the Care and Use of Laboratory Animals. All procedures in our animal research have been approved by the Animal Protection and Use Committee of Northwest A&F University (Yangling, China) (license number: 17-347, data: 13 October 2017).

Informed Consent Statement: Not applicable.

Data Availability Statement: The authors confirm that the original contributions presented in this study are reflected in this manuscript.

Acknowledgments: This study acknowledges the help and support provided by Shaanxi Aonike Dairy Goat Breeding Co. Yu Li especially wishes to thank the support and encouragement of Yue Li.

Conflicts of Interest: The authors state that the research was carried out without any commercial or financial relations that could be perceived as underlying conflict of interest.

References

1. Hunter, R.H. Fallopian tube physiology: Preliminaries to monospermic fertilization and cellular events post-fertilization. *Ernst Scher. Res. Found. Workshop* **2005**, *52*, 245–261. [CrossRef]
2. Amaral, A. Energy metabolism in mammalian sperm motility. *WIREs Mech. Dis.* **2022**, *14*, e1569. [CrossRef]
3. Owen, D.H.; Katz, D.F. A review of the physical and chemical properties of human semen and the formulation of a semen simulant. *J. Androl.* **2005**, *26*, 459–469. [CrossRef]
4. Cao, W.; Aghajanian, H.K.; Haig-Ladewig, L.A.; Gerton, G.L. Sorbitol can fuel mouse sperm motility and protein tyrosine phosphorylation via sorbitol dehydrogenase. *Biol. Reprod.* **2009**, *80*, 124–133. [CrossRef] [PubMed]
5. Goodson, S.G.; Qiu, Y.; Sutton, K.A.; Xie, G.; Jia, W.; O'Brien, D.A. Metabolic substrates exhibit differential effects on functional parameters of mouse sperm capacitation. *Biol. Reprod.* **2012**, *87*, 75. [CrossRef]
6. Du Plessis, S.S.; Agarwal, A.; Mohanty, G.; van der Linde, M. Oxidative phosphorylation versus glycolysis: What fuel do spermatozoa use? *Asian J. Androl.* **2015**, *17*, 230–235. [CrossRef] [PubMed]
7. Piomboni, P.; Focarelli, R.; Stendardi, A.; Ferramosca, A.; Zara, V. The role of mitochondria in energy production for human sperm motility. *Int. J. Androl.* **2012**, *35*, 109–124. [CrossRef]

8. Odet, F.; Gabel, S.; London, R.E.; Goldberg, E.; Eddy, E.M. Glycolysis and mitochondrial respiration in mouse LDHC-null sperm *Biol. Reprod.* **2013**, *88*, 95. [CrossRef]
9. Pasupuleti, V. Role of Glycolysis and Respiration in Sperm Metabolism and Motility. Master's Thesis, Kent State University, Kent OH, USA, 2007.
10. Losano, J.; Angrimani, D.; Dalmazzo, A.; Rui, B.R.; Brito, M.M.; Mendes, C.M.; Kawai, G.; Vannucchi, C.I.; Assumpção, M. Barnabe, V.H.; et al. Effect of mitochondrial uncoupling and glycolysis inhibition on ram sperm functionality. *Reprod. Domest Anim. Zuchthyg.* **2017**, *52*, 289–297. [CrossRef]
11. Storey, B.T. Mammalian sperm metabolism: Oxygen and sugar, friend and foe. *Int. J. Dev. Biol.* **2008**, *52*, 427–437. [CrossRef]
12. Gibb, Z.; Aitken, R.J. The Impact of Sperm Metabolism during In Vitro Storage: The Stallion as a Model. *BioMed Res. Int.* **2016** *2016*, 9380609. [CrossRef] [PubMed]
13. Varner, D.D.; Gibb, Z.; Aitken, R.J. Stallion fertility: A focus on the spermatozoon. *Equine Vet. J.* **2015**, *47*, 16–24. [CrossRef]
14. Bucci, D.; Rodriguez-Gil, J.E.; Vallorani, C.; Spinaci, M.; Galeati, G.; Tamanini, C. GLUTs and mammalian sperm metabolism *J. Androl.* **2011**, *32*, 348–355. [CrossRef]
15. Davila, M.P.; Muñoz, P.M.; Bolaños, J.M.; Stout, T.A.; Gadella, B.M.; Tapia, J.A.; da Silva, C.B.; Ferrusola, C.O.; Peña, F.J Mitochondrial ATP is required for the maintenance of membrane integrity in stallion spermatozoa, whereas motility requires both glycolysis and oxidative phosphorylation. *Reproduction* **2016**, *152*, 683–694. [CrossRef]
16. Hardie, D.G.; Hawley, S.A.; Scott, J.W. AMP-activated protein kinase–development of the energy sensor concept. *J. Physiol.* **2006**, *574*, 7–15. [CrossRef]
17. Chhipa, R.R.; Wu, Y.; Mohler, J.L.; Ip, C. Survival advantage of AMPK activation to androgen-independent prostate cancer cells during energy stress. *Cell. Signal.* **2010**, *22*, 1554–1561. [CrossRef] [PubMed]
18. Tosca, L.; Chabrolle, C.; Dupont, J. AMPK: A link between metabolism and reproduction? *Med. Sci. M/S* **2008**, *24*, 297–300. [CrossRef]
19. Huang, J.; Chen, G.; Wang, J.; Liu, S.; Su, J. Platycodin D regulates high glucose-induced ferroptosis of HK-2 cells through glutathione peroxidase 4 (GPX4). *Bioengineered* **2022**, *13*, 6627–6637. [CrossRef]
20. Ren, Y.; Shen, H.M. Critical role of AMPK in redox regulation under glucose starvation. *Redox Biol.* **2019**, *25*, 101154. [CrossRef] [PubMed]
21. Han, D.; Jiang, L.; Gu, X.; Huang, S.; Pang, J.; Wu, Y.; Yin, J.; Wang, J. SIRT3 deficiency is resistant to autophagy-dependent ferroptosis by inhibiting the AMPK/mTOR pathway and promoting GPX4 levels. *J. Cell. Physiol.* **2020**, *235*, 8839–8851. [CrossRef] [PubMed]
22. Grahame Hardie, D. AMP-activated protein kinase: A key regulator of energy balance with many roles in human disease. *J. Intern. Med.* **2014**, *276*, 543–559. [CrossRef] [PubMed]
23. Ferramosca, A.; Zara, V. Bioenergetics of mammalian sperm capacitation. *BioMed Res. Int.* **2014**, *2014*, 902953. [CrossRef]
24. Yanagimachi, R. The movement of golden hamster spermatozoa before and after capacitation. *J. Reprod. Fertil.* **1970**, *23*, 193–196. [CrossRef] [PubMed]
25. Stauss, C.R.; Votta, T.J.; Suarez, S.S. Sperm motility hyperactivation facilitates penetration of the hamster zona pellucida. *Biol. Reprod.* **1995**, *53*, 1280–1285. [CrossRef]
26. Keeley, T.P.; Mann, G.E. Defining Physiological Normoxia for Improved Translation of Cell Physiology to Animal Models and Humans. *Physiol. Rev.* **2019**, *99*, 161–234. [CrossRef]
27. Qiu, J.H.; Li, Y.W.; Xie, H.L.; Li, Q.; Dong, H.B.; Sun, M.J.; Gao, W.Q.; Tan, J.H. Effects of glucose metabolism pathways on sperm motility and oxidative status during long-term liquid storage of goat semen. *Theriogenology* **2016**, *86*, 839–849. [CrossRef] [PubMed]
28. Zhu, Z.; Li, R.; Wang, L.; Zheng, Y.; Hoque, S.A.M.; Lv, Y.; Zeng, W. Glycogen Synthase Kinase-3 Regulates Sperm Motility and Acrosome Reaction via Affecting Energy Metabolism in Goats. *Front. Physiol.* **2019**, *10*, 968. [CrossRef]
29. Zhu, Z.; Umehara, T.; Okazaki, T.; Goto, M.; Fujita, Y.; Hoque, S.A.M.; Kawai, T.; Zeng, W.; Shimada, M. Gene Expression and Protein Synthesis in Mitochondria Enhance the Duration of High-Speed Linear Motility in Boar Sperm. *Front. Physiol.* **2019**, *10*, 252. [CrossRef]
30. Lin, S.C.; Hardie, D.G. AMPK: Sensing Glucose as well as Cellular Energy Status. *Cell Metab.* **2018**, *27*, 299–313. [CrossRef]
31. Dunn, J.; Grider, M.H. Physiology, Adenosine Triphosphate. In *StatPearls*; StatPearls Publishing LLC.: Treasure Island, FL, USA, 2022.
32. Nsiah-Sefaa, A.; McKenzie, M. Combined defects in oxidative phosphorylation and fatty acid β-oxidation in mitochondrial disease. *Biosci. Rep.* **2016**, *36*, e00313. [CrossRef]
33. Heo, G.; Sun, M.H.; Jiang, W.J.; Li, X.H.; Lee, S.H.; Guo, J.; Zhou, D.; Cui, X.S. Rotenone causes mitochondrial dysfunction and prevents maturation in porcine oocytes. *PLoS ONE* **2022**, *17*, e0277477. [CrossRef]
34. Blanco-Prieto, O.; Catalán, J.; Trujillo-Rojas, L.; Peña, A.; Rivera Del Álamo, M.M.; Llavanera, M.; Bonet, S.; Fernández-Novell, J.M.; Yeste, M.; Rodríguez-Gil, J.E. Red LED Light Acts on the Mitochondrial Electron Chain of Mammalian Sperm via Light-Time Exposure-Dependent Mechanisms. *Cells* **2020**, *9*, 2546. [CrossRef]
35. Moraes, C.R.; Meyers, S. The sperm mitochondrion: Organelle of many functions. *Anim. Reprod. Sci.* **2018**, *194*, 71–80. [CrossRef] [PubMed]
36. Brownlee, M. Biochemistry and molecular cell biology of diabetic complications. *Nature* **2001**, *414*, 813–820. [CrossRef] [PubMed]

37. LaRocca, T.J.; Sosunov, S.A.; Shakerley, N.L.; Ten, V.S.; Ratner, A.J. Hyperglycemic Conditions Prime Cells for RIP1-dependent Necroptosis. *J. Biol. Chem.* **2016**, *291*, 13753–13761. [CrossRef] [PubMed]
38. Peña, F.J.; Ortiz-Rodríguez, J.M.; Gaitskell-Phillips, G.L.; Gil, M.C.; Ortega-Ferrusola, C.; Martín-Cano, F.E. An integrated overview on the regulation of sperm metabolism (glycolysis-Krebs cycle-oxidative phosphorylation). *Anim. Reprod. Sci.* **2022**, *246*, 106805. [CrossRef] [PubMed]
39. Dixon, S.J.; Lemberg, K.M.; Lamprecht, M.R.; Skouta, R.; Zaitsev, E.M.; Gleason, C.E.; Patel, D.N.; Bauer, A.J.; Cantley, A.M.; Yang, W.S.; et al. Ferroptosis: An iron-dependent form of nonapoptotic cell death. *Cell* **2012**, *149*, 1060–1072. [CrossRef]
40. Garcia, D.; Shaw, R.J. AMPK: Mechanisms of Cellular Energy Sensing and Restoration of Metabolic Balance. *Mol. Cell* **2017**, *66*, 789–800. [CrossRef]
41. Fang, X.; Xu, Q.Y.; Jia, C.; Peng, Y.F. Metformin improves epididymal sperm quality and antioxidant function of the testis in diet-induced obesity rats. *Zhonghua Nan Ke Xue Natl. J. Androl.* **2012**, *18*, 146–149.
42. Nguyen, T.M.; Seigneurin, F.; Froment, P.; Combarnous, Y.; Blesbois, E. The 5′-AMP-Activated Protein Kinase (AMPK) Is Involved in the Augmentation of Antioxidant Defenses in Cryopreserved Chicken Sperm. *PLoS ONE* **2015**, *10*, e0134420. [CrossRef]
43. Zhu, Z.; Li, R.; Ma, G.; Bai, W.; Fan, X.; Lv, Y.; Luo, J.; Zeng, W. 5′-AMP-Activated Protein Kinase Regulates Goat Sperm Functions via Energy Metabolism In Vitro. *Cell. Physiol. Biochem. Int. J. Exp. Cell. Physiol. Biochem. Pharmacol.* **2018**, *47*, 2420–2431. [CrossRef] [PubMed]

Disclaimer/Publisher's Note: The statements, opinions and data contained in all publications are solely those of the individual author(s) and contributor(s) and not of MDPI and/or the editor(s). MDPI and/or the editor(s) disclaim responsibility for any injury to people or property resulting from any ideas, methods, instructions or products referred to in the content.

www.ingramcontent.com/pod-product-compliance
Lightning Source LLC
Chambersburg PA
CBHW041344210926
51850CB00013B/2988
9783725837830

MDPI AG
Grosspeteranlage 5
4052 Basel
Switzerland
Tel.: +41 61 683 77 34

Animals Editorial Office
E-mail: animals@mdpi.com
www.mdpi.com/journal/animals

Disclaimer/Publisher's Note: The title and front matter of this reprint are at the discretion of the Guest Editor. The publisher is not responsible for their content or any associated concerns. The statements, opinions and data contained in all individual articles are solely those of the individual Editor and contributors and not of MDPI. MDPI disclaims responsibility for any injury to people or property resulting from any ideas, methods, instructions or products referred to in the content.